The United States and the Armed Forces of Mexico, Central America and the Caribbean, 2000–2014

ALSO BY RENÉ DE LA PEDRAJA
AND FROM MCFARLAND

*Wars of Latin America, 1982–2013:
The Path to Peace* (2013)

*Wars of Latin America, 1948–1982:
The Rise of the Guerrillas* (2013)

Wars of Latin America, 1899–1941 (2006)

The United States and the Armed Forces of Mexico, Central America and the Caribbean, 2000–2014

René De La Pedraja

McFarland & Company, Inc., Publishers
Jefferson, North Carolina

LIBRARY OF CONGRESS CATALOGUING-IN-PUBLICATION DATA

De La Pedraja Tomán, René.
   The United States and the Armed Forces of Mexico, Central America and the Caribbean, 2000–2014 / René De La Pedraja.
      p.   cm.
   Includes bibliographical references and index.

   ISBN 978-0-7864-9508-5 (softcover : acid free paper) ∞
   ISBN 978-1-4766-1776-3 (ebook)

   1. United States—Military relations—Latin America. 2. United States—Military relations—Caribbean Area. 3. Latin America—Military relations—United States. 4. Caribbean Area—Military relations—United States. 5. Latin America—Politics and government. I. Title.
F1418.D435 2014
355'.0310973098—dc23   2014031853

BRITISH LIBRARY CATALOGUING DATA ARE AVAILABLE

© 2014 René De La Pedraja. All rights reserved

No part of this book may be reproduced or transmitted in any form or by any means, electronic or mechanical, including photocopying or recording, or by any information storage and retrieval system, without permission in writing from the publisher.

On the cover: FAV F-16A, 2008 © Chris Lofting

Printed in the United States of America

McFarland & Company, Inc., Publishers
   Box 611, Jefferson, North Carolina 28640
      www.mcfarlandpub.com

To Beatriz and Jaroslav

Table of Contents

List of Maps	ix
List of Tables	ix
Preface	1

1. VENEZUELA: THE APRIL 2002 COUP
    The Break with the U.S. Government — 5
    The Attempt to Topple Hugo Chávez — 12
    The Coup Within the Coup — 17
    The Counter Coup — 19

2. THE EXPANSION OF THE ARMED FORCES OF VENEZUELA
    The Venezuelan Military Until December 2006 — 26
    The Military and the Presidential Election of 3 December 2006 — 34

3. UNCONDITIONAL ALLIES OF THE UNITED STATES
    Guatemala: Settling Old Scores with the Army — 38
    Rebuilding the Army of El Salvador — 43
    Improving the Armed Forces of Colombia — 49

4. RELUCTANT MEXICO
    The Role of the Armed Forces in Mexico — 55
    The Presidency of Ernesto Zedillo, 1994–2000 — 63
    The Presidency of Vicente Fox, 2000–2006, and El Chapo Guzmán — 73

5. THE MILITARIZATION OF COLOMBIA AND VENEZUELA
    Colombia: Sustaining the Momentum — 81
    The Venezuelan Military After the Reelection of Hugo Chávez in December 2006 — 88

6. SHAPING THE ARSENALS OF THE ARMED FORCES — 93
    Nicaragua: Surface to Air Missiles — 95
    Venezuela: Arms Buildup — 99
    Mexico: Fighter Jets — 105

7. New Complications
- Colombia: False Positives — 108
- Scandals in the Colombian Army — 114
- Jamaica: Trouble in Paradise — 120

8. Central America Turns to the Left
- Nicaragua: The Return of the Sandinistas — 127
- Guatemala: The First Leftist President of the Twenty-First Century — 130
- El Salvador: The FMLN in Office — 135

9. Honduras: The Coup of June 2009
- The Presidency of José Manuel Zelaya — 144
- The Path to the Coup — 150
- The Coup of 28 June 2009 — 158

10. Mexico: The Armed Forces Embrace the United States
- The Start of the Vicente Calderón Presidency — 165
- The Military Offensive Against the Drug Cartels, 2008–2010 — 171
- The Battle for Monterrey — 184
- The Hunt for El Chapo Guzmán — 189

11. Mexico: The Presidency of Enrique Peña Nieto — 194
- Pursuing the Drug Lords — 195
- The Challenge of Michoacán — 198

12. Countries Without Armies
- Costa Rica: Creeping Militarization — 206
- Panama: The Struggle to Demilitarize — 211
- Haiti: Flawed Demobilization — 221

13. An Inconvenient Ally: The Armed Forces of the Dominican Republic
- The Military of the Dominican Republic — 227

14. Defiant Cuba — 238
- Preserving the Revolutionary Legacy — 240
- The U.S. Vision for Cuba and the Caribbean — 251

15. Trying to Dismantle the Bush Legacy — 258
- Three Explanations — 267

Chapter Notes — 275
Annotated Selected Bibliography — 309
Index — 311

List of Maps

1. Venezuela — 13
2. Caracas — 15
3. El Salvador — 48
4. Eastern Colombia — 51
5. Southern Mexico — 57
6. Northwestern Mexico — 61
7. Northeastern Mexico — 70
8. Northwest Colombia — 117
9. Central America — 121
10. Nicaragua — 128
11. Guatemala — 131
12. Panama — 213
13. Haiti — 224
14. Dominican Republic — 229
15. Cuba — 239

List of Tables

1. Mexico: Military Personnel Receiving U.S.-Funded Training, 1950–2009 — 64
2. Mexico: Army and Navy Strength, 1990–2014 — 66
3. Mexico: Drug-Related Arrests, 1990–2013 — 67
4. Extraditions from Mexico, 1980–2013 — 72
5. Mexico: Desertions from Army, 1997–2012 — 173
6. Mexico: Women in Army, 2006–2012 — 174
7. Mexico: Crop Eradication and Drug Seizures, 1990–2013 — 178

Preface

"This would be like giving a 15-year old a Ferrari," was the observation a seasoned American diplomat made about giving advanced weapons to a Caribbean country. That comment revealed the fundamental assumptions behind U.S. policies toward the armed forces of Mexico, Central America, and the Caribbean, a zone often known as the "backyard of the United States." The U.S. was a distant father that knew what was best for his wards who when left without adult guidance were easily captivated by shiny new weapons. As this comment reveals, the U.S. government assumed the role of deciding both what toys countries could not possess and also of telling local rulers in a paternalistic tone what type of armed forces they should have "because it was good for you."

Actually, having armed forces in its backyard has never been the first choice for a U.S. government normally anxious about the region only when foreign trespassers appeared. American officials preferred to rely exclusively on local police forces or constabularies to preserve internal order. For over a century the United States has labored to create local police forces, but generally the results have been at best disappointing. In frustration, the U.S. government saw no option but to accept the need for armies to control these turbulent countries. American officials unenthusiastically adopted, and in some cases even created, armed forces as the only way to assure internal order in the region.

When the United States occupied the island of Cuba in 1898, this was just the first step in the long and exhausting chore to shape the armed forces in Mexico, Central America, and the Caribbean. The complex task of balancing the preference for police with the unavoidable necessity of having armed forces has continued and shows no signs of abating in the foreseeable future.

As the chapter titles of this book suggest, not all the U.S. creations turned out well. From the beginning Mexico has been the most restive, and other countries have experienced undercurrents of opposition. Three

Caribbean countries have even tried to reduce U.S. influence on their armed forces. The first and most successful was Cuba with its open defiance since 1959. Nicaragua made a first attempt from 1979 to 1990 and then again since 2006 to reduce its dependence, as has Venezuela since 2003. For the rest of the countries, either grudging acceptance of U.S. wishes or sometimes slavish adherence have been the only realistic options.

No less than U.S. officials, I have grappled with the absence of a clear dividing line between the activities of the army and of the police in these countries. It is well known that the soldiers of Mexico and the Caribbean routinely have performed functions normally assigned to police personnel in the United States and Western Europe. In a further blurring of responsibilities, the police in the Caribbean sometimes have been under the jurisdiction of the Ministries of Defense. At another extreme, the police in a few cases have routinely engaged in combat operations, supposedly the preserve of the army, as continues to happen in Colombia in its ongoing war with guerrillas. Quite naturally the U.S. government has taken a major interest in the police forces of the region, sometimes as a complement to the military and at other times as a possible alternative or substitute to an army. This book frequently delves into the distribution of functions between the police and the armies but without attempting an exhaustive examination of the police in those countries.

The above concerns emerged as I was doing research for my three volumes on the wars of Latin America from 1899 to 2013. As I was finishing the last of those three histories, I decided to examine U.S. efforts to create a special type of armed forces in the countries to the south. In spite of having accumulated considerable materials while I was conducting research on the wars of Latin America, the quality and the quantity of the information diminished after the 1950s. Secrecy restrictions prevented me from documenting the continuity in U.S. efforts to shape the armed forces from 1899 to the present. Ample evidence existed for the period up to 1960, but what about the last fifty years?

After I sadly abandoned my project, an exciting possibility came almost as a bolt from the sky when Wikileaks released the cables from U.S. embassies in the region. Now for the first time I enjoyed abundant and detailed information on recent U.S. efforts to shape and control the armed forces of Mexico, Central America, and the Caribbean. I decided to devote a single book to the twenty-first century as the best way to bring to the attention of the reading public the many exciting findings in the U.S. embassy cables. My family was equally enthusiastic about this very innovative project, and I am very happy to be able to thank my wife and my son for their support. As always, my wife drew the maps, read the many drafts of the text, and helped with the translations. I am also happy to welcome Kyra, my daughter-in-law, into my support group, and already her spontaneous offer to read the initial chapters has improved the language. I know that she and my son will make a great team in reading my future manuscripts as in all other endeavors of their lives.

It is my sincere belief that readers will find enlightening, revealing, and even at times entertaining, my account of how the U.S. government quietly and usually secretly strove to shape, not always successfully, the armed forces of Mexico, Central America, and the Caribbean during the twenty-first century.

1

Venezuela: The April 2002 Coup

> *There was a long history of Latin American resentment of America as arrogant and insensitive to their interests and problems. Whenever America reached out in genuine friendship ... we did better. [... But] when we supported the overthrow of democratically elected leaders, backed dictators, and tolerated their human rights abuses, we got the reaction we deserved.* —President Bill Clinton[1]

Building upon President Bill Clinton's bailout of Mexico in 1995 (see chapter 4), the U.S. government maintained unusually friendly relations with Caribbean countries in the final years of the twentieth century. Even the possibility of settling the long-festering disputes with Cuba no longer seemed just an illusion. But when the George W. Bush administration took office in January 2001 and promptly began to support right-wing regimes in the region, opposition to this new policy emerged in Latin America. In Venezuela, where the Bush administration collaborated in trying to overthrown a democratically elected government, not surprisingly resistance to the new U.S. policy became intense and bitter. It is often forgotten that the revolution of Hugo Chávez Frías initially was not hostile to the United States, and only after the Bush administration supported the April 2002 coup did the Venezuelan president strive to make his armed forces independent of U.S. control.

The Break with the U.S. Government

The electoral campaign of 1998 for the presidency of Venezuela revealed the key elements of Chavismo. During his electoral speeches, Hugo Chávez Frías directed the strongest attacks against PDVSA, the state-owned petroleum monopoly, and these charges made leftists uncomfortable. The business sector and foreign corporations interpreted these attacks as a prelude to a full or at least partial privatization of state-owned firms.[2] In reality Chávez was railing at those elite groups controlling institutions, whether private, public, or labor unions. Scholars sometimes refer to the "iron law of oligarchy," and Chávez found this characteristic rampant throughout Venezuela. Institutions had come to exist primarily and often exclusively for the benefit of its top management or leaders, while the rank-and-file saw their economic situation worsen or at best stagnate, as happened even inside the armed forces.

The privileges of elite groups deprived the rest of the Venezuelan population of meaningful opportunities for improving their material conditions. A deep religious feeling arising from his passionate Roman Catholic faith made Chávez value each individual, and the almost mystical qualities of many of his speeches at times seemed to be turning him into a Messianic figure. His religious principles and his many criticisms of state institutions appealed to Conservative and right-wing individuals who either supported or at least tolerated his candidacy. In reality, his agenda was neither privatization nor nationalization, because with surprising insights into the nature of organizations, he had seen how determined individuals could take over and exploit any institution. Throughout his presidency he relentlessly pursued the twin goals of increasing the popular participation of the rank-and-file in all institutions and providing economic and political opportunities for those until then left outside the existing structures of government, business, and labor.

The significance of the electoral victory of Chávez on 6 December 1998 was not immediately obvious. After his inauguration on 2 February 1999, he seemed to be following the pattern of politicians who talked of major changes but ultimately governed as moderates. Because he had a small minority in the bicameral legislature, in reality he could not enact any major legislation. He devoted most of his efforts during 1999 to deliver on his promise to adopt a new constitution for Venezuela. In a referendum voters approved convening a constitutional convention and in a separate election chose its deputies. The convention duly gathered and under the presidency of Chávez's close ally and mentor Luis Miquilena drafted a constitution in 1999. Although the Chavista coalition controlled the overwhelming majority of seats to the convention, the final text did not always reflect the wishes of Chávez who had to accept compromises to maintain his coalition intact.[3]

Under the terms of the new 1999 Constitution, Chávez ran for a new six-year term as president in what were called "mega-elections" because all elected offices were contested. In the broadest popular participation ever seen in Venezuela, 35,000 candidates competed for 6200 positions. His coalition gained 105 of the 165 seats in the newly constituted unicameral National Assembly but not the two-thirds majority needed to pass major legislation. Just when he seemed to have reached the moment of his greatest triumph, cracks started to appear in his coalition. Since the fall of 1999 he had been establishing close ties with Socialist Cuba, and his admiration for Fidel Castro generated not only predictable hostility from right-wing groups but also tensions within his leftist coalition. Chávez insisted on creating a new system that would neither be capitalist or socialist but would be uniquely Venezuelan. Not surprisingly, defining the exact nature of this system was hard, and the internal debates, increasingly acrimonious, dragged on throughout 2000. As long as the Bill Clinton administration was in office, curiosity characterized the attitude of the U.S. government toward the latest attempt to effect a transformation of a Latin American country.[4]

Official U.S. tolerance for the political and social experimentation in Venezuela abruptly ended when George W. Bush took office on 20 January 2001. After having campaigned on the promise of pushing a very aggressive foreign policy, President Bush found totally unacceptable the close ties Chávez enjoyed with Socialist Cuba. Cold War era officials eager to resume the battle against Communism returned to key positions in the Bush administration. But when the terrorist attacks of 11 September 2001 shifted U.S. efforts to Afghanistan and the Middle East, Venezuela seemed at least temporarily to have escaped Washington's anger.[5]

The lull in U.S. hostility proved short-lived, but while it lasted the many leftist factions in the Chavista coalition enjoyed the freedom to debate the specifics of the transformation for the country. The 1999 Constitution had altered the structure of government, but the implementation of social and economic changes required legislation from the National Assembly. Chávez was rightfully becoming impatient with the pace of change because he was not able to deliver on the campaign promises he had been repeating since the 1998 election. Other than a new organizational chart for the government, the Venezuela of 2001 largely resembled that of 1998. He needed to quicken the pace of transformation but the National Assembly was lagging in approving major legislation. After striving for over a year to pass new laws, he concluded regretfully that he needed to take extraordinary measures to break the logjam in the legislature.[6]

Strongly believing that a majority of voters twice elected him to improve their lives, he had recourse to a parliamentary maneuver. He obtained from an unwary National Assembly temporary authorization to issue laws, and then in November 2001 he stunned the country by issuing 49 decree laws containing the major changes previously bottled up in the legislature. The package included a land reform law and strict controls over banking as well as over many other sectors of the economy. One of the 49 decrees forced PDVSA for the first time to share its profits with the Venezuelan people. Elites, whether public or private, were outraged at the loss of their privileges and embarked on their campaign to remove Chávez from office. No less significant, the 49 decree laws irreparably fractured the Chavista coalition, as many former supporters deserted the movement. The defections of leftists who insisted on mechanically repeating the Socialism of Cuba were most ironic, because they felt that Chávez had betrayed them by failing to nationalize the private sector. Of these leftists, the most important was the old Communist Luis Miquilena, his former mentor, who although not publicly breaking with Chávez, quietly faded away and ceased to support the government. Without this fragmentation of the Chavista coalition, right wing and business groups could have never gathered a major opposition movement the next year.[7]

The 49 laws favored workers, employees, and peasants, nevertheless, labor did not enthusiastically embrace these measures putting restrictions on capital accumulation by the wealthy. This weak response was understandable, because in Latin America union leadership has often become a labor aristocracy more determined on preserving its privileges than in supporting the needs of workers. Rank-and-file union members formed part of the Chavista coalition because they obtained better responses from the government than from their senior labor leaders. As a way to help satisfy the often ignored wishes of workers, the government announced in October 2001 new measures to guarantee fair and honest union elections. The insistence on more popular participation in all sectors of Venezuelan life was a fundamental cornerstone of the Chavista program, but this principle obviously clashed with the desires of the labor aristocracy to remain in power. When in November Chávez promulgated the 49 laws, union bosses, and in particular the head of the Venezuelan Confederation of Workers (CTV) Carlos Ortega, jumped at the chance to form an alliance with business and right-wing groups. The key personality was Pedro Carmona, the director of FEDECAMARAS, the most powerful business association in Venezuela.[8]

The seemingly contradictory alliance between FEDECAMARAS and CTV became the foundation of a growing opposition movement, and these two groups carried out a 12-hour general strike on 10 December 2001 as a first demonstration against the 49 decrees.[9] Favorable TV coverage helped the opposition, because by then private media groups had turned into the most passionate opponents of Chávez and constantly clamored for his removal from office. Partly out of a sense of betrayal because most media outlets originally had endorsed his candidacy in 1998, both newspapers and TV stations became pathologically obsessed with destroying Chávez. With a crusading zeal best reserved for more worthy causes, the private media, in particular the TV stations, not only kept a steady barrage of violent attacks on the Chavista regime but also deliberately distorted news, exaggerated supposed failures, and eliminated any favorable reports. In the many low points of journalism in Latin America, the media of Caracas easily qualified among the worst and may well have ranked as the very worst in the failure to provide even a semblance of accurate reporting.[10]

Umbrella organizations emerged to provide a safe haven for the many disgruntled leftists who were abandoning the Chavista coalition, but who could not in good conscience support private business groups. And as a final frosting on the opposition cake, most high-ranking members of the Roman Catholic Church, including Ignacio Velasco, Venezuela's lone cardinal, actively supported and even blessed the campaign to remove Chávez from office. The hostility of the Church hierarchy certainly reflected its traditional links to private business and upper-class groups, but more was at stake. Throughout his seemingly interminable television appearances, Chávez incessantly talked about his two heroes, Simón Bolívar, the liberator of Venezuela from Spanish rule, and Jesus Christ, the liberator of souls. At times Chávez sounded like an avenging patriarch straight out of the Old Testament, and his inspirational homilies profoundly touched many fervently Catholic Venezuelans. In reality, Chávez was trying to place the same emphasis on popular participation from below in the Roman Catholic Church as he was trying to do in the rest of the institutions in Venezuela. The Catholic hierarchy, just like other elite groups in the country, saw its position threatened by this Messianic speaker and not surprisingly eagerly joined the growing opposition coalition.[11]

As Bush administration officials watched the array of groups lining up behind the opposition, the possibility of overthrowing a fragmented Chávez regime grew stronger with each passing day. The U.S. Embassy in Caracas chronicled in great detail the remarkable expansion of opposition groups, but U.S. diplomats neglected to report on the still formidable capacities of Chavista supporters. The task of providing accurate reporting admittedly was not easy, because as Chávez started his passionate denunciations of the Bush policy in the Middle East, many Chavista followers if not already hostile to the U.S. government were most reluctant to talk to American officials. U.S. diplomats in the Caribbean area were accustomed to being treated with sometimes servile deference in client states, and this unusual sign of independence and even defiance "was getting under their skin and the culture within the embassy was becoming increasingly anti–Chávez."[12] Without missing a beat, U.S. Embassy officials abandoned the tolerance of the Clinton administration and embraced the new aggressive arrogance of the Bush administration.

Documenting the details of U.S. involvement in the plot to overthrow Chávez has been hard simply because the Bush administration did not have to play a major role. The U.S. government enthusiastically cheered from the backseat while opposition leaders implemented the coup. The CIA was spared the costly burden of organizing major expeditionary forces as was the case for Guatemala in 1954 and for the Bay of Pigs in 1961. At the same time the U.S. government did not have to be as closely involved as happened during the overthrow of Salvador Allende of Chile in 1973. The closest historical analogy in Latin America was to the military coup in Brazil in 1964. However, in that operation the U.S. Embassy in Rio de Janeiro played the main role and generated an extensive paper trail of incriminating and embarrassing documents.[13]

For Venezuela in 2002, the Bush administration placed major restrictions on its participation. First of all, the U.S. Embassy was largely kept in the dark, thereby not just dramatically reducing the paper trail but also allowing deniability if things went wrong. Washington remained at the center of events influencing the coup, and Venezuelan opposition leaders duly paraded through the capital to obtain their assurances of U.S. support. Secondly, any removal of Chávez had to be done by constitutional means and could not be a naked power grab. In one of its rare similarities in Latin American policy with the Clinton presidency, the Bush administration correctly concluded that a wave of violent military coups across the region was not in the best interests of the United States. American diplomats insisted that the removal of Chávez from office and the creation of a provisional government required that coup plotters follow the provisions of the 1999 Constitution. Third, the U.S. government gave financing to groups hostile to Chávez. These funds, which were supposedly intended to foster "democracy" in Venezuela, went through the National Endowment for Democracy and its affiliates, with almost no coordination with the U.S. Embassy in Caracas. The CIA certainly channeled additional sums secretly, but actually this money was really not essential because so many wealthy

Venezuelans were eagerly contributing to overthrow Chávez. The modest U.S. assistance was primarily a demonstration of the Bush administration's commitment to the coup.[14]

Behind all the U.S. moves was an intelligence failure eerily similar to that for Cuba in April 1961. Just like official agencies reported in April 1961 that the Cuban population was eagerly waiting for a spark to rise up against Fidel Castro during the Bay of Pigs Invasion, so again in April 2002 the U.S. intelligence community expected the Venezuelan population to rise up and overthrow Chávez. Bolstered by the optimistic reports from the U.S. Embassy in Caracas, the Bush administration concluded that a cheap and easy victory was possible in Venezuela, where the media increasingly depicted Chávez as a mad dictator. The military stood out as the last remaining bastion of a crumbling Chavista regime. However, since the promulgation of the 49 decrees in November 2001, influential civilians had been increasing their contacts with high-ranking military officers. Gradually among the officers emerged coup plotters who sensed that the opportunity was ripe to throw out Chávez and to advance their military careers. U.S. military officers in routine contact with their Venezuelan counterparts regularly received full briefings on the coup plotting inside the armed forces. Although U.S. officials consistently repeated the formula that any change had to follow constitutional procedure, they failed to warn of any adverse response from the U.S. government in case of a coup.[15]

Quite naturally Venezuelan officers concluded that U.S. officials supported a coup, a conclusion made even clearer by the Venezuelan military attaché at Washington, D.C., General Enrique Medina. The role of this officer as a link between military plotters and the Bush administration has generally been overlooked. As military attaché in Washington, he dealt directly with CIA headquarters and effectively bypassed the U.S. Embassy in Caracas. General Medina was violently opposed to Chávez, who in one of his typical mild responses, instead of firing the general removed him from command of troops and shipped him off to the supposedly harmless post of military attaché to the United States. In addition, the military attaché was closely associated with the business elite and could count on financial support to lure wavering officers away from Chávez. General Medina could well have served also as a conduit for CIA funds to win over key individuals, particularly in the gendarme-like police force of Venezuela, the National Guard (*Guardia Nacional*), and the Caracas Metropolitan Police. General Medina did persuade military officers to join the coup and helped the CIA recruit informants among Venezuelan officers.[16]

The Attempt to Topple Hugo Chávez

If Chávez had any doubts about the discontent inside the armed forces, a series of increasingly bold statements by both active duty and retired officers confirmed that he faced a major challenge to his rule. To bolster his authority, he needed easy access to the funds of PDVSA, the principal source of revenue for the government. The management of this state-owned oil company, just like the leadership of organized labor and many other elite groups in Venezuela, enjoyed a privileged position and resented the efforts of the President to share company profits with the poor of the country. Chávez decided to get ahead of the growing opposition movement by announcing the appointment of a new director and a new board of directors for PDVSA on 21 February 2002. Opposition leaders saw this action as a sign of weakness and as a last desperate attempt to save his collapsing regime. The labor confederation CTV began to talk about another general strike around the middle of March, but then unexpectedly its leader Carlos Ortega postponed the general strike. He was already coordinating with Pedro Carmona, the director of the business association FEDECAMARAS, a much larger operation to topple Chávez in conjunction with a parallel conspiracy taking place inside the armed forces.[17]

As is common in large institutions with organized workers, not one but several labor unions represented workers and employees, and on 4 April 2002 the management union began a strike at PDVSA. In spite of all the efforts of Ortega, the head of CTV and a former oil worker himself, the other labor unions either refused to support the strike or did so only partially. The split within organized labor at PDVSA reflected the larger struggle throughout Venezuela as the elites on top of all institutions tried to defend their privileges from demands of workers and employees on the bottom. Although the management union did tremendous damage to PDVSA operations, this faltering company strike could not carry the whole weight of the struggle against Chávez. Two days later CTV set Tuesday April 9 as the date for the promised general strike and ominously stated that although it was supposed to last for 24 hours, the general strike might be extended under the right circumstances.[18]

In response to the challenge from the opposition, on Sunday April 7 Chávez used his weekly morning talk show program, *Aló Presidente*, to rally his followers. Feeling the need to solidify his support among the masses, Chávez went against the recommendation of his financial advisers and increased the minimum wage by 20 percent. On live television he fired the top seven managers of PDVSA after individually calling out each of their names and also announced the retirement of another twelve senior managers. Even he in later years admitted that this very theatrical performance had been ill-advised.[19]

No doubt existed that a major confrontation between the government and the opposition was imminent, and the rumors about a military coup spread widely. To deal with the impending threats, Chávez later that same Sunday April 7 convoked a cabinet meeting with key generals at the Miraflores Presidential Palace. Government spies already reported that the opposition intended to extend the general strike of Tuesday 9 April beyond 24 hours and to culminate the protest with a gigantic rally in front of the main PDVSA building in the morning of Thursday April 11. Because the protest site was in eastern Caracas and about 10 kilometers from downtown, at first glance the rally did not seem threatening, and legally the government could not deny the permit to assemble peacefully. But when spies also reported that the real plan was to divert the demonstrators away from the PDVSA building to the Miraflores Palace to demand the resignation of the President, the protest march took on a more sinister air.[20]

Besides the Caracas Metropolitan Police, at that time under the control

Map 1: Venezuela.

of a bitter opposition figure, the only other police unit trained and equipped to stop the crowd from reaching Miraflores was the National Guard, the gendarme-like police force. Because Chávez was already having doubts about the loyalty of most of the National Guard, he felt that he had no choice but to turn to the army. He activated Plan Ávila, a military contingency plan to secure key installations in the capital, including Miraflores Palace, by calling on army units at the large military base of Fort Tiuna (Map 1) located to the south of the city. As the meeting of Sunday April 7 came to an end, militant Chavistas entered to demand that the President authorize his followers to gather over the coming days outside Miraflores in a counterdemonstration to show the intense support he still enjoyed among vast sectors of the population.

The general strike duly began on Tuesday April 9 and was very effective in wealthy and middle class neighborhoods but barely noticeable in poor Chavista districts. On 10 April the CTV prolonged the general strike, which was fast losing steam. In a last burst of energy, the opposition rallied supporters for the giant march on the PDVSA building on Thursday April 11. At 0900 hours a crowd numbering at least 200,000 gathered in one of the largest demonstrations in Venezuela's history. Once the crowd was in a state of frenzy, at 1100 FEDECAMARAS director Pedro Carmona and CTV chief Carlos Ortega urged the crowd to march on Miraflores Palace to demand the resignation of the President.

The protesters eagerly set out, but the tiring trek under the hot midday sun and the stifling heat soon took its toll. The march of ten kilometers was the weak point in the plan, and as the demonstrators dwindled in number, all that was needed was a small National Guard picket to prevent the march from coming even near Miraflores Palace. But at the suitable choke points the National Guard failed to appear, and the orders given to deploy its units were either revoked or not followed. Police commanders in Latin America have been notoriously corrupt, and here was where opposition leaders, in conjunction with the CIA, played a decisive role in keeping National Guard units out of the path of the marching column of protestors.[21]

Chávez realized that something bigger was going on when he received word that a wealthy businessman had flown in General Enrique Medina from Washington, D.C., to Caracas. The military attaché probably finalized the details of the coup and at the very least also reported on the deployment of U.S. vessels as a concrete signal of support by the Bush administration for the plotters. The news of Medina's arrival prompted Chávez at 1030 to activate Plan Ávila, and shortly afterwards a column of tanks left Fort Tiuna for the garages near Miraflores Palace. The tanks did not immediately reach their destination, because rebel officers of both the army and the National Guard were already blocking intersections and diverting civilian vehicles into Fort Tiuna to create massive traffic jams.[22] But the column of tanks by taking circuitous routes finally reached the barracks of the presidential guard and in the trip did not hurt anybody and did not "even scratch a car."[23] The state TV station had been calling on the Bolivarian Circles to send Chavista supporters who already were starting to gather around Miraflores. Chávez felt that his loyal followers and the tank column were more than adequate to defend the palace, even without counting on his hand-picked presidential guard that was ready to die for him.

Meanwhile the march reached the grid pattern streets of downtown Caracas, and sometime around 1400 the first demonstrators were approaching the neighborhood of Miraflores Palace (Map 2), by then defended by many Chavista followers. The vanguard of the column turned into Baralt Avenue around 1500. The map shows that this route added an unnecessary step and required making a second turn to the west at Urdaneta Avenue to

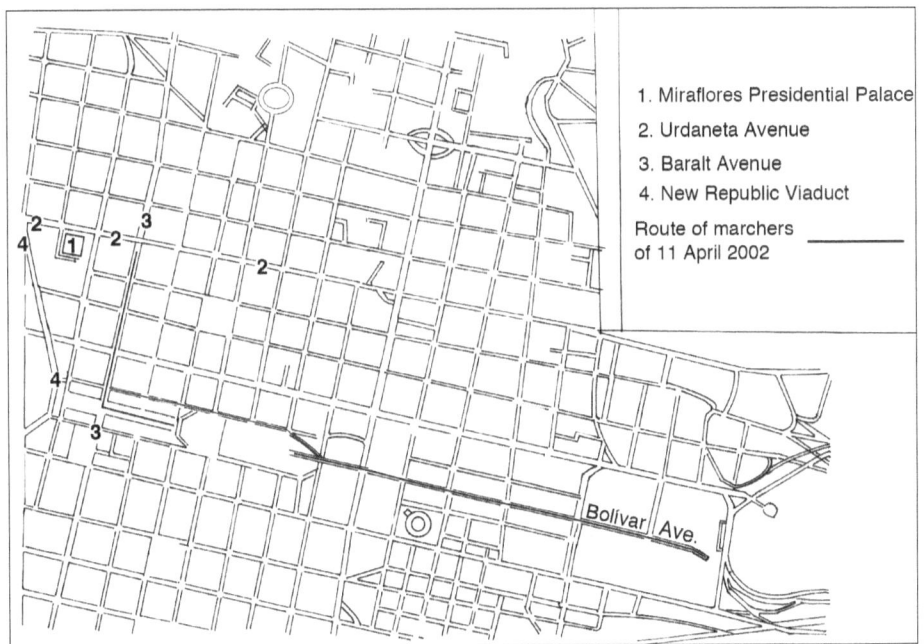

Map 2: Caracas.

reach the Palace. But even worse were the stairs at the end of Baralt Avenue; because of the uneven elevation in many parts of Caracas, streets sometimes ended in concrete stairs to allow pedestrians but not vehicles to continue. The leaders took the demonstrators into what was a dead end, and then mysteriously the key leaders vanished from the column.[24]

Sometime after the column of demonstrators marched into Baralt Avenue, Chavistas hurled rocks at the column, and then shots rang out from nearby buildings. Although the exact identity of the shooters has never been confirmed, the opposition had the most to gain from any violent confrontation. The military plotters needed a bloody incident to justify breaching their loyalty to the legally elected government. Whoever started the shooting, of the 19 persons who died that day, seven were Chavistas, seven opposition leaders, and the rest innocent bystanders, so to claim that the government had brutally massacred innocent protesters was at best an exaggerated distortion. And with the path blocked at the stairs of Baralt Avenue, the column backed up and the overflow poured into South Eight Street and New Republic Viaduct. But by then Chavistas and small detachments of police blocked these alternate and shorter routes to Miraflores Palace.[25]

Even before the first civilian deaths occurred, the military plotters were already using the bloody repression to justify their disloyalty to the Chávez regime. Because the private TV stations were repeating the most horrible accusations of a bloody repression, at 1625 Chávez ordered the private TV stations to stop broadcasting. But using alternate technologies and subterfuges, most of the private TV stations continued to broadcast even when the government took control of their principal studios. The most important announcements began at 1900 when military officers declared on TV their refusal to obey Chávez. The loss of all military control seemed to be complete when the commander of the army General Efraín Vásquez Velasco announced at 2000 that he too was abandoning the President. Gradually the opposition regained full control of the private TV stations and soon after the state television station ceased to broadcast.[26]

Chávez was trapped in Miraflores Palace and could no longer talk to Venezuelans because of the loss of the state television station. He was still in contact with officers across the country and knew that most of the regional garrisons remained loyal. At Maracay, General Raúl Baduel, the commander of the 42nd Paratrooper Brigade, the best unit of the Venezuelan military, wanted to attack Caracas, but Chávez opposed launching an operation sure to cause hundreds of civilian deaths. His preference was to lead the tanks of Miraflores to Maracay and to join General Baduel, but then the unexpected return of the tank column back to Fort Tiuna deprived Chávez of this option. He was already dressed in military fatigues, had strapped a pistol to his leg, and was lugging an automatic rifle. He seemed ready to die a martyr's death alongside the soldiers of the presidential guard who were thirsting to make a last-ditch defense. All talk of bravado ended when Fidel Castro telephoned to explain that a martyr's death was understandable for a Salvador Allende left without any support in Chile in 1973 but was pointless for Chávez who still enjoyed the support of the majority of the military forces and of his people.[27]

Shortly before 0300 on 12 April the rebel generals issued an ultimatum threatening to attack Miraflores if Chávez did not resign within 15 minutes. The President pretended to accept under certain conditions. As proof of his intentions, he authorized reading a statement at 0330 that he had resigned, and the private TV stations eagerly broadcast the news. As negotiations over the language of the resignation letter dragged on, Chávez set aside his weapons and allowed rebel troops outside the palace to take him prisoner. The rebels brought him to Fort Tiuna where TV cameras caught a glimpse of him entering the base around 0400 hours.[28] The tumultuous experiment with revolution was over even before Chávez finished his first term, and nobody seemed to worry about the final detail of securing a signed resignation letter.

The Coup Within the Coup

The imprisonment of Hugo Chávez left the coup plotters free to consolidate their grip on the country. At the same time that the removal of Chávez was taking place, General Efraín Vásquez Velasco, in consultation with other generals, was organizing a provisional government. The plotters' original plan called for the creation of a civilian-military junta to rule until new elections were held. Its chair would be Pedro Carmona, the director of FEDECAMARAS, the country's most powerful business association. But U.S. insistence on following proper constitutional procedure ruled out any military participation. General Vásquez Velasco proposed as a substitute for the military that Cardinal Ignacio Velasco become a junta member, but he declined to join without prior authorization from the Vatican. Reluctantly the generals accepted Carmona as the new president. The news came as no surprise to Carmona, who several weeks before had fitted himself out with a proper presidential sash while he was in Spain securing support for the coup from the conservative Spanish government. At 0451 on 12 April, less than an hour after TV cameras filmed Chávez entering imprisonment at Fort Tiuna, Carmona appeared on live television to announce that he was Venezuela's new interim president.[29]

Carmona's statement was inherently risky, because the resignation of Chávez automatically made Vice President Diosdado Cabello the new ruler of Venezuela. Vice President Cabello, along with many Chavista officials, had gone into hiding. Opposition deputies proposed convoking the National Assembly to confirm Carmona as interim president. Although Chavistas comprised the majority in the National Assembly, opposition leaders were convinced that the bandwagon effect of always wanting to be with the winner would convince opportunistic deputies to switch sides and ratify the selection of Carmona. However, the generals did not want a messy debate in full public view of a large TV audience, and the opposition of Roman Catholic Cardinal Ignacio Velasco, a key plotter, finished killing the proposal to consult the National Assembly.[30]

As dawn broke on 12 April the Carmona presidency seemed off to a great start. "Everyone who was anybody in the opposition gathered at the presidential palace, to celebrate, to congratulate, and to lobby for positions in the provisional government."[31] The media was ecstatic about the fall of Chávez, and private newspapers trumpeted the event in celebratory headlines while TV commentators constantly made jubilant statements. But more than genuine enthusiasm was present in the media, because the TV producers for the first time enjoyed the intense pleasure of knowing that they could make or break governments.[32]

The initial pronouncements from the U.S. government were extremely promising for Carmona. High U.S. officials, starting with the White House press secretary, saw events in Caracas as a repeat of the protests in Eastern Europe that had overthrown corrupt and decrepit Communist regimes. Sensing a resemblance to the "people's revolutions" in various countries, U.S. officials unanimously condemned the shooting of demonstrators by the Chávez government. Carmona was most pleased when both the U.S. Embassy in Caracas and the State Department in their official statements did not cast any doubts about his authority and openly welcomed the provisional government. And going beyond words, the Venezuelan Air Force reported that at 0900 hours U.S. ships entered Venezuelan territorial waters and were cooperating with military authorities on communications issues. All that remained to do was to give final shape to the provisional government, and Carmona with his advisers spent the day drafting a constituent decree and selecting cabinet members. The new government was so sure of its powers that it ordered the arrest of all prominent Chavistas, and soon small mobs were attacking homes and buildings suspected of hiding fugitive members of the fallen regime.[33]

The show of strength was premature, because before the morning of Friday April 12 was over, already the first cracks were appearing in the just constituted provisional government. Carlos Ortega, the labor leader who had been instrumental in organizing the mass rally the previous day, was outraged to hear that Carmona had become president. The labor leader understood the arrangement to be that a junta of civilian and military persons, including both Carmona and Ortega, was to head the provisional government. An angry labor leader headed for Miraflores Palace, and failing to obtain a satisfactory response in his conversation with Carmona, Ortega left Miraflores even more furious. He withdrew organized labor's support for the provisional government and returned to his home town of Coro in western Venezuela.[34]

Up to that moment the cooperation between FEDECAMARAS representing private business and CTV representing organized labor had been the foundation of the anti–Chávez coalition. Once this alliance collapsed, the immediate consequences were enormous. First, without the moderating presence of organized labor to protect the interests of workers, the provisional government was free to go far to the right in pursuing pro-business policies, and already it was stating its intention to repeal the recent increase in the minimum wage. Second, the huge crowds calling for the removal of Chávez quietly melted away. Their numbers on the previous day had not been as large as opposition groups later claimed, and accepting the clearly exaggerated figures for the number of the opposition demonstrators on April 11 makes it impossible to account for their near disappearance on April 12. Even if a majority of opposition protesters had not been union members, without the networks of organized labor it was impossible to rally and bring large numbers of anti–Chávez protesters into the street.

The Counter Coup

As the opposition presence in the streets dwindled to small groups using the search for Chavista officials as an excuse to loot homes and stores, already in the early morning of 12 April Chavista supporters were filling the streets. News of the supposed resignation of Chávez shocked his followers, many of whom doubtlessly felt guilty for not having turned out in large numbers on 11 April to protect Miraflores Palace. Their determination to save the Chavista regime was passionate, and only their burning desire can explain their recklessness in taking on directly the Caracas Metropolitan Police. Unlike the opposition march of 11 April that traveled the 10 kilometers to Miraflores Palace without encountering any violence until the last few blocks, the Metropolitan Police under the control of provisional president Pedro Carmona used brute force to repress the Chavistas. The police ended up firing on the crowds, but the Chavistas had such an intense hatred for Carmona that they continued to risk their lives in confrontations. Hundreds of Chavistas were wounded on 12 and 13 April in street battles against the Metropolitan Police, and at least 56 Chavistas were killed. It was amazing to see the strange reversal of opinions by the Carmona government that claimed its legitimacy from the supposed Chavista repression of demonstrators on 11 April but now on 12 and 13 April showed no hesitation or even remorse about ruthlessly hurling the Metropolitan Police to kill the loyal followers of the fallen Chavista regime.[35]

As the demonstrations increased in size on 12 April, a demoralized General Jorge García Carneiro went outside to talk to Chavistas who had gathered in front of Fort Tiuna. Their passionate devotion to Chávez revived his hopes and prodded him to do everything possible to liberate the imprisoned President. General García Carneiro knew that inside Fort Tiuna the majority of senior officers supported Carmona, but that junior officers and soldiers whether Chavistas or not refused to accept the coup as valid. Looking over the list of generals who had publicly disobeyed Chávez on the night of 11 April, General García Carneiro noticed that missing from the list was General Raúl Baduel who was the commander of the 42nd Paratrooper Brigade. At Maracay (Map 1) General Baduel commanded the best trained and best equipped unit in the Venezuelan army, and only he had sufficient stature to neutralize the influence army commander General Efraín Vásquez Velasco enjoyed in the military. If General Baduel came out strongly in favor of the Chavista regime, then General García Carneiro felt that a real chance existed of retaking control of Fort Tiuna from the opposition generals, but it would not be easy.[36]

The Carmona provisional government was in trouble but not necessarily doomed. It had lost organized labor and was facing growing rifts in the military, but it still could count on support from the business sector, elite groups, the Roman Catholic hierarchy, and the middle class. And the private media, in particular the TV stations, besides broadcasting false news and hiding information, was constantly and fanatically defending the provisional government. Carmona desperately needed to retain the support of the remaining elements of the opposition coalition to assure the survival of his provisional government. Everything depended on making a successful formal presentation of the provisional government at Miraflores Palace beginning at 1700 hours on April 12. Swarms of office seekers demanding plum jobs crowded the palace since the early morning hours, but for the public inauguration the provisional government invited only the most prestigious members of Caracas society. The first impressions of the gala event were far from reassuring: "For the first time in the past three years, the faces of the basement media room of the Miraflores Palace were almost exclusively white, the suits Armani, the accents those of the Country Club district of Caracas."[37]

At 1730 hours the new attorney general Daniel Romero began reading the 11 articles of the decree establishing the provisional government. After the reading of each article, elite members cheered enthusiastically and shouted slogans. When Romero read article 3 abolishing the National Assembly, the audience enthusiastically acclaimed the measure ending the country's elected legislature with rapturous shouts of ¡*Democracia! ¡Democracia!* The decree continued abolishing all the state institutions of Venezuela including the Supreme Court and in effect made Carmona a dictator with powers vastly exceeding anything Chávez had ever enjoyed. Not only could Carmona remove any government employee (and not just appointed individuals) from any national office, but the authority extended also to all state and local officials. Clearly, the provisional government intended to carry out a major purge of the entire bureaucracy. The decree revealed the real cause of the bitter opposition to Chávez when article 9 suspended the 49 decrees Chávez issued in November 2001. As mentioned in the first section of this chapter, elite groups regarded the 49 decrees as a fundamental threat to their privileges. But the provisional government had invited the high-society members to the ceremony not just to be cheerleaders, and once the reading of the decree concluded, they all agreed to affix their signatures to show the entire world that all of Venezuela was solidly behind the changes. And throwing all discretion away, the first to sign the decree was Cardinal Velasco, who had been instrumental in organizing the opposition and guiding the coup.[38]

Shortly after reading the infamous decree on live TV, Pedro Carmona swore himself in as the interim president of Venezuela, but he may not have realized that by going so far to the right with his pro-business positions, he had struck a possibly fatal blow to his new regime. As news of the impending decree leaked even before the formal reading at 1730 hours, support from two key players, the middle class and the U.S. government, quietly evaporated. The middle class saw too many of its benefits threatened by the rightist lurch of Carmona and found repulsive the dictatorial powers of the provisional government. And although its dislike of Chávez remained as strong as ever, the realization was obvious that the remedy was turning out to be worse than the ailment. Any possibility that the middle class would turn out in large numbers to save the crumbling Carmona regime vanished.

No less significant was the withdrawal of U.S. support. First and most obvious, after U.S. diplomats for months reiterated *ad nauseam* the official position line that any removal of Chávez from power needed to follow constitutional procedures, the U.S. government failed to practice what it preached. In the excitement at the fall of Chávez, U.S. officials forgot to mention in the celebratory statements in the morning of 12 April that the change had not followed proper constitutional procedure. The constitutional bar was high, because Chávez, his vice president, and the chair of the National Assembly all had to resign formally, and then the legislature needed to ratify Carmona as president. But without these indispensable steps the U.S. not only risked bringing a dictator into power in Venezuela but also encouraged the launching of additional coups throughout Latin America. Much as the Bush administration sympathized with the right-wing ideology of the Carmona provisional government, the U.S. government placed a higher priority on stability in the region. And with the deepening U.S. involvement in the Middle East, the last thing Washington wanted was to have to put out fires in its own back yard. Quietly the U.S. government tampered its support for Carmona, and soon U.S. ships in Venezuelan territorial waters returned to the high seas, while the aircraft carrier *George Washington* resumed its routine exercises and was no longer on stand-by status. U.S. military officers who had been visiting Fort Tiuna throughout the crisis quietly faded away. CIA agents likewise stopped supporting the provisional government.[39]

Carmona was left with civilian support only from business groups, the upper class, and the media. The TV stations in their delusions of grandeur believed that they could prop up the provisional government indefinitely, but their arrogance kept them from realizing that their shrill cries against Chávez were increasingly falling on deaf ears. And with Chavista crowds growing in number and in intensity as night fell on April 12, the Metropolitan Police was increasingly overwhelmed and could no longer guarantee control over Caracas. More than ever Carmona needed the army to clear the streets by force and to crush all popular resistance to the provisional government. Realizing that he was falling into a dangerous dependence on the army, he began to fear the political ambitions of General Vásquez Velasco. In what seemed at the moment a masterful political stroke, Carmona eased out General Vásquez Velasco by appointing as defense minister an admiral of lower rank. As in standard practice in armed forces, when a lower-ranking officer was appointed to command senior officers of higher rank, traditionally the latter retire from the institution. Carmona believed that he had eliminated a political rival when in reality he turned Vásquez Velasco into a bitter foe. An outraged general saw his power and influence diminish, and in retaliation he abandoned all efforts to keep the army loyal to the provisional government.[40]

Whether General Vásquez Velasco was planning still another coup to seize power remains unclear, but obviously the bungling by the provisional government created an opportunity. A temporary void in authority appeared inside the armed forces, and into this gap stepped in General Raúl Baduel who at Maracay commanded the 42nd Paratrooper Brigade. The overthrow of Chávez had been largely a Caracas affair, and Maracay was one of many places outside the capital remaining firmly under Chavista control and where state television continued to broadcast pro–Chávez messages. The spontaneous Chavista demonstrations in Maracay and elsewhere in the country emboldened Baduel to draft a manifesto calling for the overthrow of Carmona, and the General publicly communicated his intentions by TV in Maracay at 1335 hours on April 13. His control of the paratrooper brigade, which many observers considered to be the only efficient combat unit in the Venezuelan armed forces, gave his words particular weight and began to sway opinion inside the army. Whether hostile or indifferent to Chávez, officers and soldiers had no intention of going into the battle against the powerful 42nd Paratrooper Brigade.[41]

As news of the open defiance spread by word of mouth among Chavista followers, General García Carneiro ordered the presidential guard to strike. The opportunity came when Carmona scheduled a second ceremony to swear in his new cabinet in the afternoon of April 13. The presidential guard planned to seize him and his cabinet members at one sweep, but the plot was leaked. Panicked officials of the provisional government fled in their luxury cars, while TV cameras filmed this shameful escape as upper class women in high heels scrambled awkwardly to reach their chauffeured vehicles. Apparently Attorney General Daniel Romero was a slow runner and fell prisoner. Among those who escaped was Carmona who sought refuge in Fort Tiuna where he hoped to rally the flagging troops to support his provisional government.[42]

At 1637 hours General Vásquez Velasco went on the air to state that the army could support the new provisional government only if it complied with 12 conditions, in particular the reestablishment of the National Assembly and other state institutions. Carmona, seeing his provisional government crumbling by the minute, promised on television at 1711 hours to reinstall the National Assembly, but it was already too late, and shortly afterwards troops loyal to Chávez secured Fort Tiuna and arrested Carmona. About the same time Chavistas regained control of all the TV stations.[43]

The provisional government ceased to exist, but with so many rumors swirling about the fate of Hugo Chávez, it was not clear whether he was still in the country, had resigned, or was even alive. To show the determination to bury forever the Carmona regime as a blatant usurpation, Chavistas swore in Vice President Diosdado Cabello as interim ruler until the fate of Chávez could be determined. While all this frenetic activity was going on, the captive Chávez had not been seen since going into Fort Tiuna at 0400 hours on April 12. Rebel generals had shuttled him across the country until on 13 April he ended up at the military base at La Orchila Island about 100 kilometers off the coast of Venezuela (Map 1). During his captivity he was kept largely in the dark about all the news.[44]

A series of strange coincidences stopped Chávez from signing his resignation as President of Venezuela every time he was ready to affix his signature. The provisional government was trying to break his will to obtain his resignation but did not want to use force such as torture to compel him to sign. By April 13 it really did not matter whether he signed the resignation or not, but in a desperate attempt to bolster its crumbling position, the provisional government sent Cardinal Velasco to persuade Chávez to sign the resignation as the best way to bring peace to Venezuela. Sending the Cardinal was a masterful maneuver on the part of the provisional government, because the Cardinal was ideally suited to manipulate the deep religious spirituality of Chávez to obtain the resignation. The Cardinal took Chávez aside to pray together in long walks on the beach to seek divine guidance, and apparently the President experienced a profound spiritual conversion.[45]

The captivity at La Orchila was turning into a religious retreat, but before an impatient Cardinal could persuade Chávez to sign the resignation, attack helicopters carrying the best Venezuelan commandos landed on La Orchila to rescue the imprisoned President. The commandoes encountered no resistance, because the majority of the island garrison sympathized with Chávez; indeed, one officer at a crucial moment whispered to the President not to sign anything because help was on the way. The mission of the Cardinal was not a complete failure, however, because the spiritual experience prevented Chávez from returning to power as an avenging angel and instead he turned the other cheek on the coup plotters and all escaped without severe punishment.[46]

Chávez left La Orchila at 0200 hours on 14 April, and at 0400 he was greeted at Miraflores Palace by enthusiastic followers joyfully celebrating his triumphant return. He made a conciliatory statement thanking his supporters and stated that he did not want any witch hunts against the opposition because he was committed to preserving the liberties of all citizens.[47] But the coup experience had left him profoundly shaken, and over the next years he strove to prevent the Venezuelan military from ever again trying to overthrow him. Not able to take out all his anger at the opposition, he increasingly put all of the blame on the United States for instigating the coup. As he solidified his base of support inside the country, he concluded that he could not really be safe until he neutralized U.S. influence. Chávez labored hard to build an anti–American alliance among countries in Latin America and even in other parts of the world as the best way to assure that the U.S. government could never again attempt to block the transformations he was trying to carry out in Venezuela.

2

The Expansion of the Armed Forces of Venezuela

> *Because the Romans did in these instances what all prudent princes ought to do, who have to regard not only present troubles, but also future ones, for which they must prepare with every energy, because, when foreseen, it is easy to remedy them.*—Niccolò Machiavelli[1]

After his inauguration on 2 February 1999, President Hugo Chávez Frías ordered the Venezuelan armed forces to provide assistance to many citizens suffering from poverty. This radical change in mission was revolutionary for a military accustomed to repressing rather than helping the population. As this cooperation intensified, the armed forces forged a close union with the Venezuelan masses. But an additional mission also appeared after the coup attempt of April 2002, when Chávez imposed on the military the major obligation of defending the country from a U.S. invasion. Not surprisingly the dual missions of both defending Venezuela and also helping its people resulted in a major expansion in the armed forces, as this chapter explains.

The Venezuelan Military Until December 2006

Hugo Chávez Frías, as a former Lieutenant Colonel in an elite paratrooper unit, had great pride in the army and wanted to tap its resources to carry out what he named the Bolivarian Revolution. The reputation of the military had been very poor ever since President Carlos Andrés Pérez sent troops to crush demonstrations in Caracas on 27 February 1989. Thousands of protestors were injured and at least 534 died in the bloody repression subsequently called the Caracazo. The close ties of President Pérez to Washington persuaded the U.S. government to downplay the event. For its part the international news media did not feature the massacre as prominently in its coverage as the Tianamen Square crack-down in Beijing, China. But inside Venezuela the Caracazo created in the public the image of the army as a savage and destructive force.[2]

Chávez was determined to restore the prestige of the military, and just weeks after his inauguration on 2 February 1999, he announced his intention to send 70,000 soldiers, or nearly three-fourths of Venezuela's military personnel, into poor neighborhoods and rural areas. The soldiers carried out Civic Action programs such as building roads, repairing buildings, and removing garbage; they also provided medical services and sold foodstuffs at discounted prices from trucks. Besides giving the program the patriotic name of Plan Bolívar after Simón Bolivar, the founder of the Venezuelan nation, Chávez also launched the operation on the symbolic date of 27 February 1999, exactly on the same day as the tenth anniversary of the Caracazo. At first the residents did not know how to react, but when they saw soldiers for the first time coming to provide help rather than repress them, the popular attitude to the military drastically changed. Plan Bolivar continued later under different phases, and its functions became permanent responsibilities of the armed forces.[3] In fact, "the government has prioritized poverty reduction over traditional military expenditures."[4] Not surprisingly, during the rest of Chávez's years in office the military enjoyed a popularity comparable to that of Mexico's armed forces (see chapter 4).[5]

Chávez personally contributed to enhancing the prestige of the military by often appearing in uniform at many public events in the cities, and when visiting the countryside he almost always wore combat fatigues. Chávez knew that he needed to draw on the military to be able to carry out his Bolivarian Revolution. Under the previous regime, soldiers and officers were completely disenfranchised from political life. Inside the military a privileged elite used the soldiers to repress the population while depriving all members of the armed forces of any voice to determine the direction of the country. In a first correction, the 1999 Constitution gave military personnel the right to vote in elections at all levels of government in Venezuela. And the new Constitution restored the right of retired officers to run for elected office and even left open the possibility that active duty officers could be candidates in elections with the authorization of their superiors.[6]

In another major innovation, the 1999 Constitution placed all decisions about promotions in the hands of senior officers and of the president. Under the previous regime, promotions to the ranks of colonels and generals required legislative approval and entangled candidates in a web of partisan intrigue. Instead, the new Constitution opened the door for competent candidates without political connections to reach high rank. Chávez used his new authority to incorporate personnel passionately committed to the social dimensions of his Bolivarian Revolution.[7]

Chávez did not wait for the approval of the 1999 Constitution in December to institute the practice of appointing officers to civilian positions. In his first year in office, one-third of cabinet secretaries were retired or active-duty officers. Military personnel also held key positions at other levels of government. The opposition denounced these appointments as part of a plan to convert Venezuela into a military dictatorship, but the reality was more complex. A shortage of civilian managerial talent in Venezuela crippled Chávez's hopes to carry out many of his reform proposals. Sometimes because of incompetence or corruption, and usually because of inadequate qualifications, suitable civilian candidates were in short supply. To overcome the obstacle of inadequate human resources, Chávez turned to the military. Because of his long career in the army, he knew many officers and preferred to appoint acquaintances he assumed would be loyal.[8] But as the coup attempt of April 2002 showed (see chapter 1), he was not always a good judge of loyalty.

The many unexpected defections during the April 2002 coup attempt shocked Chávez who decided to carry out a much more thorough purge of the senior officers than he had done when he took office in 1999. Predictably he fired all those involved in the coup against him and duly promoted in their places the many officers who remained loyal.[9] But as details of U.S. complicity in the coup attempt came out, he realized that he needed much more than changes inside the Venezuelan military to safeguard the Bolivarian Revolution. As the views of Chávez coalesced during 2003, he came to the conclusion that he was facing in the international arena the same problems he faced inside Venezuela. Chávez in effect extended his original diagnostic to cover not just Venezuela but the entire world. And the problems seemed to be remarkably similar, as privileged elites exploited the masses throughout the globe. Just as in Venezuela elites benefited from policies detrimental to the rest of the population, so in the international arena the U.S. government was the elite power depriving the rest of the countries in the Western Hemisphere of opportunities to prosper.

When the Bush administration launched the most aggressive and militaristic foreign policy in the history of the United States, Chávez regarded the U.S. invasions of Afghanistan and then Iraq as simply the prelude to preemptive strikes against other countries including Venezuela. To Chávez the April 2002 coup attempt had been just the opening gambit of an imperial power determined to use its military might to crush any dissent in the Western Hemisphere. Starting in 2003 Chávez regularly used the term "imperialism" when referring to the United States, because he felt that Venezuela was in grave danger of U.S. invasion. However, the Venezuelan military was woefully unprepared to face a foreign attack. According to the role U.S. officials originally assigned to the Venezuelan military, its duties were counterinsurgency and crushing popular protests. When the last Venezuelan guerrillas petered out in the late 1970s, counterinsurgency became irrelevant, while under Chávez the immense popularity of his regime precluded any internal revolts against the government.[10]

The military needed a new mission, and partly to fill this void Chávez sent soldiers to carry out Civic Action programs throughout the country. However, in his initial years in office he did not foresee any increase in the military, and actually defense expenditures declined in 2001 and 2002 as he shifted government spending toward poverty reduction. The trend continued into 2003, but in 2004 military expenditures rose when he realized that Washington was determined to overthrow his regime. In a final attempt to avoid the need for military expansion, on 29 February 2004 Chávez tried to use petroleum as a weapon when he threatened to interrupt the export of petroleum to the United States in case of any invasion or threat of an invasion. A sudden halt to oil deliveries would certainly cause momentary disruptions in the United States, but in reality because of a lack of comparable markets for its petroleum exports, ultimately Venezuela would suffer the most from any petroleum boycott. Although for political purposes Chávez continued to talk about stopping petroleum exports, Washington accurately dismissed the threat as silly theatrics.[11]

When the Bush administration responded to the bluff by increasing funding for the political opposition, Chávez concluded that he had no choice but to prepare for the inevitable U.S. invasion. By 2004 the internal debates inside his coalition reached the conclusion that the Bolivarian Revolution needed to adopt Socialism as a fundamental component. Chávez talked increasingly about Socialism in his speeches, and in one of his *Aló Presidente* Television talk shows, he appeared next to a big book bearing in large letters the word "*SOCIALISMO.*" Actually his paramount consideration had been the need to use nationalization as a political weapon to weaken powerful enemies in the private sector. More out of political necessity than from ideological commitment to Socialism, Chávez steered his Bolivarian Revolution to the left as the government expropriated many private businesses. But no wholesale nationalization of the entire economy took place, because Chávez remembered all too well how state-owned enterprises could easily fall under the control of privileged managerial elites. Although he continued to seek alliances with those business leaders staying away from the opposition, at least the threat of nationalization helped to keep many private investors from turning into bitter opponents. But because previously he had not announced Socialism as a formal policy, his declaration in January 2005 that he was a socialist became a rallying cry for opposition figures who for years had repeatedly accused him of wanting to establish Cuban-style Socialism in Venezuela. To the opposition this January 2005 declaration merely followed and duplicated Fidel Castro's similar statement in April 1961, but in reality Chávez continued to talk about creating a different type of Socialism appropriate for the twenty-first century.[12]

The relentless accusations about Socialism from the opposition and the threat of a U.S. invasion made Venezuela change its national defense policy. Once Chávez increased military expenditures in the budget, he implemented four main measures: (1) the creation of a large force of reservists; (2) arms purchases abroad; (3) the adoption of the new military doctrine of asymmetrical warfare; (4) fostering the loyalty of the military to his regime.

The Creation of a Large Force of Reservists

Chávez publicly launched the new anti-imperialist phase of the Bolivarian Revolution with a major rally in Caracas on 16 May 2004, and in his televised speech he announced the need to create citizen soldiers in the country. Clearly Chávez had done his math, and he realized that expanding the army to a size capable of repulsing a U.S. invasion would place a crushing burden on the economy. The cost-effective solution was to create militias, but he avoided using the term militia for fear of antagonizing career officers. Defining the exact nature of the new force consumed nearly a year, and only on 4 April 2005 did he issue two decrees organizing army reserves. He appointed retired General Julio Quintero Viloria to be in charge of the reserves and to report to the President directly rather than through the Ministry of Defense.[13]

On his *Aló Presidente* Television talk show Chávez claimed on 3 April 2005 that "the new military reserves will number more than 1.5 million and the rest of the population will be recruited to help defend the country in guerrilla warfare."[14] The astronomical figure seemed beyond reach, and even General Quintero Viloria in that same Television program explained that he was beginning with only 80,000 reservists, mostly former service members. A confidential estimate calculated the entire reserves at a maximum of five thousand and one army colonel explained "that if the military could not keep regular troops trained, it would seem unlikely that it would be able to prepare millions of reserves for combat."[15] In his *Aló Presidente* of 3 July, Chávez claimed that the reservists were reaching 500,000 in number, although their more modest commander General Quintero Viloria announced the training of only 50,000 during the second half of 2005. The government "by offering modest payments and free lunches, does not appear to be having difficulty recruiting reserves" and pollsters reported "that one-third of the Chávez supporters interviewed said they were prepared to join the reserves to fight the United States."[16] Independent of the actual number of reservists, the campaign was giving Chávez a very favorable boost in popularity as he headed toward the presidential election of December 2006. And for that election, the armed forces were able to deploy 19,000 reservists to guarantee normality in the electoral process.[17]

Arms Purchases Abroad

Because both the enlargement of the reserves and the improvement of the military required additional weapons, Venezuela starting in 2004 went on a buying spree and by 2007 became one of the largest purchasers of Russian weapons. Chapter 6 discusses the efforts of Venezuela to acquire armament abroad in defiance of U.S. obstacles. From the perspective of the individual soldiers, the most important acquisition was that of 100,000 AK-103 automatic assault rifles. Although some believed these new arms would go to the units of citizen soldiers, in reality they were needed to replace the worn out Belgian FAL rifles of the regular army. The discarded weapons temporarily equipped the citizen soldiers until Venezuela started to manufacture its own rifles at a new factory. In March 2008 machinery from Russia arrived to set up a production facility for weapons.[18]

These arms purchases were also the most visible sign of the rising military expenditures. By 2004 when Chávez finally gained full control over the revenues of PDVSA, the state petroleum monopoly, he was able to obtain funds to cover the costs of both his social programs and military expenditures. Because of serious fiscal constraints, the amounts spent on weapons were never as great as some American observers believed, and many contracts and letters of intent were never fulfilled. Chávez, by announcing very large arms purchases, hoped to deter the United States and its proxy Colombia from trying to invade Venezuela. New weapons did arrive, but just enough to keep the image of a large arms buildup credible.

New Military Doctrine of Asymmetrical Warfare

In the face of hostility from the Bush administration, the Venezuelan military needed a new doctrine to replace the outdated counterinsurgency concepts imposed by the United States in the 1960s. The best way to introduce the new doctrine was to break completely with the old one, and Chávez sought to end all bilateral military relations with the U.S. government. Besides refusing to continue the routine visits of U.S. officers to Venezuela, in a dramatic public gesture he expelled the U.S. military training mission from its offices in Fort Tiuna in May 2004. Likewise he cancelled exchange programs for training Venezuelan officers in the United States in April 2005. He sought to isolate Venezuelan officers from their American counterparts and even those in diplomatic positions learned not to appear too friendly with Americans.[19] He put Venezuelan officers to study and apply the doctrine of "asymmetrical" warfare, which refers to the strategy a small country has to follow to defeat an invasion by a large country. "The armed forces have held several conferences to plot strategies for war against an overwhelming conventional force such as the U.S. military."[20]

But for the Venezuelan armed forces, asymmetrical warfare did not exist in a vacuum and formed an integral part of the military campaign to improve the social conditions of the people. Only by combining the population and the military in a war to maintain national liberation could asymmetrical warfare have a chance of wearing down the invader. Reservists were the key link between the population and the Venezuelan military, and they started to participate in regular army maneuvers.[21]

The doctrine of asymmetrical warfare dated back to the Vietnam War, and Socialist Cuba had published many texts in Spanish explaining and elaborating this doctrine. Cuba provided textbooks and limited training to a small number of Venezuelan officers, but reports of extensive Cuban involvement in the Venezuelan military turned out to be vastly exaggerated. As the U.S. Embassy commented, "Post has received no credible reports of extensive Cuban involvement in the Venezuelan military, despite the Venezuelan Armed Forces' attempts to imitate Cuban military doctrine and uniforms."[22] The replacement of American-style uniforms with attire similar to Venezuelan uniforms of the eighteenth century had sparked the unfounded rumors about Cuban involvement because the new clothes resembled those used by the Cuban military.[23]

Fostering Military Loyalty to the Regime

After the traumatic experience of the April 2002 coup Chávez was particularly sensitive to any reports of discontent within the military. He devoted considerable efforts to guarantee the loyalty of the army and to persuade military personnel to become enthusiastic backers of the Bolivarian Revolution. He intensified the practice dating back to his first year in office of appointing senior officers to civilian positions. Whether retired or active duty, and the distinction was often left deliberately vague, 25 percent of his cabinet secretaries, 30 percent of the vice ministers, and 39 percent of the governors came from the military in 2006. Not only these high-level appointments but many more for junior officers carried the immense benefit that officers received both their military and civilian salaries, a particularly valuable perk because for five years military pay scales had been frozen. For officers of lower rank, whenever deciding upon promotions, Chávez gave more weight to their performance in civilian jobs than to traditional military criteria. Because in 2006 Chávez completed approving four promotion cycles since the 2002 coup, he already had ample opportunity to shape the officer corps according to the principles of his Bolivarian Revolution.[24]

This ideological transformation of the military generated discontent among mid-level personnel passed up for promotions. These were the officers who held the ranks equivalent to captain, major, and lieutenant-colonel in the U.S. army. Their subordinates were undergoing intense indoctrination in the social principles of the Bolivarian Revolution, while their peers received promotions to higher rank because of loyalty to Chávez. Older lower-ranking officers were the ones who resented the most the new Cuban-style uniforms and preferred their previous U.S.–style uniforms. A more serious source of discontent was that in five years the government failed to readjust military salaries to keep up with the cost of living increases. In a calculated maneuver, in May 2005 the government announced that salaries would remain frozen for the rest of the year. This announcement sparked considerable grumbling; disgruntled officers later claimed that loyalists used reports about who had complained to determine assignments and promotions.[25] And then suddenly in a speech on 24 June 2005, "Chávez announced a 60 percent salary increase for cadets and junior officers, and a fifty percent increase for officers and senior enlisted personnel."[26]

Chávez needed to justify the changes inside the armed forces and the shift toward Socialism to the population if he was to overcome the bitter backlash from the opposition. The urgency was great, because as the year 2005 was winding down, Venezuela was fast approaching presidential elections on 3 December 2006. Chávez could not allow the opposition to destroy his Bolivarian Revolution at the polls. And the electoral tactics he used often had a strong military component.

The Military and the Presidential Election of 3 December 2006

Depicting the opposition as unpatriotic has always been a reliable campaign tactic in electoral systems across the world. The opportunity came early in 2006 when the public learned that the opposition candidate was regularly meeting with U.S. diplomats including the American ambassador. This revelation gave Chávez the opportunity to depict his rival as an unconditional ally of the United States. The President created the impression that the opposition candidate Manuel Rosales, who was the governor of the traditionally conservative and petroleum-rich state of Zulia (Map 1), was receiving orders from the U.S. Embassy at Caracas. To add fuel to the fire, Chávez in the 5 March 2006 television broadcast of his *Aló Presidente* depicted Rosales as part of a movement seeking independence for Zulia from Venezuela. Chávez even called on PDVSA employees in Maracaibo to join the army reserves in order to protect the refineries in that city from sabotage. The accusation about a Republic of Zulia proved very effective in discrediting the opposition, even though concrete evidence to back the charges never appeared.[27]

Easy as it was to question the patriotism of his opposition candidate, Chávez, a master politician, was striving to create a much more significant choice for voters in the December election. "Chávez continues his efforts to treat the United States as the opposition candidate on the Venezuelan presidential ballot."[28] Reducing the election to voting either for the United States or Venezuela was the ideal situation for Chávez, but even he knew that he could not pull off such a bold maneuver without decisive help from Washington. The Republic of Zulia accusations could go only so far, he needed direct responses from the U.S. government to stir up nationalistic feelings in Venezuela.

The U.S. government took a first timid step when the amphibious assault ship USS *Saipan* visited the Dutch island of Curaçao (Map 1) on 27 February 2005 as part of routine maneuvers. Chávez and the official Venezuelan media described the ship as an aircraft carrier and soon depicted the visit as a provocation and as a prelude to an imminent invasion. However, the USS *Saipan* incident came too early in the election cycle to be remembered for long by voters with short memory spans, and photographs soon confirmed that this U.S. Navy ship was too modest to mount a credible invasion threat. Into the fray stepped the Dutch ambassador to Caracas who was delighted with the visit of the U.S. Navy ship and on 21 June 2005 urged both his government and Washington to make a "Port call by U.S. Navy Carrier Battle Group" because this "would send a very clear signal to Chávez, a man who does not catch subtle messages. Chávez accused us of sending a carrier four months ago during the USS *Saipan* visit; why not send a real one this time?"[29] The proposal of gunboat diplomacy and saber rattling reflected the nineteenth-century era of colonial empires, and the U.S. Embassy in Caracas wisely did not go beyond forwarding the Dutch ambassador's suggestion to Washington.

The Bush administration did feel at home in the age of colonial empires and decided to flex its muscles in early 2006, just as the electoral campaign was heating up in Venezuela. The U.S. government planned to blanket the Caribbean islands with warships supposedly on goodwill and humanitarian missions. The Bush administration heartily endorsed the ideas of the Dutch ambassador and deployed a carrier battle group to the Caribbean Sea. And the flagship of the battle group was none other than the USS *George Washington*, the aircraft carrier that appeared intent on finishing in 2006 what it had started to do during the April 2002 coup attempt. Either U.S. officials missed the symbolism of deploying this carrier already well-known to the Venezuelan public or more likely wanted to send a crystal clear message of the serious consequences that criticizing the U.S. government always brought.

The U.S. Embassy at Caracas had the obligation to oppose the deployment of this aircraft carrier because it played right into the hands of Chávez. Instead, perhaps afraid of not being considered a team player or fearing reprisals over promotions and assignments, the U.S. Embassy did a remarkable *volte-face* from its earlier position and twisted the gunboat diplomacy into a great opportunity to reinforce the relentless propaganda message that Chávez was a mad dictator. To prompt him into taking ridiculous behavior, the Embassy believed that "timely notification that the Ambassador will visit the *George Washington* may contribute to Chávez paranoia....

If Chávez takes the bait, the deployment will expose the international community to apocalyptic statements and Venezuela-centric views of the region that make Chávez appear at best silly and at worst clinically paranoid in the eyes of many observers."[30] Resoundingly the U.S. Ambassador added his own personal endorsement of the carrier battle group: "This is a win-win for us. The Strike Group does its operational training; we advance drug and humanitarian interests in the region; and we give Chávez opportunities to make a mistake by feeding his paranoia."[31]

The one that fell into its own trap was not Chávez but the U.S. government. Without the need for any histrionics or theatrical performances, the Venezuelan media reported in detail the deployments of U.S. warships. Venezuelan officials had little to do to finish framing the presidential election of December 2006 as a referendum on voting either for Venezuela or for the United States, and in personal terms the election was between voting for Chávez or for Bush. As the poll numbers reported on the plummeting popularity of the opposition candidate and the growing victory margin of Chávez, it was too late to correct the blunder U.S. officials had committed in propelling the Venezuelan President to a landslide victory.

All that remained was to arrange the final details of the 3 December 2006 elections. Chávez counted on the members of the armed forces not just to vote for him as they had done in the mega-elections of 2000 but also to display enthusiastic support for the Bolivarian Revolution. Chávez wanted to show the Venezuelan people that the entire armed forces, in the fulfillment of their patriotic obligations, were solidly behind Chávez as the only candidate willing and able to defend the independence of Venezuela from U.S. invasion. The ostensible role of the military was to safeguard polling stations and to transport voting equipment and materials in a secure manner, but their presence conveyed the subliminal message that the soldiers were all behind Chávez. To accomplish its polling duties, the military deployed 128,000 soldiers, but because active-duty personnel did not exceed 100,000 in number, the high command completed the quota by drawing on reservists. Regulations prohibited military personnel from checking the identification cards of voters, but the traditional laxness and inefficiency of the civilian bureaucracy gave soldiers an opening in rural regions and poor neighborhoods. Soldiers available right there with nothing to do in every one of the 11,000 polling stations inevitably participated in the voting process as replacements for the many civilian employees who came late or not at all, went for leisurely meals, or left before the polls closed at the end of the day.[32]

Chávez wanted the military to do more than simply guarantee a smooth electoral process, but not all officers shared his vision of a wider role for the military. The divergence came out in public when a leak to the press first disclosed that all executives of PDVSA had been told to support the reelection or else lose their job. Chávez weighed in by saying that both PDVSA and the military were red, very red (*rojo, rojito*) in allusion to the campaign colors of his candidacy. This was the cue for Defense Minister General Raúl Baduel, one of the saviors of Chávez during the 2002 April coup, to follow the official line on the eve of the presidential election in December 2006. However, the Defense Minister publicly equivocated in expressing full support. The rift inside the military was now out in full view, but it was too late in the electoral cycle to take drastic remedial action, and Chávez tolerated the ambivalence at least for the moment.[33]

General Baduel was really behind the curve in supporting the Bolivarian Revolution, when already Chavista sympathies were rampant throughout the military and not just among the reservists. Many soldiers and officers were wearing red t-shirts instead of the usual white ones under their combat fatigues and uniforms, a practice sure to impress voters at the polling stations. The National Military Academy, Chávez's alma mater, placed in front of its main building the slogan "Cradle of the Bolivarian Revolution." Even the National Guard was trying to improve its image sullied by the participation in the 2002 April coup; among other actions, the National Guard inflated a huge image of Chávez over its barracks at the Maiquitía International airport serving Caracas.[34]

Chávez tried to leave nothing to chance in the election and hurried the completion of major public works. When told that the 15 percent increase in the minimum wage of February 2006 had not been adequate, he announced an additional increase of 10 percent in late September. Not surprisingly, he coasted to a landslide victory in his reelection bid of 3 December 2006 and in many ways this moment marked the high point of the Bolivarian Revolution.[35] After Chávez successfully concluded his first full six-year term, the U.S. government could only watch anxiously the start of his second six-year term.

3

Unconditional Allies of the United States

Diplomacy is the police dressed up in gala clothes.—Napoleon Bonaparte[1]

The three unconditional allies of the U.S. government during the first decade of the twenty-first century were Colombia, El Salvador, and Guatemala. The continuous alliance of Colombia with the United States dated far back into the nineteenth century, while the alliance of El Salvador was slightly less old. Instead the relations of Guatemala with the United States had suffered major interruptions during the twentieth century. Before 2000 the Guatemalan army had blocked close relations, but as the influence of this institution declined, the Guatemalan government became an unconditional ally of the U.S. government.

Guatemala: Settling Old Scores with the Army

The need to block an insurgency dating back to 1961 forced the upper class to tolerate a large military establishment in the country during the second half of the twentieth century. But as the guerrilla threat receded into insignificance in the mid–1990s the momentum to end the war and to reduce the size of the military grew. Peace negotiations with the guerrillas dragged on, and as one way to speed up the peace process, the army reluctantly accepted the abolishment of the military commissioners (*comisionados militares*) and the Civil Patrols under their control in September 1995. For decades these armed civilian officials, usually veterans, had provided a crucial link for maintaining state authority over Guatemala, particularly in the countryside. Human rights activists hailed the demobilization of these 24,000 officials and the disbandment of the Civil Patrols as decisive steps toward peace and justice in Guatemala. Historical perspective shows that the abolishment of military commissioners turned out to be not just the first major step in the breakdown of state authority but also the decisive step in the descent of Guatemala into savage criminal violence.[2]

As long as the Guatemalan military remained intact the government still had the possibility of preserving order and stability throughout the country. Unfortunately for Guatemala, the upper class, human rights activists, and the U.S. government pressed hard to reduce and preferably to eliminate the Guatemalan military. The upper class resented the arrogance of the haughty army officers and wished to reduce the burden of financing a large military establishment. Human rights activists sought punishment and vengeance against the military because of the genocide committed during the 1980s when over 200,000 persons died. The U.S. government claimed to share this outrage but in reality was motivated by the desire to destroy an institution that had openly defied Washington during the campaign against guerrillas of 1975–1996. Not only had the army refused to conduct the counterinsurgency according to U.S. instructions but had brazenly carried out the worst genocide against Indian peoples since the independence of Latin America. Shocking as were the human rights atrocities, it was the open defiance that most angered U.S. officials. The U.S. government, the upper class, and human rights activists united to obtain in negotiations what guerrilla groups had not been able to win in combat. The Peace Accords of 29 December 1996 ended the insurgency and more significantly for the later history of the country also stipulated a 33 percent reduction for the military. The army, accustomed to wielding enormous power and influence over the country's life, resisted bitterly these mandated cuts. Only two reductions took place, and the army still numbered 31,423 soldiers in 1997.[3]

A weakened army deprived of military commissioners and Civil Patrols could only watch as the country slid into lawlessness. No sooner had the war ended than lynching and other acts of spontaneous violence became frequent in Guatemala. Highway robberies previously unknown in the countryside under military rule multiplied as criminals enjoyed their new freedom. Theft of crops and even new crimes such as the wanton destruction of coffee and corn plants spread in rural areas. Already in the late 1990s drug traffickers and armed gangs were filling the void left by the receding military presence. The worsening crime situation did not worry civilian politicians, U.S. officials, and human rights activists who actually welcomed the decline of military control as an essential first step to provide the political space for the spread of democratic practices. The assumption was that a revitalized police could maintain order, but in the history of Guatemala the police had been the most corrupt and inefficient institution. It did not seem clear how the new National Civilian Police created under the Peace Accords of 1996 could replace the military in the task of controlling crime throughout Guatemala.[4]

The growing crime wave persuaded the majority of citizens of the need for a strong military presence, and voters rejected the May 1999 constitutional reforms that would have prohibited the army from engaging in any police functions. Until a solid and well-functioning police force could emerge, everything argued for preserving the armed forces as at least a temporary bulwark to prevent Guatemala from falling into anarchy. But the Guatemalan government, under pressure from U.S. officials, human rights groups, and the upper class, relentlessly pursued the goal of eliminating the army. As soon as newly elected President Alfonso Portillo took office in 2000, he promptly fired all the generals, a reasonable measure given that the army had too many. This firing also made possible the promotion of some colonels into the rank of general. The number of colonels was still too large, and the government intended to dismiss most of the remaining colonels. But the officer corps issued warnings that such a drastic reduction risked triggering a military coup unless the government offered generous compensation packages for the dismissal of officers in all ranks below general. A scared government backed off and contemplated alternative solutions. The ranks of major through colonel were excessively bloated and, in an anomaly for Latin America, so was the number of Non-Commissioned Officers (NCOs). Retirement was mandatory only after 33 years of service, so attrition was too slow a mechanism to reduce the large number of officers and NCOs. Lack of funds to cover the severance packages delayed any substantial reduction until 2003, and only in that year did the treasury succeed in floating bonds in foreign markets to cover the costs of compensating the military personnel.[5]

In 2003 the Portillo administration offered severance payments and pension benefits to military personnel who took early retirement. Many officers below the rank of colonel accepted the offer, and in the best response of any rank, nearly half of NCOs took the early retirement packages. The army automatically granted the requests for early retirement to all officers except for pilots. The biggest disappointment came at the rank of colonels, because most were so close to mandatory retirement age that staying in the ranks for a couple of more years gained them virtually the same benefits. The Portillo administration succeeded in reducing the size of the military from the 31,423 to 27,000, of whom about 3,000 were civilians.[6]

The U.S. government, human rights activists, civilian politicians, and the upper class were in agreement about the need to reduce and preferably eliminate the Guatemalan military. To be an unconditional ally of the United States, a country could not just share the same goals; client states also needed to seek and follow instructions from U.S. officials. The defiance of the Guatemalan army between 1975 and 1996 required punishment both on its own merits and also to send a strong message to the other countries of the region about the dangerous consequences of opposing U.S. instructions. With the election of Óscar Berger to the presidency in 2004 the possibility appeared of completing the task of disbanding the army of Guatemala. As U.S. Embassy officials exclaimed "now that we have a government attuned to our interests ... we have in hand a perhaps one-time opportunity to partner with the Berger administration to positively influence the future of one of [the] hemisphere's most insular and stodgy militaries."[7]

The military was on the defensive because already in the 2003 presidential campaign candidate Berger had been making statements about replacing the army and the police with a constabulary or a gendarmerie-style force. The army bitterly opposed this proposal to merge with the despised police, but it was not just military opposition that blocked the proposal. The inefficiency and corruption of the Guatemalan police were so well known that abolishing the police and recruiting a brand-new force was the only realistic solution. When it was learned that the new police-army constabulary would be called the National Guard, the red flag was too obvious even for human rights activists who feared a body reminiscent of the hated National Guard of Dictator Anastasio Somoza in Nicaragua. As much as U.S. officials wanted the Guatemalan army to disappear, even they questioned the advisability of merging the army with the police. In a final attempt to preserve something of the merger, President Berger proposed abolishing the Defense Ministry and placing both the army and the police under the civilian Ministry of the Interior, but the army blocked even this scaled-back plan.[8]

The U.S. government wanted drastic reductions in the military and strongly supported the Berger administration in its negotiations with the army. When the President proposed that the number of soldiers drop from 27,000 to 14,869, an outraged Defense Minister threatened to resign. Even milder disputes in previous decades had ended with a military coup d'état, but this time the public opposition of the U.S. government to coups in Latin American countries made the generals hesitate. Realizing where his real strength came from, President Berger requested and obtained categorical support from the U.S. Embassy in his struggle with the generals. The Defense Minister wavered in the face of U.S. pressure, and he made a counter proposal of reducing the army to 16,000. Seeing that the army was giving a signal to negotiate rather than to stage a coup, the President agreed to the figure of 15,500. The Defense Minister, preferring to live to fight another time, accepted the compromise.[9] For President Berger this seemed to be a substantial accomplishment, because he had succeeded in reducing the large expenditures of "the most expensive military in Central America."[10]

By 30 June 2004 the army declined in size by 67 percent in comparison to the force existing at the end of the war in 1996. Actually, this drastic cut was twice as much as the 33 percent reduction stipulated in the Peace Accords of 1996.[11] The deterioration affected all aspects of the armed forces:

> The Guatemalan military suffers from an inventory of aging and obsolete equipment in its ground, air, and naval forces.... The military also has little funding for procurement of spare parts, which in many cases are expensive or even non-existent because of the age of the equipment. For example, of the six UH-1H and Bell 212 helicopters in the Guatemala air force inventory, only two are currently operational. All three of the air force's A-37 ground attack aircraft are currently grounded due to a lack of parts. The Guatemalan navy shares some of the air force's readiness problems but is even more handicapped by a lack of funds to pay for fuel costs. The lower-tech ground forces are less constrained but the army's truck fleet is dilapidated because of a lack of spare parts and maintenance funds.[12]

The decline in the military was so drastic that a nationalistic public feared that the real aim was to abolish the army altogether; some critics even blamed the U.S. government for wanting to destroy this institution. But even U.S. officials were realizing by 2006 that they may have miscalculated in slashing the Guatemalan military as punishment for its past defiance. Reducing the size of the military resulted in a collapse of law and order as the police proved incapable of dealing with the crime wave already reaching epidemic proportions across the country. President Berger began to soften his opposition when he explained that the knowledge and experience of former army privates made them excellent candidates for incorporation into the police. However, he still remained opposed to recruiting former army officers into the police force.[13]

The President had to abandon his opposition to army officers because ex-soldiers were willing to return to duty only under the command of army NCOs and officers. At a moment when the crime wave was intensifying, the National Civilian Police could not attract sufficient candidates and suffered from personnel shortages. The example of Haiti collapsing into anarchy because of the failure to organize an adequate police force (see chapter 12) served as a warning to U.S. officials. As the country seemed on the verge of collapse because of the crime wave, gang violence, and drug trafficking, U.S. officials softened their attitude toward the Guatemalan army from one of strong opposition to grudging acceptance. When the government fired 2,500 members of the National Civilian Police because of corruption and criminal activities, President Berger had no choice but to turn to the military to fill the gap. In a first guarded step, he announced the creation of a special force of 3000 former soldiers. Only reservists with at least a high school degree could join this force, and its commanders were 72 active-duty officers plus 70 officers recalled from retirement. These reservists received additional training on human rights and on dealing with civilians and were deployed in six units of 500 men each. Although all the officers were from the army, this special force was under the operational control of the National Civilian Police.[14]

The partial rehabilitation of the military came too late to save the reputation of the Berger administration, which was unable to stop the rampant crime wave afflicting the country. The collapse of authority gave leftist groups the opportunity to run Álvaro Colom as candidate for president. Campaigning on crime as one of his key issues, the moderate leftist Colom won the elections and for the first time since the overthrow of Jacobo Arbenz in 1954, the country had a president who did not come from the right wing of the political spectrum. And the arrival of a leftist president promised great changes for the military, as chapter 8 will show.

Rebuilding the Army of El Salvador

El Salvador along with Guatemala and Colombia were the trio of the unconditional U.S. allies in the Caribbean. Unlike Guatemala whose close ties to the U.S. government only blossomed after 2000, El Salvador's alliance dated back to the 1960s. Already on 1 January 2001 El Salvador reinforced the close links by adopting the U.S. dollar as its currency. The high point of the relationship with the United States came with the inauguration of Tony Saca in 2004 who headed "an unabashedly pro–American administration."[15] Quite naturally, El Salvador enthusiastically joined CAFTA-DR, the Central American Free Trade Agreement with the United Sates.[16] By

2007 the U.S. Embassy was ecstatic about the relationship "simply put, El Salvador is arguably our closest friend in the Western Hemisphere,"[17] even though the administration of Álvaro Uribe Vélez in Colombia could make an equally convincing claim to that title.

While the army of Colombia continued to increase in size to try to defeat guerrillas, the end of the war in El Salvador brought a substantial reduction in its discredited army. The Peace Accords of 1992 slashed the army to 12,000 soldiers and 2,500 civilian personnel. The Salvadoran army slowly regained its prestige in the 1990s because of its participation in relief programs to help areas devastated by natural disasters. In the initial years of the twenty-first century polls showed that the Salvadoran army enjoyed high approval ratings.[18]

Collaboration with U.S. troops on humanitarian missions and for the distribution of relief supplies were activities boosting the popularity of both Salvadoran and American soldiers. The presence of U.S. military personnel on foreign territory required signing a Status of Forces Agreement (SOFA) between the U.S. government and the host country. The agreement granting immunity to American soldiers usually was simply a contractual letter between both governments valid for one or several years. As one more indication of its close alliance with the United States, El Salvador became the second country in the Western Hemisphere (after Trinidad and Tobago) to give formal legislative approval to SOFA. With this blanket legislative permission, the U.S. government could then automatically meet any request by the Salvadoran government for U.S. troops to participate in humanitarian missions.[19]

The George W. Bush administration had bigger plans for the Salvadoran army than merely participating in relief functions during natural disasters. In a desperate search for allies to join what was then called "The Coalition of the Willing," the Bush administration was frantically trying to persuade countries to contribute troops to the Iraq war. El Salvador was one of only four Latin American countries that provided modest detachments. Nicaragua maintained a unit for less than a year in the Middle Eastern country. And after the departure of the units from the Dominican Republic and Honduras in 2004, El Salvador remained as the only Latin American country with troops in Iraq.

The deployment of an annual contingent of nearly 400 Salvadoran soldiers was extremely unpopular in El Salvador. The government was always able to secure legislative approval to keep the battalion deployed for one more year in Iraq by claiming that it needed that leverage to secure favorable consideration on other vital matters affecting the country. Although U.S. officials disingenuously claimed that the participation of the Salvadoran battalion was separate from the rest of the issues in the bilateral relationship, the Salvadoran troops earned considerable good will in the Bush administration. And when the White House had to decide each year whether to extend the Temporary Protected Status (TPS) to hundreds of thousands of otherwise illegal Salvadoran immigrants, invariably the Bush administration signed the decree renewing TPS for another year. In a cruel reality, those Salvadoran soldiers wounded and dying in combat were paying with their blood so that fellow citizens could stay in the United States for another year.[20]

U.S. officials were so pleased with El Salvador's unconditional support that in 2006 a Department of Defense official inquired to find out whether the government would be interested in acquiring the status of "Major Non-NATO Ally" (MNNA), a prestigious classification the U.S. Congress created in 1989.[21] The U.S. government ranks foreign countries in a continuum beginning with NATO ally as the closest and ending with hostile states at the other end. If the executive branch wanted a closer alliance, then usually the next step was either to incorporate that country into the treaty system of NATO or to negotiate a separate bilateral treaty for congressional ratification.

Legislative approval in both countries could be cumbersome, and as an alternative, granting MNNA status was an ingenious way to reward countries already working closely with the United States. Among other benefits, MNNA expedited the shipment of defense materials and facilitated the training of military personnel. In reality, MNNA status made the local government more submissive to the United States and was more demanding than NATO membership. So while the U.S. government could not expel member states, such as Britain and France, from NATO, Washington could easily strip an uncooperative country of MNNA status. President Tony Saca was fascinated with the possibility of MNNA and sought earnestly to gain that status as much as a dedicated student seeking to receive the highest grade from a teacher, but El Salvador never received the coveted prize.[22]

Despite the unbounded enthusiasm of the Saca administration for intensifying ties with the U.S. government, all was not well in El Salvador. Right-wing business elites traditionally controlled economic policy in the country, and they enthusiastically embraced free market ideas of the Bush administration such as the CAFTA-DR trade agreement. The result was a concentration of income in the upper class and the impoverishment of the mass of Salvadorans. The worsening income inequality fueled the migration of over a million citizens who sought to ease their economic hardship by working as illegal laborers in the United States. As a partial solution to the plight of many Salvadorans living illegally in the United States, the government of El Salvador insistently sought and obtained the renewal of Temporary Protection Status (TPS) for its citizens.

Another consequence of the deteriorating economic situation in El Salvador was the crime wave sweeping the country. Drug traffickers found in El Salvador a convenient place to make transshipments of cocaine coming from Colombia to the United States, and soon rival street gangs were fighting for control of drugs, while many other criminal activities multiplied. TV broadcasts and newspaper headlines were full of accounts about "the epidemic of gang violence" and "in 2005 El Salvador eclipsed Colombia as the country with the highest homicide rate per capita in the Western Hemisphere."[23] Because of the army's human rights violations during the insurgency of 1979–1992, the Peace Accords of 1992 sharply reduced the army's contact with the civilian population to avoid a repetition of the atrocities committed during the civil war. Responsibility for controlling crime passed solely to the new National Civilian Police. But when the Peace Accords of 1992 created this body, political considerations prevailed over crime prevention. The Peace Accords stipulated that 20 percent of the members of the new institution had to come from former guerrillas, 20 percent from former government security personnel, and the rest would be new recruits. Filling the guerrilla quota proved difficult because many former fighters could not meet the minimum educational requirements, and by 2006 their percentage stood at 14 percent. The recruitment of former security personnel was more successful, but their percentage declined to 18 percent by 2006.[24]

In the new professional force all new hires had to graduate from the police academy. Meeting the recruitment quotas proved impossible, and the National Civilian Police was never able to fill its ranks and thus could not patrol the country effectively. Precisely in the rural areas and neighborhoods lacking a police presence the gangs first took root and then gradually spread across El Salvador. Actually, the National Civilian Police was shrinking. For example, in an eight-month period in 2006, the academy graduated 534 cadets while the police force lost 575 individuals. The reasons why the officers left were very revealing: "307 resigned, 221 fired for corruption, 39 incarcerated, 8 died in the line of duty."[25] And of those who resigned, a substantial number became illegal immigrants in the United States.

3. Unconditional Allies of the United States

The U.S. government had the obligation to provide extensive assistance to the National Civilian Police if only for the purpose of reducing the migration of former policemen to the United States. Instead the U.S. government worsened a bad situation by accelerating the deportation of Salvadoran inmates back to their native country. The return of these hardened gang members usually from California increased the violence in El Salvador. One way the U.S. government could have helped El Salvador was by keeping these violent individuals in American prisons, because otherwise without any charges pending against them in El Salvador, they were free to join the gangs and participate in the drug trade to the United States. In contrast to the steady increase in the number of gang members, the National Civilian Police was slowly shrinking in size.[26]

The impact of the sharp reductions in both military and police strength after the war was immediately felt in the countryside. As early as 1992 citizens called on the army to patrol the countryside, and beginning in 1993 the government routinely deployed soldiers to guard highways and to protect the coffee harvest. *Plan Guardián* in March 1995 instituted joint army-police patrols as a permanent practice in the countryside. Human rights activists were quick to denounce these army patrols as a violation of the Peace Accords. In contrast, the population largely welcomed the soldiers and even clamored for greater military involvement in stopping the growing crime wave.[27]

The start of the twenty-first century witnessed a major expansion of violence in urban areas. Street violence reached unheard of proportions, with gangs robbing passengers aboard public transport and sometimes burning the buses. The National Civilian Police created special anti-narcotics and anti-gang units, but because those detachments removed policemen from street patrols, inevitably criminal activity increased in the less policed neighborhoods. Seeing the ineffectiveness of the police, the government used a constitutional loophole to adopt its "iron fist" plan (*Plan Mano Dura*) on 23 July 2003 against crime. Under this tough anti-crime policy, the soldiers patrolled permanently not just the countryside but also all urban areas. Supposedly the army would join with the police in carrying out joint patrols, but commanders stated that even when patrolling alone the soldiers were ready to make arrests of citizens as necessary.[28] In a startling reversal "soldiers are thus back on the streets of San Salvador for the first time since the end of the civil war."[29]

The Saca presidency was not going to fall behind on a "tough on crime" policy and announced with great publicity a "Super Iron Fist" policy (*Plan

Super Mano Dura) in August 2004. In 2006 the government elaborated the policy of concentrating soldiers in those areas suffering the highest crime rates. For example, a violent crime wave in the city of San Miguel (Map 3) led to the deployment of 500 soldiers in 2006 to conduct operations in that city. Because not enough policemen were available for the joint army-police patrols, over 500 cadets from the police academy took part in these operations throughout the country. Soon more soldiers joined the police to carry out foot patrols, and in other instances troops took up guard duty at installations and thus released policemen for street patrols. And the most dangerous neighborhoods of San Salvador were off-limits to policemen unless they were accompanied by soldiers able to counter the automatic weapons of the gangs.[30]

With El Salvador virtually disintegrating around him, President Saca still did not consider the crime violence to be the most serious threat facing the country. Instead, he was "especially worried about Central America now that the Ortega administration in Nicaragua has been added to the tide of the leftist populism in the region."[31] President Saca, like most upper-class Salvadorans, was still trapped in the Cold War mentality and considered leftists, or Communists as they were still called in El Salvador, to be the biggest problem facing the country. But the population of El Salvador, scared for decades about the horrors that a leftist takeover would bring,

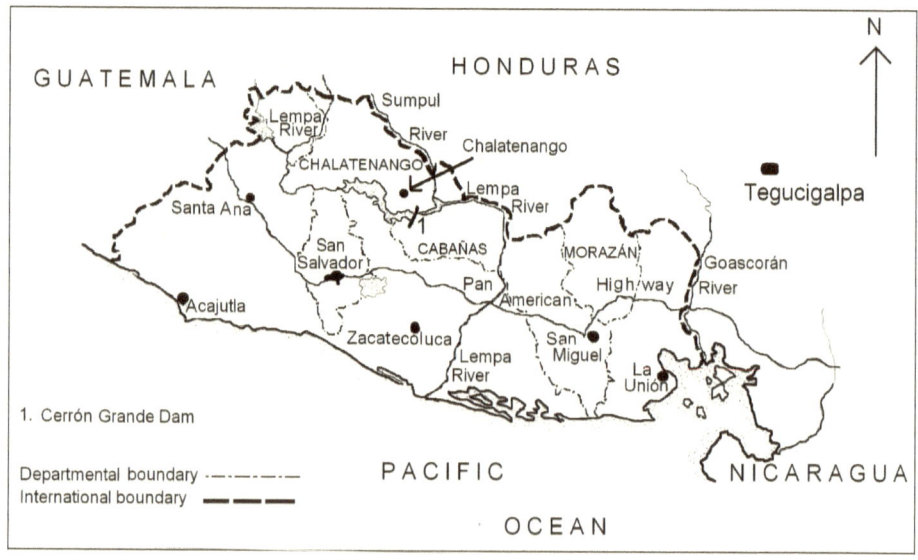

Map 3. El Salvador.

was increasingly disillusioned with the right-wing clichés and was becoming more receptive to new ideas, as chapter 8 will show.

Improving the Armed Forces of Colombia

The military involvement of U.S. officials with Colombia was the largest in all Latin America from 2000 to 2013. Not since the U.S. government tried to prop up the military of El Salvador during the insurgency of 1979–1992 had Americans been so deeply and constantly involved in the transformation of the armed forces of a Latin American country. The guerrilla war ongoing in Colombia since the late 1960s prompted the U.S. involvement, but a major commitment to rebuild the armed forces of Colombia came only at the start of the twenty-first century.

Since the 1960s the U.S. government supported the efforts of Colombia to destroy guerrilla groups, but in the initial decades Colombian guerrillas did not seem to pose a major threat. With the Cold War raging in many parts of the world and not just in Latin America, Washington devoted scant attention and limited resources to Colombia. Yet already in 1986 it was becoming clear that the Colombian armed forces were in no condition to crush the guerrilla groups. To begin with, the Colombian army of about 70,000 soldiers was too small to face the threat. "Given the geographic spread of the guerrilla forces, the inhospitable terrain, the need to defend many fixed high-value targets, the army is already stretched thin."[32] The army needed to increase its size at the very least by an additional 15,000 soldiers.

Another major obstacle was poor intelligence on the guerrilla groups, and in the case of operations near Cali, "at least in this theater, the army appears bogged down and unable to bring its concentrated strength to bear."[33] The army was poorly equipped, and while in an exaggeration officers claimed that guerrillas possessed better weapons, the shortage of quality arms certainly hindered military operations. To increase the mobility of troops, the government purchased helicopters in 1987, and this new capacity improved deployment and contributed to some successes on the battlefield in the early 1990s. But by 1994 the increased size of guerrilla columns necessitated deploying even larger army units, and officers clamored for more helicopters.[34]

A series of battlefield defeats persuaded the Ernesto Samper Pizano administration (1994–1998) that the most serious problems were inside the officer corps. President Samper Pizano and his successor Andrés Pastrana relentlessly searched and rooted out officer incompetence. This determination to transform the officer corps into an effective combat force avoided the nonchalance characteristic of the Salvadoran officer corps during that country's insurgency. The U.S. government found in Colombia civilian officials determined to demand quality performance from army officers, a situation totally lacking in El Salvador during its insurgency. Results would have come earlier in Colombia had not accusations of illegal campaign contributions from drug lords tarnished the image of the Samper Pizano administration. The U.S. government took a holier than thou attitude to Colombia and refused to provide the military assistance necessary to stop the guerrillas while Samper Pizano was in office.[35]

By the time President Andrés Pastrana was inaugurated in August 1998, the Revolutionary Armed Forces of Colombia (*Fuerzas Armadas Revolucionarias de Colombia*) or FARC had spread across the country and seemed poised to challenge the Colombian government. The expansion of FARC was hard to understand for U.S. officials who held the naive assumption that once the Soviet Union collapsed then any remaining Marxist guerrillas would wither away. Contrary to conventional wisdom FARC continued to increase, and it was clear than only a massive influx of U.S. assistance could stem the tide of victory then favoring the guerrillas. Washington was sympathetic to the request, but the Bill Clinton administration was worried that public opinion might not welcome an open-ended commitment to a war in a South American country. Instead, the drug epidemic in the United States was too obvious to deny, and the Clinton administration, by shifting the focus to anti-drug campaigns, was able to obtain congressional approval for Plan Colombia, a military and police assistance program for Colombia.[36]

Because in Colombia the police have primary responsibility for the anti-drug effort while the military have been in charge of the war against guerrillas, initially the military received only a small part of the funds from Plan Colombia. In trying to explain the survival and remarkable expansion of FARC, American and Colombian officials came to believe that the root cause was the drug trade. If the police and military stopped the drug trade, then FARC without drug money was destined to wither away.[37]

As U.S. funds poured into Colombia starting in 1999, training and equipment improved the capabilities of the Colombian armed forces. The new electronic surveillance equipment proved invaluable in detecting FARC movements, and the tipping point came with the Seventh of August oper-

3. Unconditional Allies of the United States 51

Map 4. Eastern Colombia.

ation of 2001 in the headwaters of the Guaviare River (Map 4). This ground campaign began early in August and lasted until late 2001, and for the first time the Colombian military was able to make a preemptive strike against a FARC offensive. The army, by rapid deployment of troops, was able to disrupt the guerrilla plans but failed to capture or destroy large numbers of guerrillas. The damage sufficed to persuade FARC Secretariat to abandon its offensive strategy and shift to a defensive posture.[38]

U.S. assistance and training put FARC permanently on the defensive, but it remained to be seen whether the Colombian military could smash the insurgency. The Colombian government very cleverly used the wave of anti-terrorism feeling in the United States after the 9/11 attacks to secure a reduction on the restrictions on using funds from Plan Colombia. By labeling FARC a terrorist organization, Colombia could freely spend U.S. funds either for the police or the military without restrictions. And the George Bush administration, delighted to have found an ideological collaborator in the incoming President Álvaro Uribe Vélez, actually increased funding for Plan Colombia.[39] And for its part the Uribe Vélez administration obtained approval from the Colombian congress to collect a wealth tax. Thus with the combination of local revenue and U.S. assistance the Colombian military for the first time enjoyed adequate resources to carry out its transformation and take the offensive in the war against FARC.

The U.S. presence spread throughout the Colombian armed forces. The most enthusiastic embrace of U.S. methods came from the Colombian Marines. To align with U.S. best practices, Admiral Luis Fernando Yance Villamil, the Commandant of Marines, raised the status of non-commissioned offers (NCOs) and in particular of sergeant majors. "Yance directed that Battalion and Brigade Commanders will turn over as much as possible of the day-to-day-running of the battalion to the sergeant major so that they can focus on operations and training."[40] The expansion of the Marine Corps to 23,000 beyond its original 20,000 required additional officers and NCOs, and to provide this new leadership a special program sought the best conscripts for promotion to NCO and officer rank as a way to entice them to reenlist. The Marines were increasing in number to assume additional responsibilities patrolling the many rivers and tributaries in eastern Colombia in new riverine battalions. The Colombian Marine Corps "continues its progressive transformation and is setting the standard for change in the Colombian military."[41]

Upgrading the Colombian air force was an essential element in improving the military's capabilities against FARC. U.S. assistance provided training, sophisticated equipment, and advice to this branch of the military. A little-noticed function of U.S. officials was to expedite the shipment of new equipment and replacement parts for the Colombian air force. Although the U.S. government gave blanket approval for the Colombian government to purchase military equipment and parts in the United States, in actual practice delivery proved very slow and enormously complex. The lack of replacement parts for maintenance kept many planes grounded and only 60 to 70 percent of airplanes were operational. The need to secure permissions and authorizations delayed delivery, and many times U.S. personnel could not train or arm airplanes until all the components had arrived from the United States. A particularly complex task was trying to equip the old airplanes of the Colombian air force with the new smart bombs. Different systems were available for using smart bombs in combat, and after extensive consultation U.S. military advisers recommended using a laser-guided system as the one easiest to fit to existing aircraft.[42]

Perhaps inevitably the Colombian government regarded the acquisition of new warplanes as a necessary measure in the war against FARC. U.S. military advisers did not agree with this idea, and they proposed instead that the Colombian air force improve and increase its airlift capacity, particularly of the C-130 cargo planes, which also were eminently suitable for flying humanitarian missions and could land in the many short runways scattered across the country. However, the lure of shiny supersonic jets was hard to resist, and only with reluctance did the air force accept the compromise formula of Tucano light attack aircraft manufactured by the Brazilian firm of Embraer. The U.S. government readily gave Brazil permission to use American components in the manufacture of the Tucanos destined for Colombia. And the U.S. continued to assist in outfitting both the new and old planes with the latest aeronautical technology.[43]

Cooperation with the air force and the Marines proved a lot easier for U.S. officials than with the army, where old-time officers still resisted the overwhelming preponderance of American power throughout Latin America. Yet even the army, under the prodding of President Uribe Vélez, was becoming even more pro–American than it always had been. With the appointment of General Reinaldo Castellanos Trujillo as army chief, "Colombia perhaps now has the most professional, internally cooperative, U.S.-friendly high command in its national history, and for the first time is experimenting with a joint operational structure in the traditional divisions to mimic the success of U.S. combat commands. The military created a joint Caribbean Command and a joint intelligence coordination center, both of which are headed up by U.S.-friendly generals with proven track records."[44]

These joint commands were a major innovation for the Colombian military, and their creation responded to the need to have all three services operate as one when confronting guerrilla groups. Having a joint commander give orders to units of other branches in one area was not an easy sell, and only under wartime conditions did the proposal gain partial acceptance.[45] As the commander of Colombian armed forces explained, "the toughest part of his job is changing the mindset of the services and getting joint force commanders to report directly to him, while still answering to their respective service chiefs. The services see this arrangement as a loss of power and control over their subordinates."[46]

The Caribbean Command was largely trouble free, but the second command created, Joint Task Force Omega, ran into difficulties. The problems started in December 2004 when General Carlos Alberto Fracica Naranjo became its commander replacing General Castellanos who assumed command of the entire army. Unlike General Castellanos who was engaged in extensive friendly cooperation with U.S. officials, General Fracica "is not willing to work with U.S. trainers."[47] The new general was also very temperamental, antagonized fellow officers, and failed to obtain results in the war. In the more than a hundred operations General Fracica conducted against FARC in the areas of the Meta, Guaviare, and Caquetá rivers (Map 4), he failed to capture or kill any high guerrilla official. The Uribe Vélez administration placed the greatest emphasis on the elimination of "High Value Targets," but the inability of General Fracica to hit any of these targets ultimately ruined his career, and in November 2005 he was removed from command of Joint Task Force Omega.[48]

As Washington increased its involvement in Colombia, the 800 American soldiers and 600 private contractors stationed in the country in 2004 needed better facilities to provide military assistance to the armed forces. U.S. officers chose Apiay, to the east of Villavicencio (Map 4), as the place for a Forward Operating Site. The location was appropriate to the task of supporting Joint Task Force Omega in its campaign against FARC. At Apiay, U.S. troops instructed Colombian military personnel and provided training on the use of new weapons and sophisticated equipment. In one of many examples, a counter guerrilla unit received marksmanship training on the proper use of new M-16 rifles equipped with telescopic sights. And Joint Task Force Omega needed all the help it could get, because far away from its supply bases in the highlands, the logistics of keeping its 15,000 soldiers well supplied was formidable.[49]

More than logistics was at play because many officers were reluctant to take risks in tracking down FARC. Deployment in battalion and brigade-size units remained the rule to avoid falling into ambushes as happened during the 1990s. Joint Task Force Omega, deep in the region of Meta, Guaviare, and Caquetá rivers, faced "the effects of low morale, a hostile environment, disease, resource shortages, and an inability to drive FARC out of the area."[50] U.S. support transformed the Colombian military and made possible blunting the guerrilla offensive, but the complete destruction of FARC still seemed beyond reach as 2005 came to an end.

4

Reluctant Mexico

> *The Mexican military is key to many counterterrorism and law enforcement functions here.*—U.S. Ambassador Pascual Garza[1]
>
> *The Mexican army customarily keeps its distance from the United States.*—Anthony DePalma, *New York Times* correspondent[2]

During World War II Mexico briefly opened up to the United States and became a staunch ally in the global struggle to destroy fascism. The Mexican people wholeheartedly supported the war effort, and the government closely collaborated with the Franklin D. Roosevelt administration. In a striking reversal from past suspicion, the Mexican armed forces eagerly accepted American equipment, training, and assistance. U.S. influence was at its peak after the war when the Mexican army acquired considerable amounts of surplus weapons and adopted U.S. training manuals for instructing its troops. But after World War II the U.S. government lost interest and diverted its attention to Europe and Asia.[3] The window of opportunity to affect a fundamental change in relations between the U.S. and the Mexican armed forces rapidly closed, and Mexican generals reverted to their traditional distrust of the powerful northern neighbor. But beginning in the last two decades of the twentieth century, new forces gradually drove a reluctant Mexican military into the orbit of U.S. influence, as this chapter explains.

The Role of the Armed Forces in Mexico

Unlike the rest of the Caribbean countries, Mexico since its 1917 Constitution has operated under a fully structured federalist system with separate layers of local, state, and federal authorities. First Venezuela in the 1970s and then Colombia with its 1991 Constitution have adopted some trappings of a federal system, but in reality power and money have remained overwhelmingly concentrated in the country's capital. Instead in Mexico state and municipal authorities assumed many duties, including the crucial police functions. But the realization was nearly universal that the state governments were far from ready to assume these responsibilities, and in a precautionary move the drafters of the 1917 Constitution granted broad powers to the president much greater than those enjoyed by the U.S. counterpart. The Constitution also limited all officials in Mexico to serving only one term in an elected position as a way to prevent any official from becoming a dictator for life.

Until today, presidentialism has remained a permanent characteristic of the Mexican political system to govern the country in spite of its federalist structure. A second major institution to bridge the gap between centralism and federalism has been the military. In many ways the army was the primary agency keeping the country together in the wake of the Mexican Revolution (1910–1929). Ruthless behavior earned a ferocious reputation for the army, and until the late 1980s soldiers commanded healthy respect if not fear among large segments of the population, while at the same time enjoying almost continuously the highest popularity ratings among Mexican institutions.[4]

Unlike the rest of the Caribbean countries, Mexico has two ministries for its military. In spite of frequent demands from U.S. officials, Mexico has refused to place all its armed forces in a single cabinet department. The government established a Ministry of Defense for the army and air force in 1937 and then a separate Ministry for the Navy in 1939; by law only uniformed personnel may head these two ministries. The Marines are under the Navy Ministry, and this force numbering 8000 elite soldiers in 1992 has performed difficult missions, sometimes at the order of the president. A higher percentage of recruits from the middle classes, greater opportunities for promotion, and frequent testing for merit have made the navy a much more efficient service than the army. The separation of the navy from the rest of the military reduces the possibility of any coup, because rivalry and competition between the navy and the army has been constant and frequently bitter; indeed the navy seems keen on "maintaining its near state of war with the army."[5] Of the military budget, about 70 percent went to the Defense Ministry and 30 percent to the Navy Ministry during the last decades of the twentieth century.[6]

The military zones of the Ministry of Defense have been instrumental in maintaining the authority of the central government throughout Mexico. The 35 military zones created in 1923 corresponded roughly to the number of geographical entities in Mexico, so that each state had its own military

Map 5. Southern Mexico.

zone. Only the long and narrow Veracruz State had three military zones while the traditionally turbulence-prone states of Chiapas and Guerrero each had two military zones (Map 5). Although the general commanding the local garrison was the head of the military zone, his primary duty was political, and he was supposed to work closely with the civilian governor on all local matters. If any crisis arose in the state, Mexico City often appointed the zone commander to be the interim governor until new elections could be held. But the practice of appointing zone commanders as governors declined gradually and was last used in 1946. The zone commanders remained as the final defense of the central government's authority, but their political role was largely assumed by the new official party.[7]

The military prevented and put down revolts in the Mexican states, but the civilian officials of each of the states could still challenge the authority of the central government. Because governors, mayors, and deputies to state assemblies and municipal councils were popularly elected, Mexico City in theory had no voice in their election. The creation of the official political party in 1929, shortly after known as the PRI (*Partido Revolu-*

cionario Institucional) avoided any possibility of challenges to Mexico City. Although other political parties were free to exist, Mexico in reality had a single-party system until the late twentieth century. The president of Mexico could not appoint the governors and mayors of the country, but as head of the PRI he picked the candidates for those positions. And since the PRI controlled the electoral machinery throughout the country, the election of those candidates was guaranteed.[8] By means of this official party, Mexico completed bridging the gap between a federalist and centralist system. Almost as if having squared a circle, Mexico enjoyed the appearances of a federal system and the reality of strict control from the national capital.

As Mexico transitioned starting in the 1920s from armed revolts to civilian opposition, the benefits of having a secret police became evident to the new rulers of the country. The fusion of several police forces into the Federal Directorate of Security (*Dirección Federal de Seguridad*) or DFS in 1947, directly responsible to the president, marked the culmination of efforts to create a national intelligence agency. Supposedly patterned after the FBI in the United States, the DFS quickly reverted to spying and reporting on any threats to the existing regime. Student movements, dissident groups, labor unions, other political parties, and suspicious individuals were the targets for this agency. The DFS also established a firm control over local and state police forces and then used them to collect information and to carry out operations. The existence of local and state police perpetuated the image of federalism, while the DFS gave the central government the reality of a single national police force.[9]

From the 1920s until the 1980s the four institutions of presidentialism, the military, the official political party, and the secret police, all operating under a veil of secrecy, governed Mexico. Other institutions supported this structure, and at least one, PEMEX, the state-owned petroleum company, should be mentioned. The government created PEMEX after the nationalization of petroleum in 1938, and this vast monopoly came to have a presence in every corner of Mexico. In remote regions PEMEX officials often are the only official representatives in the area, and even when local or state authorities exist, PEMEX operates in splendid independence and responds only to the president of Mexico.

By the 1960s the gap in economic prosperity between Mexico and the United States was reaching disturbing proportions, and in response many Mexicans fled the country to find work as illegal immigrants in the rich northern neighbor. The faltering growth inside Mexico led to demands for changes in the governing structure of the country, and the political pressure became very strong after the 1968 Tlatelolco Massacre in Mexico City. The single-party system weathered the crisis, but the cracks were too obvious to deny, and the government had to appease the growing opposition movement. In a fateful decision with many unforeseen consequences, President Miguel de la Madrid abolished the DFS in 1985 as a way to show tolerance to opposition groups and greater openness in the political system. Without this secret police, the government not only was left in the dark about events in the states but also lost direct control over state and local police forces. Although most of the files of the DFS seem to have survived, the entire investigative and spying apparatus disappeared. And the network of informants, local contacts, and control so laboriously constructed over decades could not be replaced. As late as 2009 U.S. officials were bemoaning the absence of real intelligence capabilities among Mexican agencies both civilian and military, while the lack of cooperation with local and state police has remained a recurrent problem.[10]

One of the reasons behind the abolishment of the DFS was its failure to control the drug trade emerging in Mexico in the late 1970s. Accustomed to spying on political dissidents and guerrilla sympathizers who were notoriously poor, the DFS faced a much harder task when trying to track down wealthy drug traffickers easily able to bribe officials.[11] As a partial solution to this weakness, the U.S. government secured a secret agreement in 1974 with the Mexican government authorizing the Drug Enforcement Administration to operate agents in the country. DEA, just created the year before in the United States, soon recruited many informants throughout Mexico. The DEA network complemented the many spies already operating under the control of officials from the CIA, the FBI, and other agencies of the U.S. government such as ATF; at times it seemed that Mexico was flooded with U.S. informants and agents. And when the DFS was abolished in 1985, the DEA agents through their networks became the main source of information for the Mexican government on the activities of drug traffickers.[12]

The Mexican Constitution prohibited all foreign officials from carrying arms inside Mexico, a prohibition frequently ignored. Whether armed or not, DEA officials needed local authorities to disrupt the activities of drug traffickers and to conduct operations on the ground. A rough division of labor took place, with state and local police carrying out searches in urban areas, while the army partnered with the Attorney General's Office to conduct operations in the countryside during the 1980s. Large numbers of soldiers were necessary to uproot the marihuana and poppy plants growing across vast stretches of farmland in northern Mexico near the Pacific Coast.

Aerial eradication by chemical herbicides has constituted only a small part of the labor, and only after 2003 has its share increased from the previous average of 10 percent to 20 percent. Mexican soldiers, usually of peasant origin, became expert in the back-breaking hard toil of uprooting, piling up, and burning the illegal plants. Officers earned promotions and decorations in proportion to the number of hectares cleared and the amount of illegal crops burned. And while the eradication of these crops has remained a routine activity of the Mexican military until the present, the drug traffickers by the late 1980s were diversifying their sources and increasingly turning to Colombia to obtain large quantities of cocaine.[13]

As U.S. law enforcement sharply disrupted the air and sea lanes from Colombia to Florida, the drug cartels in Colombia turned to Mexico as an alternate route to get cocaine into the United States. The modus operandi was for airplanes to fly shipments from Colombia and land in deserted air strips usually in northern Mexico prior to delivery of the cocaine across the U.S. border. But because army units frequently were near those air strips, the Mexican drug traffickers paid bribes to generals in exchange for allowing the shipments to pass undisturbed. Reports reached Mexico City of soldiers even carrying and guarding the cocaine. The most dramatic incident occurred on 7 November 1991 at Llanos de la Víbora (Map 5) in Veracruz State. When agents of the Attorney General's Office were about to confiscate a cocaine shipment from a just landed airplane, troops opened fire, killed seven agents, and then left with the drugs. The cocaine was never recovered and even though the scandal caused an international sensation, all attempts to bring the perpetrators to justice failed. Even worse, army officers insistently denied the incident ever occurred and tried to block all press coverage, because "the army, used to being portrayed, and portraying itself, as incorruptible, remains extremely sensitive about the incident."[14] The army ignored the wake-up call that was the Llanos de la Víbora massacre and, by refusing to carry out intense inquiries, left the door open for even more embarrassing revelations in the future.

And for their part, the drug traffickers were mastering their trade and became very adept at bringing cocaine from Colombia and sending it by airplane and by tunnels across the land border into the United States. By 1990 two large cartels controlled the drug trade in Mexico. The first was known by a variety of names and is easiest to remember as the cartel of the Lord of the Skies (*el Señor de los Cielos*) or Amado Carrillo Fuentes. He began his drug empire in the key frontier city of Ciudad Juárez (Map 6) and from there expanded until he controlled virtually all the Pacific Coast region of Mexico and also established outposts in Central America. He also was the first to develop the domestic market by selling low-quality drugs at the local corner stores (*narcotiendas*). The second was the Gulf Cartel, which operated in the Mexican states bordering the Gulf of Mexico and relied on Caribbean routes from Colombia. The Gulf Cartel smuggled

Map 6. Northwestern Mexico.

cocaine from Colombia into Texas and to other destinations in the eastern United States. Under the Carlos Salinas de Gortari presidency, these two cartels grew tremendously, and observers agree that only thanks to protection from high-level officials could this remarkable expansion have taken place. The obsession of President Salinas with securing the NAFTA free trade agreement for Mexico shifted his attention to keeping negative news out of the media and made him neglect the rapidly expanding drug cartels in Mexico.

His desire to hide any bad news led President Salinas to ignore the warnings the army was sending about an uprising in Chiapas State (Map 5); even bordering Guatemala sent a file full of reports about the planned uprising by a new group calling itself the Zapatista National Liberation Army (*Ejercito Zapatista de Liberacion Nacional*) or EZLN. The army had been involved in counterinsurgency activities since 1968 and had been carefully monitoring the activities of likely subversives in southern Mexico and in particular of the Zapatistas as they were afterwards referred to. Only 1,055 soldiers were stationed in Chiapas, the poorest state in Mexico, and the military recommended increasing the garrisons to act as a deterrent. President Salinas refused to act because he wanted to portray the image of a modern Mexico free of any violent past. Thus by default he left the initiative to the poorly armed Zapatistas who on 1 January 1994 captured four towns in Chiapas, including San Cristóbal de las Casas, the third largest town in the state. The date was also full of symbolism, because on the first day of that year the NAFTA agreement with the United States went into effect.[15]

The Chiapas uprising turned into a debacle in many ways for the Mexican army. First, to provide political cover to President Salinas who claimed not to know anything about this movement, the army had to state that the Zapatista revolt also caught it by surprise. In fact the army was aware of the imminent revolt, although it did not know the exact day. Secondly, when the army sent units to put down the uprising, the deployment proved difficult and time consuming. The slow response was particularly embarrassing when TV camera crews waited for days to record the arrival of the first troops trickling into Chiapas. And after the army laboriously deployed troops and heavy weapons in preparation for pounding the rebels into submission, the possibility of slaughtering the almost unarmed rebels provoked a backlash in public opinion against the army. Right when the soldiers were in hot pursuit and ready to smash the fleeing rebels, President Salinas halted the military offensive and announced the opening of negotiations with the Zapatistas as the only way to bring the dispute to a peaceful end.[16]

Just when the creaking military machine was about to crush the rebels, the President's decision snatched victory away from the army, and not surprisingly officers felt spurned and humiliated. The army also felt betrayed by the U.S. government. As soon as the Zapatista revolt broke out, Mexico sought the immediate purchase of equipment and weapons abroad. In a manner reminiscent of Álvaro Obregón's urgent appeal to the U.S. government to make rush deliveries of weapons during the De la Huerta Revolt of 1923–1924, the army in 1994 turned to the U.S. government for immediate delivery of vehicles, equipment, and weapons; Mexico also wanted specialized training for its officers. The U.S. refusal to satisfy these requests deeply angered Mexican officers who during many years "continue to lose no opportunity to remind us that they see a deep wound and an insult."[17] The army turned to European suppliers who had no hesitation about selling not just vehicles and equipment but also a vast array of heavy weapons to Mexico; the Mexican navy even took advantage of the government's unusual largesse to acquire Russian helicopters.

The Presidency of Ernesto Zedillo, 1994–2000

The refusal of the Bill Clinton administration to provide weapons and equipment to Mexico at its time of need was a serious blow to the efforts to bring the Mexican army into the orbit of U.S. influence. The Mexican generals vented their suspicions of U.S. motives and intensified their traditional practice of deliberately avoiding even the most casual contact with U.S. officials, both military and civilian.[18] The new president, Ernesto Zedillo, initially was most reluctant to overrule his generals. The worst economic collapse in Mexico since the Great Depression finally forced him to seek closer relations with the United States. Shortly after Zedillo took office on 1 December 1994, the Mexican peso started to drop in value, businesses began to fail, and investors refused to extend loans to Mexico. Millions lost their jobs, and the inability to pay the foreign debt was driving Mexico toward an economic meltdown.[19]

President Bill Clinton immediately realized the danger that a disintegrating Mexico right on the border posed for the United States, but Congress was too paralyzed to act. Relying on presidential authority over U.S. Treasury funds, he authorized a bailout of 20 billion dollars in January 1995. This massive infusion of funds saved the Mexican economy, and already the next year prosperity returned to the country as unemployment dropped dramatically. The U.S. Treasury not only recovered all the money it had lent to Mexico but also made a nice profit on the transaction. Even President Zedillo was surprised at the speed and magnitude of the U.S. bailout, and in a major policy reversal, he concluded that the traditional Mexican opposition to U.S. requests for cooperation could not all be automatically rejected as in the past.[20] And high on the list of things to do was forcing the Mexican armed forces to cooperate with the United States for the first time since World War II. As an initial step, U.S. officials insisted on bringing the Mexican military into the counter-drug campaign as the only force capable of stopping the spread of the drug cartels.[21]

TABLE 1. MEXICO: MILITARY PERSONNEL RECEIVING U.S.–FUNDED TRAINING, 1950–2009

Year	Number of Personnel
1950–1978	906
1984–1992	512
1996	440
1997	236
1998	693
1999	1,271
2000	282
2001	N.A.
2002	225
2003	207
2004	162
2005	185
2006	184
2007	N.A.
2008	177
2009	517

Sources: 1950–1978, Lars Schoultz, *Human Rights and U.S. Policy Toward Latin America* (Princeton: Princeton University Press, 1981), p. 215; 1984–1992, International Consortium of Investigative Journalists, "U.S.-Trained Forces Linked to Human Rights Abuses," 12 July 2001, Internet; 1996–2009, Secret/No Forn Cable from U.S. Embassy, Mexico City, 21 August 2009, Wikileaks. A very different set of figures appears for 1999 and 2001–2006 in U.S. Congress, House, Committee on Foreign Affairs, *U.S. Security Assistance to Mexico* (Washington, D.C.: Government Printing Office, 2007), p. 70; I have followed the numbers in the Secret Embassy cable.

A first meeting of the minds occurred when Zedillo had lengthy conversations with Clinton during a White House visit on 10 October 1995. The two leaders agreed to strengthen relations between their two countries not just on a wide range of economic matters but also for the first time on military cooperation. In exchange for Zedillo's promise to bring the Mexican military directly into the campaign against drug cartels, the U.S. government agreed to provide extensive assistance. The United States resumed arms sales to Mexico and even provided many types of weapons and equipment as military aid. The U.S. government allocated large sums to train Mexican officers in the United States, and a reluctant Mexican military sent candidates to receive this training.[22] Table 1 indicates that the number of military personnel receiving U.S. training vastly increased during the Zedillo years and reached the all-time high of 1,261 in 1999. In three years of his presidency a larger number of Mexican officers received U.S. training than in the preceding thirty-year period.

As a follow-up to the Zedillo visit to the White House, U.S. Secretary of Defense William J. Perry came to Mexico for a two-day visit on 23–24 October 1995. This was the first time in history a U.S. Secretary of Defense had ever come to the country. Over the decades Pentagon officials repeatedly proposed reciprocal visits by high-level defense officials of both countries, but Mexican generals relied on the traditional Mexican courtesy to wiggle out of almost all the invitations. Sometimes Foreign Ministry personnel filled in for generals who cancelled at the last moment their participation in bilateral meetings. But now under direct orders from President Zedillo, the generals had to meet their uniformed American counterparts in both official and social functions.[23]

U.S. support allowed the armed forces to intensify their efforts against insurgencies in southern Mexico. After the outbreak of the brief Zapatista revolt of 1 January 1994, troops had been pouring into Chiapas until the Mexican army came to have 12,000 soldiers deployed in that state in addition to a small number of Marines. The military installed 111 posts (*partidas*) throughout Chiapas, and this high density of occupation served both as a deterrent against any revival of Zapatista operations and also as a containment to prevent the revolt from spreading to bordering states. As a boa constrictor, without having to fire a shot, the large number of troops slowly squeezed the life out of the Zapatistas who dared not challenge this military might. The government also poured money into Chiapas, spending more federal funds there than in any other Mexican state. International agencies also sent economic assistance, while those inhabitants who could still not find satisfactory employment departed for the United States in a new migratory wave. The combination of economic assistance and massive military pressure undermined the Zapatista movement, which faded away in the last years of the twentieth century.[24]

Table 2. Mexico: Army and Navy Strength, 1990–2014

Year	Army Personnel	Navy Personnel
1990	151,178	42,105
1991	155,218	44,087
1992	157,142	45,449
1993	162,169	46,809
1994	165,463	48,166
1995	172,753	48,955
1996	179,038	53,128
1997	182,328	54,247
1998	182,329	53,566
1999	182,329	54,972
2000	182,329	55,223
2001	185,143	49,165
2002	188,143	50,026
2003	191,143	47,301
2004	191,143	47,316
2005	191,143	47,644
2006	196,767	47,471
2007	196,710	50,032
2008	202,355	51,680
2009	206,013	52,979
2010	206,013	53,224
2011	209,716	53,997
2012	209,716	54,466
2013	212,916	54,314
2014	212,916	54,704

Sources: Mexico, Instituto Nacional de Estadística y Geografía, *Anuario estadístico 2001* (Mexico: Instituto Nacional de Estadística y Geografía, 2002), Table 8.28; George W. Grayson, *The Impact of President Felipe Calderón's War on Drugs on the Armed Forces* (Fort Carlisle, PA: U.S. Army War College, 2013), pp. 44, 56; Jorge Luis Sierra Guzmán, *El enemigo interno: Contrainsurgencia y fuerzas armadas en Mexico* (Mexico: Universidad Iberoamerciana, 2003), p. 279; Jesús López González, *Presidencialismo y fuerzas armadas en México, 1876–2012* (Mexico: Gernika, 2012), p. 239; Presidencia de la República, *Segundo Informe de Gobierno, 2013–2014* (Mexico: Talleres Gráficos, 2014), p. 40. For some years the sources show slight variations in the figures, but the differences do not affect the broad trends.

Historical perspective makes obvious the disappearance of the Zapatista armed threat but for the military of the late 1990s the victory was not immediately clear. The Zapatista danger remained latent mainly because a separate guerrilla movement emerged in the state of Guerrero on 1 July 1996 (Map 5). Mexico City feared that this new movement, the People's Revolutionary Army (*Ejército Popular Revolucionario*) or EPR, was just one more in a series of guerrilla groups set to rise up. And because Guerrero, the home base of EPR, was separated only by Oaxaca from Chiapas, a linkup between the Zapatistas and EPR seemed a real possibility. And unlike the more pacifist Zapatistas who never actually clashed with the army after January 1994, EPR guerrillas engaged occasionally in raids and ambushes.[25]

The deployment to Chiapas strained the resources of the army, and in response Zedillo authorized annual increases in the military budget during the rest of the twentieth century. Already in 1997 army strength rose to 182,000 (Table 2), and with this additional manpower the military was gradually able to suffocate the revolt of EPR in Guerrero. The operations were bloodier than in Chiapas, and by the time the campaign came to an end in late 1999, 68 soldiers and 15 policemen had died. Deep divisions within ERP and fragmentation into splinter groups finally made most guerrillas abandon the obviously hopeless struggle against a numerous and well-armed military in Guerrero.[26] However, the claims about the demise of EPR proved premature, as chapter 10 will show.

By the end of the twentieth century neither EPR nor the Zapatistas posed any threat to the internal security of Mexico but their lingering presence still channeled resources away from the struggle against the drug trade. This concentration on leftist groups was a major reason why the army failed to develop comparable intelligence on the drug cartels. Already in 1995 observers believed that Mexico was neglecting its anti-drug efforts because of the Chiapas revolt, and the distribution of troops did reveal a major concentration in those states with guerrilla groups. In 2004 the army still stationed 25 percent of its soldiers in areas of previous guerrilla activities and 34 percent in areas of the drug cartels. Meanwhile the army also increased the number of military zones, its basic organizational unit. In 1983 the army created its 36th zone, and the number of military zones went up from 36 in 1992 to 44 in 2004.[27] Although the army had a deeper and more visible presence throughout the country than before, its enlargement was still inadequate to face the combined threats from insurgency and drug cartels.

TABLE 3. MEXICO:
DRUG-RELATED ARRESTS, 1990–2013

Year	Number of Persons Arrested
1990	18,115
1991	8,621
1992	27,369
1993	17,551
1994	8,545
1995	17,397

Year	Number of Persons Arrested
1996	18,335
1997	18,010
1998	17,533
1999	15,632
2000	16,035
2001	16,581
2002	14,231
2003	13,571
2004	23,459
2005	28,651
2006	18,914
2007	29,381
2008	28,630
2009	42,142
2010	32,319
2011	41,675
2012	28,813
2013	19,723*

*Preliminary figures

Sources: Presidencia de la República, *Segundo Informe de Gobierno, 2013–2014* (Mexico: Talleres Gráficos, 2014), p. 51; Instituto Nacional de Estadística y Geografía, *Anuario Estadístico de los Estados Unidos Mexicanos 2002* (Mexico: Instituto Nacional de Estadística y Geografía, 2003), Table 8.27.

The struggle against the drug trade received a major boost from the strong personal ties between Clinton and a grateful Zedillo. The greater involvement of the Mexican military in the struggle against the drug trade increased the number of drug-related arrests (Table 3). But Mexican prisons did not seem capable of restraining high-profile inmates. As a major solution and in a complete break with past practice, President Zedillo authorized for the first time the extradition of Mexican citizens to the United States in April 1996. Although the extradition treaty with Mexico had gone into effect in January 1980, its clauses left final decision in the hands of Mexican officials who traditionally authorized only the extradition of Americans and other foreigners.[28] Zedillo's decision unleashed an uproar among nationalists and drug dealers, and the massive public outcry counseled the President to authorize the extradition of only small numbers of Mexican criminals during his presidency (Table 4).

As mentioned earlier, the visit of U.S. Defense Secretary William J. Perry to Mexico in October 1995 strengthened ties between the armed forces of both countries to the point that the Mexican navy scheduled joint maneuvers with U.S. warships.[29] As part of the negotiations during the visit, the

U.S. government also donated 72 UH-1H helicopters (the Hueys of Vietnam War fame). The Mexican military considered these helicopters unsuitable for the struggle against the drug trade and previously had refused to receive them. But U.S. officials insisted on imposing an anti-drug strategy that would allow Mexican soldiers to reach promptly remote airstrips just when airplanes were landing with drug shipments from Colombia. Under prodding the Mexican army reluctantly accepted the donated helicopters.[30]

The Perry visit had been the high point of military cooperation under President Zedillo, but the amicable feelings started to crumble after the Mexicans put into operation the donated helicopters. The machines had a low flight ceiling and could not operate in the high elevations of northern Mexico, the region of the greatest drug activities. The helicopters were not supposed to fly for more than 15 hours a month and were not designed for near continuous use against drug traffickers. These "were thirty-year cranky machines that needed constant care and spare parts that were either unavailable or too costly,"[31] and soon the helicopters were grounded and in 1999 Mexico returned all of them to the United States. Mexican officials gloated: "when gringos offer you the hand of cooperation, you have to be careful [or] they may take everything you have."[32] This attempt to foster a close relationship between the armed forces of both countries ended in resounding failure. Worse yet, the experience confirmed in the minds of Mexican generals their traditional distrust of the United States and made them wary of Americans bearing gifts.

Actually, it was not inadequate numbers of helicopters but the lack of intelligence on the drug cartels that crippled most army interdiction efforts. The drug cartels awash in money bribed officers to secure safe passage for shipments into the United States. DEA agents with their network of informants partially filled the deficiencies of both army and civilian intelligence. As will be recalled, the Gulf Cartel and the cartel of the Lord of the Skies were the two principal drug trafficking operations in Mexico during the 1990s. DEA agents scored one of their greatest successes when they obtained the arrest of the head of the Gulf Cartel, Juan García Ábreu, in Monterrey (Map 7) on 14 January 1997. DEA agents had been tracking this drug lord for years and finally found his location. Agents of the Attorney General's Office carried out the raid, and because evidence suggested that Ábreu was an American citizen, his extradition to the United States proved an easy matter for President Zedillo who promptly handed him over to U.S. officials. The arrest and deportation of Ábreu weakened the Gulf Cartel and undermined its position in the drug trade for years.[33]

Map 7. Northeastern Mexico.

A second blow against the drug trade came at a very high price to the Mexican army. As proof of his firm determination to destroy the drug cartels, President Zedillo appointed General José de Jesús Gutiérrez Rebollo on 9 December 1996 to the newly created post of anti-drug czar; the expectation was that by combining civilian and military efforts for the first time, this senior official would be able to strike major blows against the drug cartels. General Gutiérrez Rebollo supposedly had been very successful in weakening the drug cartels, and his seemingly ideal record also gave U.S. officials an opportunity to introduce U.S. influence. Since the Mexican Revolution the U.S. government lacked any voice over the choice of individuals to high-level positions, and U.S. officials saw the appointment of General Gutiérrez Rebollo as a favorable opening wedge to demonstrate effusive admiration for an appropriate choice of candidate; at a later stage the U.S. government could take the next step of expressing disapproval over future appointments. "Since Gutiérrez's appointment as Mexico's drug czar in December [1996] U.S. authorities had embraced him without reservation. His North American counterpart, retired General Barry McCaffrey, had gone out of his way to praise him as 'a serious soldier, a guy of absolute unquestioned integrity.'"[34] The General had even "traveled to Washington, D.C., where he had received a detailed briefing full of sensitive information on U.S. narcotics strategies, priorities, and operations."

However, his elevation to this high position brought him under close scrutiny, and soon the DEA discovered that the General's successful drug busts were all against the rivals of the Lord of the Skies. As DEA agents dug deeper, they soon confirmed that this senior officer had been on the payroll of the Lord of the Skies since the late 1980s. The evidence was so overwhelming, that President Zedillo finally authorized the arrest of General Gutiérrez Rebollo in his own office inside the Defense Ministry on 6 February 1997. It was hard to tell which government came out looking worse in this sordid episode, but no doubt existed that the first attempt of the U.S. government to influence the appointment of high Mexican officials had turned out to be a fiasco.[35]

The arrest of General Gutiérrez Rebollo was the second major scandal to rock the army after the massacre of Llanos de la Víbora in November 1991. But while the army had been trying to downplay the 1991 incident as a series of confused events, in this case no doubt existed that General Gutiérrez Rebollo had devoted most of his army career to the service of the Lord of the Skies. Public outrage was so great that even the possibility of execution loomed over the disgraced general. The Mexican Constitution prohibits the death penalty in all cases except for active-duty military personnel. For a while it seemed that General Gutiérrez would be the first army officer executed since the Mexican Revolution. Cooler heads prevailed and the general received a sentence of 40 years in prison.[36]

The army hastily learned public relations techniques to present this corrupt general as a loner or as the proverbial bad apple. Damage control seemed to be working until 31 August 2000 when two other generals were also arrested on charges of helping the drug cartels. The passage of three years since the Gutiérrez Rebollo incident gave the army an argument that the incidents were not related, and since these two generals did not hold high positions, the damage to the institution did not seem so serious. But the worry that the drug cartels had infiltrated the officer corps became pervasive.[37] The steady drip of arrests and dismissals of army officers of rank lower than general suggested that drug-related corruption was rampant inside the army.

TABLE 4. EXTRADITIONS FROM MEXICO, 1980–2013

Year	Number of Extraditions to the United States	Total Number of Extraditions
1980–1994	8	N.A.
1987–1994	N.A.	13
1995–1999	56	80
2000	12	N.A.
2001	17	N.A.
2002	25	29
2003	31	35
2004	34	34
2005	41	42
2006	63	72
2007	83	86
2008	95	N.A.
2009	107	N.A.
2007–2011	N.A.	505
2012	117	N.A.
2013	54	N.A.

Sources: 1980–1994, Yvonne M. Dutton, *U.S.–Mexico Extradition and Cross-Border Prosecution* (San Diego: Trans-Border Institute, 2004), p. 5; 1995–2005, Cable from U.S. Embassy, Mexico City, 25 October 2006, Wikileaks; 2000 and 2006, U.S. Congress, House, Committee on Foreign Affairs, *U.S. Security Assistance to Mexico* (Washington, D.C.: Government Printing Office, 2007), p. 28; 2007, *Los Angeles Times* 30 November 2008; 2008, France24.com, 11 February 2009; 2009, *Washington Post*, 22 May 2010. Total number of extraditions 2002–2007, International Monetary Fund, *Mexico: Detailed Assessment Report on Anti-Money Laundering and Combating the Financing of Terrorism*, Country Report no. 09/7, January 2009, p. 288; 1987–1999 and 2007–2011, Ruth Roque and Rogelio Velásquez, "Gobierno extradita a medio millar de mexicanos," *Revista Contralínea*, 31 January 2012; 2012–2013, CapBiz, 18 February 2014.

The drug cartels, however, were running into difficulties. As mentioned before, the arrest of García Ábreu plunged the Gulf Cartel into turmoil. Everything seemed to indicate that the Lord of the Skies was coming close to achieving his dream of being the boss of the entire drug business in Mexico, once he finished knocking out the weakened Gulf Cartel. But in a manner never satisfactorily explained, the Lord of the Skies died under mysterious circumstances while undergoing plastic surgery to change his face. After his death on 3 July 1997, his empire over the drug trade on the Pacific side of Mexico collapsed, because his highly individualistic style of running the drug business could not be duplicated or transferred.[38]

Succession struggles erupted after his death, and the inevitable result was the fragmentation of his drug empire. His followers set up their own mini cartels for specific regions in the Pacific Coast. The drug trade suffered more from these internal disputes than from law enforcement. Most significantly, the Mexican cartels were starting to lose ground to Asian and Colombian criminal organizations. The entry of Mexico into the drug business had been a rather accidental affair, and unless the drug cartels became efficient, Mexico risked losing most of its market share in the United States.

As the drug cartels struggled to restore their former position, the Mexican government with the help of its armed forces seemed to have succeeded in containing the drug trade. The military also had been instrumental in subduing the two insurgencies in Chiapas and in Guerrero. However, these successes, a returning prosperity, and the high popularity of outgoing President Zedillo were inadequate to prevent the opposition candidate Vicente Fox from winning the elections of 2 July 2000. For the first time since its creation in 1929, the PRI lost control of the presidency. The public had high expectations about the president-elect who announced his intention to take Mexico in a new direction.

The Presidency of Vicente Fox, 2000–2006, and El Chapo Guzmán

When in a surprise upset Vicente Fox of the conservative PAN party (*Partido de Acción Nacional*) won the presidential election of 2 July 2000, this was the first time an opposition candidate defeated the PRI, the official party until then. The victory seemed unexplainable, because Ernesto Zedillo enjoyed very high approval ratings as president of Mexico. The economy was recovering, the threat of an insurgency in Chiapas and Guerrero largely vanished, and the drug cartels were in deep trouble by 2000. The death of the Lord of the Skies (*el Señor de los Cielos*) or Amado Carrillo Fuentes and the arrest of Juan García Ábreu left their cartels in disarray, and the power struggles among drug traffickers distracted their attention from their core business of exporting Mexican marihuana and Colombian cocaine. Slowly the Mexican cartels were losing market share in the United States, and as the money and influence of the drug traffickers diminished, Mexico came to enjoy a pause in drug-related violence. Yet in spite of all the accomplishments of Zedillo, the PRI could not transfer his prestige to its candidate, and voters instead believed Vicente Fox who promised major improvements in Mexico under a PAN presidency.

Most of the promises of Fox revolved around his attempts to bring democratic practices to Mexico. In a manner reminiscent of his predecessor Francisco Madero nearly a century before, Fox offered an idealized democracy as the principal solution to Mexico's vast structural social and economic problems. While he was sure that he could avoid a martyr's fate as befell Madero, whether Fox could solve Mexico's problems seemed highly unlikely. He wisely left the armed forces alone mainly because tampering with their privileges would have meant igniting a political firestorm. However, shortly after his inauguration on 1 December 2000, he could not resist acting upon a long-held PAN core belief. Opposition figures traditionally blamed the highly politicized Ministry of the Interior for undermining democratic practices by channeling its police forces into political repression at the cost of neglecting efforts against criminal activities and the drug cartels. The solution seemed simple, and President Fox duly created a separate Ministry of National Security (*Secretaría de Seguridad Nacional*) in charge of the majority of the federal police forces. The Ministry of the Interior remained as a rump agency grouping an assortment of motley bureaus.[39]

In reality these civilian bureaus in the Ministry of the Interior had often been instrumental in providing leads to the police forces, and the new institutional separation from the police diminished the efficiency of the new Ministry of National Security. The only other large federal police force was in the Attorney General's Office, and with great fanfare Fox baptized this force with the name of Federal Agency of Investigation (*Agencia Federal de Investigación*) or AFI. The name reflected the attempt to copy the FBI of the U.S. Department of Justice, the same goal behind the establishment of the Federal Directorate of Security or DFS in 1947. And just as quickly as the previous DFS had lost all resemblance to the FBI, the new AFI resumed the function of spying on political suspects. The attempt to replicate the FBI failed miserably; indeed, it seemed impossible to clone any U.S. institution on Mexican soil.[40]

With these at best meaningless and at worst harmful bureaucratic changes, the Fox presidency completed making the innovations for the campaign against the drug cartels. In a much more effective manner, the drug traffickers had been carefully preparing a master plan to restore their declining influence. The fragmentation of the drug trade inside Mexico was costing the country market share in the United States, as Asian and Colombian criminal organizations delivered a growing percentage of the illegal drugs directly into the U.S. market. Only large cartels could successfully repulse the threat from foreign competitors, but none of the drug traffickers seemed capable of reconstituting the drug cartels existing in Mexico before 1997. Imprisoned inside a jail cell was one individual who had the abilities and the vision to restore the Mexican drug cartels to their former wealth and influence.

In one of the paradoxical occurrences so typical of the Mexican justice system, Joaquín Guzmán Loera, universally known as El Chapo Guzmán, was serving a prison term for perhaps the only crime he had not committed during his already long and fearful criminal career. He was charged with participating in the shocking assassination of Roman Catholic Cardinal Juan Jesús Posadas Ocampo in Guadalajara (Map 6) on 24 May 1993. After his extradition from Guatemala and arrest in June 1993, El Chapo Guzmán was kept in a maximum security facility near Mexico City, perhaps the only secure prison in the country. Through circumstances probably impossible to uncover, he obtained in November 1995 a transfer to Puente Grande in a suburb of Guadalajara. Although this penitentiary was supposedly also a maximum security facility, in reality its standards were not as strict. Soon El Chapo Guzmán had put the warden and most of the guards in his payroll, and the question became why he stayed so long in the prison.[41]

As long as the Lord of the Skies was alive, it did not make sense for El Chapo Guzmán to escape from Puente Grande, but once his boss died in July 1997, new opportunities appeared. The Colombian drug lord Pablo Escobar already was a legend, and El Chapo Guzmán knew that the Colombian had enjoyed a great life until the escape from prison led to a fatal end. El Chapo Guzmán was determined to avoid the mistakes of Pablo Escobar and settled down to enjoy the many amenities that prison life offered to wealthy inmates in Mexico. He became a voracious reader and accumulated an extensive collection of books on history, particularly on the history of Mexico. El Chapo Guzmán lived in spacious suites with every possible convenience, and outside individuals regularly came to provide additional services. While the press has focused on his many affairs with females, he also enjoyed princely attention ranging from private chefs to exclusive medical treatment. Needless to say, he was in constant communication with his associates and operated a lucrative drug business out of the comfort of his luxurious suites. And most important for a drug trafficker, he was safe, because the government no longer was looking for him, and the entire prison staff served as his private army to protect him from attacks by rival drug traffickers.[42]

After the election of Vicente Fox in July 2000, things began to change in Puente Grande, and rumors about the imminent escape of El Chapo Guzmán began to circulate. For El Chapo Guzmán an escape from the penitentiary was a momentous decision in his life. He could stay running his fiefdom safely from inside, but if he wanted to run a drug empire he needed the freedom to operate outside prison walls. And although escaping from prison was usually easy for recent inmates who could bribe guards, for long-term residents such as El Chapo Guzmán the same was no longer true. He had become a cash cow, and although prison officials would obey him in everything else, they did not want to lose this steady source of income. At least complicity if not collusion from high officials seems to be behind the escape of El Chapo Guzmán.

After Fox was inaugurated on 1 December 2000, the preparations for the escape of El Chapo Guzmán went into high gear. He even had many acquaintances come to visit him as if they were saying the last farewells to a person permanently departing to another country or to the next life. After hosting spectacular Christmas and New Year's parties openly celebrated as the end of an era in Puente Grande, El Chapo Guzmán was ready to make his move. On 19 January 2001 around 2200 hours a guard reported that the prisoner was not in his suite. The warden immediately sounded the alarm for a lock-down and also summoned units of the AFI. According to a prearranged plan, this police force was supposed to take full control of the penitentiary. Soon AFI agents with their faces partly or completed covered and all wearing caps flooded the place. With so many people entering and leaving the buildings in complete confusion, nobody noticed when the next day on 20 January 2001 one of the persons boarding a departing vehicle was an individual in AFI uniform who later turned out to be El Chapo Guzmán. Out of his many valuable possessions and abundant cash he had left behind in his suites, he had made arrangements to retrieve only his collection of books on the history of Mexico.[43]

The official complicity in the escape, with even claims that the DEA had suggested the idea, received additional validation by the weak pursuit after his flight. And this was not just the prison escape of one drug lord but a calculated move in the long-term strategy to bring order and rationality to the chaotic drug business. Through constant traveling across the country (also helpful to elude any pursuers), El Chapo Guzmán reestablished personal relations with most drug traffickers and persuaded them to attend a drug summit in Cuernavaca (Map 5) and Mexico City in October of 2001. Cuernavaca with its warm climate provided an ideal site for more relaxed vacation-style gatherings of the drug traffickers, but the pervasive role Mexico City has played as the locale for the majority of operations for drug traffickers has always amazed journalists and scholars. Not only have drug traffickers paraded with impunity in front of the headquarters of law-enforcement agencies in the country's capital, but the Benito Juárez International airport has always operated as the most reliable entry point for cocaine shipments from Colombia.[44]

The drug summit of October 2001 at Mexico City and Cuernavaca was a complete success for El Chapo Guzmán. At least 25 major drug traffickers attended, and after intense negotiations they agreed to establish *La Federación*, as their consortium. The drug traffickers, as was their tradition, doubtlessly sealed these agreements with wild orgy-like parties involving liquor, music, and prostitutes. A council with representatives from the main drug organizations ruled *La Federación* under the guidance of El Chapo Guzmán. That so many drug traffickers agreed to an apparent power grab by him reflected the desperate nature of the business, because all knew that unless Mexico regained market share in the United States, the drug business was doomed to decline. A key component of the agreements was that each trafficker would make available his routes to other drug traffickers on a temporary basis; the strategy was to be always keeping border officials guessing as to the entry points of the next cocaine shipments. U.S. officials as soon as they blocked one route suddenly encountered a flood of drugs appearing in another part of the border.[45]

However, not all the drug traffickers became members of *La Federación*. The Gulf Cartel did not join but agreed to collaborate in the interest of maintaining peace and restoring prosperity to the business. In particular, the Gulf Cartel agreed for a moderate payment to allow members of *La Federación* to use its routes passing from Monterrey and Nuevo Laredo to the Texas border. In the extreme north west of Mexico, the Tijuana Cartel (Map 6) of the Arellano Félix brothers failed to join *La Federación*, probably because from the start El Chapo Guzmán had made other plans for the Tijuana Cartel.

He knew that profit alone could not make *La Federación* succeed and that the new cartel could only impose discipline by instilling a healthy fear among drug traffickers. He also needed to show the Mexican and U.S. governments the benefits of collaborating with him. Barely had the agreements been signed when El Chapo Guzmán launched an offensive against the Tijuana Cartel. At first it seemed just a war between rival drug cartels, but soon *La Federación* was providing AFI and DEA with excellent intelligence on the location and the operations of the Tijuana Cartel. Following up these leads, AFI racked up an impressive number of drug seizures and arrests. The decision of President Fox to establish this new police agency inside the Attorney General's Office seemed vindicated. And the Fox administration earned even more goodwill with U.S. officials by sharply increasing the number of extraditions.[46] Already in the first three years of the Fox presidency, Mexico extradited more criminals to the United States than in the entire Zedillo presidency (Table 4). Even though a favorable Mexican Supreme Court ruling eased the requirements to extradite individuals to the United States, the government always retained the right to refuse extradition for Mexican citizens.[47]

By early 2002 the Tijuana Cartel had been crushed, and the Fox administration saw this success as proof that its new anti-drug policy was bringing positive results. The U.S. government was very pleased with the Fox presidency, and it seemed that the change from rule by the PRI to the PAN political party had helped the campaign against the drug cartels. Actually, the amount of drugs confiscated stayed within the range of traditional figures. The publicity surrounding the confiscation of quantities of drugs belonging to the Tijuana Cartel fooled U.S. officials who only years later realized that in spite of these seizures, the volume of cocaine entering the United States barely declined. And indeed, while the Tijuana Cartel no longer could get its cocaine into the United States, *La Federación* quietly picked up the slack and provided more than adequate replacement cocaine for American consumers. El Chapo Guzmán by the middle of 2002 controlled the drug trade in the entire Pacific Coast region of Mexico and was reestablishing the bases the Lord of the Skies had enjoyed in Central America.[48]

With good luck still on his side, El Chapo Guzmán decided to take the next big step of trying to create a monopoly cartel over all of Mexico. Even the Lord of the Skies had preferred in the 1990s not to challenge his rivals in the Gulf Cartel, but El Chapo Guzmán was determined to surpass his former boss. Late in 2002 he decided to hurl the victorious *La Federación* against the Gulf Cartel, which, however, was ready and waiting for him.

During the late 1990s this cartel succeeded in recruiting army deserters to form security forces and death squads later famous in Mexico under the name of the Zetas. Another opportunity to find recruits came when the Guatemalan government went too far in reducing the size of its army after the end of the guerrilla war. Soon the Gulf Cartel was recruiting unemployed personnel from the Kaibiles, the Special Forces of the Guatemalan army.[49]

When El Chapo Guzmán sent his followers to invade the territories of the Gulf Cartel, the Zetas were waiting to repulse the attacks. When AFI agents came to deliver the knockout blow against the Gulf Cartel, the police usually could not match the weapons and the combat skills of the Zetas whose savage brutality rapidly gained national notoriety. The agents did intercept some cocaine shipments and secured enough arrests and extraditions to keep U.S. officials satisfied with the progress of the drug war in Mexico. But the government knew that the firepower of the Zetas required calling in the army to counter this powerful armed band. The army already had played a supporting role in destroying the Tijuana Cartel and now struck a greater blow against the Gulf Cartel by capturing its chief, Oziel Cárdenas Guillén, in Nuevo Laredo on 16 March 2003. Police forces did not know anything about this strike, and doubtlessly El Chapo Guzmán provided the army with the information to track down the rival drug lord. In spite of this success, the offensive of *La Federación* proved very costly, and El Chapo Guzmán decided to withdraw in order to regroup for a second try at destroying the Gulf Cartel, and he did return in 2008 as chapter 10 explains. And without the intelligence *La Federación* provided, arrests and drug seizures declined.[50]

The Gulf Cartel, however, did not resign itself to waiting for the next blow and sent Zetas to penetrate the heartland of *La Federación* in Guerrero state in May 2005. But when the Zetas arrived in Acapulco (Map 5) and started to follow the movements of members of the rival cartel, it turned out that they were being watched by the AFI. When on 14 May 2005 the Zetas were preparing to attack, AFI agents struck and wiped out many of them. In hot pursuit and helped by *La Federación*, AFI finished killing the rest of this advance team of the Zetas on 15 May 2005. For his part El Chapo Guzmán sent his own armed followers to pursue and kill any survivors of the Zetas in Guerrero state.[51]

The AFI had proven effective against the Zetas when they operated outside their home territories but had not been very successful against them in their traditional bases such as in Nuevo Laredo. Seeing the violence spiral out of control, President Fox called in the army to help the AFI in the struggle against the Zetas in Nuevo Laredo. He announced this army deployment as part of his new program labeled "Safe Mexico" (*Mexico Seguro*) to stop the fighting between the drug cartels. The real purpose of the army deployment was to gain time to tide over his administration until the next presidency. The country was already immersed in the campaign leading up to the presidential elections on 2 July 2006. Fox wanted to maintain the impression that his administration had made real progress in the campaign against drug traffickers, and the external display of military patrols and convoys brought a temporary decline in fighting between drug cartels.[52]

But all that President Fox had done was postpone dealing with the problem in Mexico. As chapter 10 shows, it was up to his successor to decide what to do about the growing power of the drug cartels, and as actions by Fox already anticipated, the armed forces were going to be central in any war against the drug lords in Mexico.

5

The Militarization of Colombia and Venezuela

> So what enables an intelligent government and a wise military leadership to overcome others and achieve extraordinary accomplishments is foreknowledge.—Sun Tzu[1]

The expansion of the armed forces of Colombia and Venezuela intensified after 2005. Colombia was enlarging its military to try to end the guerrilla insurgency existing in the country since the 1960s. The mission of the Venezuelan armed forces was radically different, because their primary obligation was to defend the Bolivarian Revolution from a U.S. invasion. As both countries increased their armed forces, tensions appeared in their relations. A war scare even came close to starting hostilities between Colombia and Venezuela.

Colombia: Sustaining the Momentum

As the year 2006 began, the military offensive against Colombian guerrillas was faltering. The Revolutionary Armed Forces of Colombia (*Fuerzas Armadas Revolucionarias de Colombia*) or FARC remained on the defensive, and the armed forces seemed incapable of crushing the Colombian guerrillas. Attention shifted from the guerrilla war to brutality inside the army when the press revealed that 21 recruits had been savagely tortured by noncommissioned officers. The physical abuse of recruits had been traditional practice in the Colombian army since the nineteenth century, but the torture of soldiers was a new low. When the reports of torture reached the press, President Álvaro Uribe Vélez took steps to prevent the scandal from growing, and he promptly fired army commander General Reinaldo Castellanos in February 2006.[2]

The firing of General Castellanos caused tremendous disarray among senior officers who did not know whether they were next in line to lose their positions. The anxiety eased when the new Defense Minister, Juan Manuel Santos, indicated that the reshuffling of senior commanders and the dismissal of several generals in June 2006 marked the end of the housecleaning.[3] The new Defense Minister once again made the defeat of FARC the primary mission of the armed forces, but he soon realized that the current approach was not working. What happened over previous years was that "each military victory requires the Colombian army to occupy more territory, leading it to convert mobile brigades into units of fixed area control and reducing offensive capacity."[4]

Defense Minister Santos concluded that Colombia needed a new four-year plan for the next stage of the war against FARC, but in a tacit admission that final victory was still not within reach, the military gave the name of *Consolidación* to the new plan. As its name suggested, the aim was to preserve existing gains rather than to finish extirpating the last guerrillas. The strategy assumed that once the armed forces reduced FARC to very remote regions of the country, the demoralized guerrillas would eventually abandon their struggle to overthrow the government. But for *Consolidación* to succeed, the military had to resume the practice of sending small-unit patrols as the way to keep the guerrillas under constant pressure.

Out of fear of ambushes, in the late 1990s the army shifted to columns of battalion size or larger and thus gave the guerrillas ample warning of troop movements. An outer perimeter of sentries sufficed to alert the main guerrilla force that a large army detachment was noisily approaching. American advisors and the U.S. Embassy had been urging the army to take a more aggressive if riskier approach against the guerrillas "involving an expanded layout of forces with platoons sent further afield to rotate around the main camp. Widening each battalion's radius of operations will expand its footprint and multiply offensive contact with the guerrillas."[5] To implement this change in tactics, Defense Minister Santos ordered the deployment of veteran soldiers to the front line and the posting of new conscripts only to garrisons in peaceful regions. In an interesting attempt to find the most suitable soldiers for *Consolidación*, the army stepped up its recruitment of former guerrillas into the ranks. Other rebel groups had demobilized in the 1990s, and because many former guerrillas were having a hard time finding jobs in the civilian sector, entry into the army gave them a chance to apply their military skills.[6]

Obviously both President Uribe Vélez and Defense Minister Santos realized that the armed forces needed to increase in size, and to finance this expansion, the government obtained congressional approval in 2006 to renew the wealth tax for another four years. Just like for the first wealth tax, the majority of these funds went to cover the cost of adding 14,000 recruits to the armed forces and 20,000 to the police. The number of recruits was larger for the police because the U.S. government provided additional funds for the reestablishment of a police presence in areas recaptured from guerrillas. For the first time in the history of Colombia, the number of military personnel, not counting the police, came to the record-high figure of over a quarter of a million. And the figure of nearly six soldiers per 1000 inhabitants also reached the highest number in the history of the country. Under President Uribe Vélez the militarization of Colombia was becoming a reality, yet in spite of constant high unemployment, the armed forces struggled to meet their quotas of recruits.[7] The violence and brutality traditionally suffered by recruits had made service in the army a last option for unemployed Colombians.

Somehow recruiters met their quotas for privates, but finding adequate numbers of non-commissioned officers (NCOs) and officers required drastic measures to find individuals with the requisite abilities. One of the structural defects of the army in the war against FARC was that soldiers left the army after their tour of duty. This weakness meant that veteran guerrillas generally faced raw conscripts in clashes. The army had tried to correct this error by giving incentives for soldiers to reenlist for at least another tour of duty. However, promotion from private to NCO normally could not be an incentive for reenlistment, because the prevailing practice was that they should come from a separate career track and not from the ranks of privates. NCO candidates underwent a training program of 18 months and then received a posting to a peaceful zone for two years before being finally sent to combat in the front lines. The prevailing system obviously could not provide large numbers of NCOs to meet the sudden expansion of the army. As a temporary measure and a radical innovation for class-rigid Colombia, the army for the first time opened promotion to NCO rank but only by invitation to the most experienced combat veterans. If these candidates passed a NCO training program lasting only six months, then they were immediately sent to combat units, and in this way the army covered its shortage of 5,143 NCOs.[8]

Colombia's rigid class system held firm for entry into officer rank, and nobody ever dreamed of promoting even the most gifted NCOs to second lieutenants. The military academy graduated a maximum of 600 officers each year but this number was vastly inadequate for the expanding army.

Merely admitting more high school graduates into the academy was not a solution, because the army could not wait three years until the first cadets graduated. At the academy half of the courses were non-military and helped provide officers with a well-rounded education. The army needed an additional 1,425 officers, and as an ingenious alternative, the academy started an accelerated program for university graduates. These individuals already had mastered the academic component and could graduate from the army academy after completing their purely military courses in one and a half years. Since the 1960s Colombia has suffered from a surplus of unemployed college graduates, so finding citizens with college degrees who were willing to enter into a military career was not too hard.

The graduates from this accelerated program could expect to rise in rank to captain and possibly major, but higher rank seemed highly unlikely. In Colombia as in other Latin American countries, the members of a particular graduating class of the army academy remained a tightly bound group, and they inevitably supported the attempts of their fellow class alumni to rise to high rank. Aware of this inherent bias, the civilian government instituted a system to try to make sure than only the best officers rose to the rank of general or to the rank of admiral in the case of the navy. Once a year full colonels and navy captains filed applications for promotions and had their names posted on billboards at military installations for public comment. A committee of generals and admirals reviewed the responses and studied the files and service records of the applicants. Then all generals, admirals, and the Defense Minister voted on the candidates. To qualify, an applicant had to receive two-thirds of the votes, and this meant that the number of promotions varied each year from as low as none to a maximum of twenty. Final authority, however, rested on the president who could reject any and all candidates. And success in the voting gained only admission to the Advanced Studies War College, because graduation from that institution was the final requirement for promotion to general or admiral.[9]

The shortage of officers and NCOs was most acute not in the army but in the naval Marines. The Marine Corps of Colombia was one of the largest in the world, and graduates from the naval academy in Cartagena provided the officers for the Marines. Naval officers preferred to serve aboard ocean warships and regarded commanding the eight battalions deployed in the river wars against FARC as a highly undesirable posting. Marines routinely rode aboard patrol craft and were frequent targets of FARC ambushes. The Marines suffered casualties higher than those of army units, and not surprisingly, two-thirds of officer and NCO positions went unfilled. In spite of these odds, the Marines developed novel strategies to patrol the rivers. Specially built mother ships (*nodrizas* or wet nurses) served as floating repair platforms and provided logistical support to the boats patrolling the rivers. By shifting locations, these mother ships deprived FARC of easy targets and also allowed rapidly sending reinforcements to threatened areas.[10]

As far as acquisitions of new weapon systems, the most important were the jet fighter bombers. In 2005 Colombia wanted to purchase Super Tucano light attack jets from Brazil, but U.S. approval was needed because these planes contained many U.S. components, including the jet engines. The U.S. government preferred that Colombia acquire additional cargo airplanes as more urgent for deploying troops to guerrilla zones, but because these Brazilian subsonic planes proved very effective for bombing guerrilla targets, the U.S. government unenthusiastically approved the purchase.[11]

Colombia made a strong case for counterinsurgency with the Super Tucano purchase, but this argument was not as convincing when seeking replacements for its aging supersonic Mirage fighters. In 1989 the air force acquired 11 Kfir fighters manufactured in Israel, but these too were in need of an upgrade. Israel proposed updating the planes and also throwing in at give-away prices an additional 13 Kfirs of the same model on condition that Colombia also pay for their upgrades. Although commercial considerations predominated in the purchase, the deal seemed too good to pass up. Because the Kfirs also contained U.S. components, both Israel and Colombia pressured the U.S. government to authorize the sale. The possibility that FARC might obtain Surface to Air Missiles (SAMs) able to shoot down the Super Tucanos finally tipped the balance in favor of the deal with Israel because the upgraded Kfirs could overcome SAM attacks. And not mentioned was the fact that the advanced electronics of the Kfirs easily accommodated the latest model of U.S. air-to-air missiles in case the U.S. government decided to use Colombia as a proxy in a war with Venezuela.[12]

A second weapon acquisition for Colombia was a new combat system for its two Diesel submarines. These two vessels were in very good condition, but rapid growth in technology made obsolete their 1970s electronics. The Colombian navy had been requesting this upgrade for years, but the U.S. Embassy refused to approve the purchase as unnecessary. But when both Colombian police and the Drug Enforcement Administration (DEA) confirmed that drug traffickers were using semi-submersible craft to smuggle drugs to the United States, the attitude of the diplomats changed. The existing sonar of the submarines could not track these semi-submersible craft, but the new technology possessed that capability. By disrupting the flow of drugs to the United States, the new digital combat systems would help to reduce the profits drug lords were sharing with FARC and thus in this way the vessels could contribute to the war against the guerrillas. This new evidence persuaded the U.S. Embassy to authorize the installation of the new combat systems in the two Colombian submarines.[13]

U.S. military assistance was an integral part of the government's *Consolidación* Plan for the war against FARC. Officers and technicians in all branches of the armed forces and the police received extensive training both in Colombia and in the United States. As mentioned earlier, U.S. funds went directly to finance the expansion of the Colombian police in areas recently reclaimed from FARC. The number of helicopters in the armed forces and in the police rose from about two dozen in 1998 to 285 in 2008, with half of the new helicopters coming from the United States. The most important contribution of the U.S. government was providing technology to the armed forces, and most of the training went to prepare Colombian personnel to learn how to handle the new digital and electronic combat systems. Rapid and secure communications were essential for rapid deployment, and U.S. assistance provided the Colombian military with radio equipment to create an extensive network. U.S. officials also provided and installed an intelligence communications computer system so that officers could have easy on-line access to sensitive information.[14]

The computer system for intelligence activities turned out to be indispensable for tracking the information for what became the most visible application of U.S. technology in the war against guerrillas: drones. These unmanned aerial vehicles or UAVs proved to be of immense value in hitting the FARC with both air and ground attacks:

> UAVs have proven useful before, during, and after strikes against FARC. In the planning stage, they provide updated reconnaissance imagery of target sites previously identified by human and signals intelligence. Prior to launch, the Colombian Air Force deploys UAVs to target sites to assess weather patterns before putting bombers in the air. The UAVs then keep their cameras trained on targets to ensure identification until planes arrive. Loitering UAVs give commanders a real-time, bird's-eye view of Colombian military assaults, enabling them to review pilots' accuracy and to provide battle damage assessments. When inserting troops at enemy sites, the Colombian Army uses the UAV for force protection and as a lookout in case of anti-aircraft threats to its helicopters. After strikes are over, UAVs can continue site surveillance, watching for enemy survivors and their escape routes.[15]

Either by guiding airplanes for bombing runs or for directing commando teams to the hiding places of guerrilla leaders, the drones seemed capable of inflicting irreparable and deadly damage on FARC. As U.S. officials proudly stated, 70 percent of the successful strikes against guerrilla targets resulted from U.S. participation.[16] The drones had their limitations, however, "they are meant to supplement but not substitute for human intelligence/signal intelligence leads."[17] More significantly, drones could only be effective in hastening the end of the war if they formed part of a larger effort directed against the guerrillas. The crucial requirement for victory was relentless patrolling on the ground, and here drones could only partially fill the void. Drones fed the illusion that the government could finally achieve victory in the guerrilla war by targeting senior commanders of FARC.

Colombian officials regarded the aerial vehicles as a great opportunity to end the war with victory, and every time drones helped to eliminate a major FARC leader, the official propaganda repeatedly announced victory in the war. But the guerrilla resistance continued and the military offensive lost steam because of two main reasons. In 2008, U.S. military assistance declined by $40 million or a nearly 40 percent drop. And the U.S. Congress specified a shift from military to primarily economic and social assistance. Secondly, when the world economic recession hit Colombia, the slowdown in the economy reduced the revenues of the government in 2008. The treasury imposed a cut of 14 percent in the budget, including military expenditures.[18]

The sudden drop in funds from the United States and from Colombia abruptly halted the expansion of the Colombian military. Only by organizing new mobile units to throw into battle could the military hope to shatter the remaining FARC zones, but that alternative became financially impossible. FARC's "deep historic roots" and "detailed knowledge of the terrain as well as protective family ties and militia networks" gave the guerrillas tremendous capacity to resist indefinitely.[19] The estimates for its strength remained between 8,000 and 12,000 fighters.[20] But resiliency did not mean success, and when FARC tried in February 2009 to deploy units for an attempt to infiltrate the areas near Bogotá, the Colombian armed forces were ready with the new U.S. technology. As drones tracked the movements, the new airplanes bombed and shattered the guerrilla columns.[21]

A chastened FARC retreated again to its core zones in the headwaters of eastern rivers such as the Meta, Guaviare, and Guayabero (Map 4). Skirmishing and mutual raids continued over the next years, but it was becoming clear that neither side could destroy the other completely. The Colombian military controlled the most valuable territories in the country, but FARC was safe in its hidden redoubts in the southern and eastern parts of Colombia. The guerrillas lacked the means to increase their capabilities in any significant way, and the Colombian government could not raise additional revenue through taxation to enlarge the army. Whatever chance existed that the Uribe Vélez administration might collect additional tax revenue from the citizens ended when the Colombian army fell into the worst scandal in its history, as chapter 7 will show. And without additional resources, the Colombian military could only promise an endless stalemate in the war with the guerrillas.

The Venezuelan Military After the Reelection of Hugo Chávez in December 2006

As expected, Hugo Chávez in his inaugural speech on 10 January 2007 reaffirmed the socialist phase of his Bolivarian Revolution. Even before his inauguration he announced a Five Year program to make the Venezuelan military more revolutionary. He began by adopting the title of Bolivarian Armed Forces as the military's official designation and intensified the indoctrination of officers. All military personnel had to take a new oath pledging loyalty to "the country, socialism, or death." Because not everyone might accept enthusiastically the new direction, Chávez planned to incorporate 57,000 new military personnel into the armed forces to replace any discontents. On 24 June he gave a 30 percent pay raise to the military. On 5 July 2007 on annual promotion day he announced that 3,300 officers were advancing in rank; he also proclaimed that the most important criteria in selecting the individuals were fervor and loyalty to the Bolivarian Revolution.[22]

The next step for Chávez was to make sure that the military establishment solidly embraced his movement. As mentioned in chapter 2, Defense Minister General Raúl Baduel was uncomfortable with the new Bolivarian direction Chávez was giving to the military. The opposition failed to use this split inside the military, and this public divergence of views actually helped to dispel the accusations that Chávez was a despot. Very astutely Chávez postponed housecleaning until after the presidential election had passed. He waited until 6 July 2007 to announce the replacement of General Baduel as Defense Minister by General Gustavo Rangel Briceño, the commander of the reserves, who previously had reported directly to the President outside the normal chain of command. General Baduel did not depart quietly from office, and in a final speech he emphasized the need to put restrictions on the 21st century socialism of Chávez to prevent Venezuela from falling into a Soviet style regime.

To the media's surprise, Chávez thanked Baduel for his many years of service and for his thoughtful remarks. The President even urged his followers to study the speech because he too did not want Venezuela to turn into a Soviet regime. His gratitude at Baduel for having been one of the saviors of Chávez during the 2002 April coup was not infinite, however. When the opposition goaded a gullible Baduel into making anti–Chávez statements rather than expressing only policy concerns, this was too much for the President. In a year Baduel faced criminal charges and was imprisoned for having engaged in corruption while defense minister.[23]

The opposition and the media tried to exploit the Baduel episode to the maximum, but inside the military the fall from grace of the previous Chávez insider fell rather flat. "Few, if any, military officials have echoed Baduel's concerns in recent days. The deafening silence of military officials in the face of sweeping changes to the FAN indicates that in the Bolivarian Republic of Venezuela there is much more to gain from appearing loyal to Chávez than to lose in opposing Chávez radical program."[24]

Loyal as the military seemed to be, Chávez continued to reduce the remaining contacts with the U.S. military. One of the last links was the Conference of the Armies of the Americas, a two-year cycle of technical conferences on topics of interest to the armies of the Western Hemisphere. The Secretariat of the Conference rotated among the member armies, but when it was Venezuela's turn to be the host for 2008, Chávez cancelled Venezuela's participation, because the Conference meant extensive contact with U.S. officers stationed in San Antonio, Texas. In his first term Chávez strove to isolate Venezuelan officers from their U.S. counterparts and now in his second term he extended his efforts at isolation to the multilateral arena.[25]

In many ways 2007 marked the high point of Chávez's efforts to transform the military of Venezuela. The armed forces and the reserves seemed to be increasing in number, and the wholesale transfer of officers into civilian jobs did not seem to have diminished the efficiency of the military, although warnings to the contrary were not lacking. A major step toward military independence came when in late October 2007 construction began for a new rifle factory equipped with Russian machinery. Once Venezuela, enjoying its own supplies of steel and aluminum, manufactured its rifles and ammunition, its vulnerability to foreign suppliers diminished substantially. And with the long-promised arrival of many new advanced Russian weapons, Venezuela seemed poised to become a regional power.[26]

Because Chávez in his statements frequently confused weapons ordered with those delivered, the U.S. Embassy in Caracas was always skeptical about the real threat posed by the new military capabilities of the Venezuelan armed forces. "As there remains a significant disconnect between the emerging asymmetric doctrine of the Bolivarian Republic of Venezuela and these arms purchases, this arms build-up will not upset the balance of power in northern South America unless training and logistics greatly improve."[27]

Confirmation of the skepticism came during the March 2008 crisis with Colombia. The war scare began on 1 March 2008 when a Colombian strike team attacked a guerrilla camp of the Revolutionary Armed Forces of Colombia (FARC) and killed one of its leaders, Rafael Reyes. The raid became an international incident because the camp was inside Ecuador, and Colombia had attacked without seeking authorization. Ecuador was a close ally of Venezuela, and Chavez sensed in the raid an opportunity to show off his new military establishment. In his first responses, he ordered mobilization for war with Colombia and closed the Venezuelan embassy in Bogotá.[28] The war scare frightened many observers, but the first comforting news came when the U.S. Embassy in Caracas reported that "nearly 48 hours after Chávez ordered ten battalions to the Colombian border, there has been only small scale movements of troops or equipment from major combat units … a lack of training, rehearsal, and failure to invest in unglamorous military equipment like trailers and cargo planes has slowed the Venezuelan military's mobilization."[29]

As the world fearfully expected the outbreak of a war, it seemed that Chávez was waiting only for his army to complete the deployment to the border to order the attack to begin. But as the days turned into a whole week, gradually it became clear that the Venezuelan army was having major difficulties trying to reach the Colombian border. For unclear reasons, most soldiers had not received the new AK-103 rifles from the Russian shipment of 100,000, and officers scrambled to distribute the new rifles to their soldiers. Sending soldiers to war without any training with new weapons, no matter how reliable and simple, did not seem a very good idea.[30] Most of the Venezuelan armed forces normally were stationed far from the Colombian border, and their deployment was proving to be slow and exhausting. "The Venezuelan army remains far from reaching its stated deployment of 10 battalions since numerous tanks and armored vehicles are still awaiting transport."[31]

When Chávez learned on 9 March 2008 that barely a third of the ten battalions had reached the border, he realized that he had to abort the deployment to avoid embarrassing publicity. In a dramatic gesture, that same day he "abruptly changed roles from Field Marshal to Peacemaker" when he announced his decision to recall the battalions from the border and to reopen the Venezuelan embassy in Bogotá.[32]

No indication exists that Chávez had ever wanted to start a war with Colombia, and indeed denouncing the neighboring country as a proxy or an ally of the U.S. government was an inexhaustible source of propaganda for domestic politics. But he did want a show of force to let Colombia and the United States know that Venezuela was a power to be reckoned with. Instead "the Venezuelan Army's embarrassing deployment in the 2008 purported war with Colombia" took the winds out of his political movement.[33] The poor performance of the army in what could at most be considered routine maneuvers confirmed all the reports and rumors about the diminished capabilities of the military. In previous years Chávez took a big gamble by spreading the scarce managerial talent inside the military across the government, and the result was that he lost an efficient army and did not transform the country.

He had tried to grasp too much all at once and could not deliver on his promises. Chavismo was in no danger of imminent demise, but after the fiasco of the March 2008 mobilization against Colombia, the upward ascent of the Bolivarian Revolution ended. Chávez had to fight just to maintain the gains he had made, and in a recognition of his predicament, he made an attempt to reach out to the new Barack Obama administration early in 2009. But after hopes of a restoration of friendly relations with the United States faded, he decided to manufacture a second war scare in 2009 against Colombia as a way to bolster internal political support. A second mobilization in 2009 was only slightly more successful than that of 2008. And the abilities of the military continued to decline when the U.S. government, in reprisal for the temporary expulsion of the U.S. ambassador to Venezuela on 11 September 2008, stopped training Venezuelan officers.[34] In 2009 "Chávez's political project faces diminished political support due to deteriorating public services and a crime problem that the Government of the Bolivarian Republic of Venezuela has been unable to contain."[35]

In one major example, the internal struggles over the political composition of the Venezuelan police had weakened its crime-fighting capabilities. By 2009 a crime wave swept the capital and gave Caracas one of the highest homicide rates in the world. The deterioration was in sharp contrast to the situation existing before 1999 when the new Chavista penal code released 12,000 criminals into the streets. The overstretched resources of the military could not take on this new responsibility, and in response to the crime wave, in April 2008 the National Assembly created a new Bolivarian National Police to complement the efforts of local and provincial police forces. In January 2010 the first graduates from the training programs of the Bolivarian National Police began to patrol dangerous neighborhoods of Caracas, and they seemed to be making a dent in crime. But the reality remained that only very costly programs of training, education, and enforcement could make possible the reduction of crime in Caracas.[36] And crime was just one of many problems hampering the Bolivarian Revolution as it entered the second decade of the twenty-first century.

6

Shaping the Arsenals of the Armed Forces

> *Today the United States is practically sovereign on this continent and its fiat is law upon the subjects to which it confines its interposition.*—Secretary of State Richard Olney[1]

The U.S. government has exercised its influence to determine those weapons the countries of the Caribbean and Mexico can possess. The efforts of the U.S. government to shape the arsenals inside its informal empire have fallen into five main categories. The first is determining the armament that U.S. officials consider suitable for a specific country, as section 1 on Nicaragua and section 3 on Mexico demonstrate. Sometimes the arms happen to be the best choice but usually they satisfy U.S. objectives rather than the needs of a country. The most obvious way for the U.S. government to shape the arsenals of governments has been military assistance and training. Officers trained in the United States are more likely to prefer American weapons over European or Chinese alternatives. Countries can obtain arms at below market prices and sometimes for free from the U.S. government. The weapons usually consist of surplus and discarded items or at best of stripped down models without the newest technology. But if U.S. interests dictate it, countries may receive the latest versions. At the end of the Cold War the U.S. government dumped many surplus weapons on countries throughout the world at bargain-basement prices and sometimes for just the shipping costs, but by the start of the twenty-first century the great giveaway was over. Since 2000 the U.S. government has taken a very narrow approach to military assistance. The two most important goals have been the struggle against the drug cartels in the region and counterinsurgency in Colombia and, to a lesser degree, in Peru.

A second category of control was limiting the quantity of a particular weapon in the hands of one country. Thus, small quantities of a weapon may well fit with U.S. interests, but large amounts might be menacing. Limiting the numbers could also be a fallback position for the U.S. government, as happened in Nicaragua. Unable to persuade the Nicaraguan government to destroy all its Surface-to-Air Missiles (SAMs), the U.S. government reluctantly shifted to a policy of reducing their numbers.

The third category was the commercial motivation to make profits by selling American weapons to governments and individuals in those countries. In the sales of small arms to individuals, U.S. embassies closely watched the private arms trade to make sure that American weapons did not fall into the hands of criminals or drug traffickers. Because all the Latin American countries have very severe restrictions on private gun ownership, the task of U.S. diplomats was not particularly intrusive and coincided completely with the goals of local governments. U.S. diplomats carried out inspections to make sure that merchants were following proper security procedures to keep criminals from buying or stealing guns at gun shops and retail outlets. In reality, moral, ethical, and even legal considerations obligated the U.S. government to help control the trade in small arms throughout Latin America. The open arms bazaar in the United States was the source of the weapon smuggling rampant throughout the Caribbean, Central America, and Mexico. The problem of arms smuggling originated in the United States and just as for the drug trade, the U.S. government had the responsibility of helping to stop the illegal flow of weapons into the hands of criminals throughout Latin America.

Much more exciting was the competition for advanced strategic weapons such as missile systems and jet fighters. As expected, U.S. diplomats stood ready to use every tool at their disposal to persuade Latin American governments to purchase American instead of European versions of the same weapons system. Dramatic as was the intense lobbying to land weapons contracts for American companies, this phenomenon primarily concerned the large countries of South America, in particular Argentina, Brazil, Chile, and Peru.

Commercial competition over arms sales was a very infrequent occurrence in Central America and the Caribbean because so many of the governments could not afford expensive hardware or even bulk purchases of cheap items. The small arms and simple equipment these countries usually preferred for internal security were often available at knockdown prices from European and Asian suppliers. And in many cases American companies no longer manufactured those obsolete arms or did not have comparable items. Local competition existed, for example Mexico and Colombia operated their own munitions factories and frequently produced pistols and small arms to satisfy the needs of their army and police. Lobbying to secure U.S. military assistance rather than fighting to obtain contracts for American manufacturers was usually the only realistic activity for U.S. diplomats stationed in the poor Caribbean and Central American countries.

The fourth category consisted of imposing an arms embargo either on a particular country or on one combatant as happened during the Mexican Revolution. The best example of this policy was the arms embargo the U.S. government placed on the Fulgencio Batista tyranny in Cuba in March 1958. The fifth category was much more invasive than the fourth because the U.S. government not only placed an arms embargo on one country but also brought massive pressure on all other countries to halt arms sales to a particular Caribbean government. The U.S. used this policy very astutely to contribute to the overthrow of the Jacobo Arbenz regime of Guatemala in 1954 and less effectively since 1959 to try to weaken the revolutionary regime in Cuba.[2] For the twenty-first century, the most dramatic application of the policy of denying weapons took place in the Venezuela of Hugo Chávez, the topic of the second section.

Whatever the specific tool, U.S. officials have quietly and constantly pressured local governments to acquire only those arsenals most suitable to meeting U.S. policy goals. At a time when Washington shifted its attention and the bulk of its resources to Asia and the Middle East, complete success was not possible as this chapter shows. Nevertheless, U.S. diplomats, having only crumbs of military and economic assistance to use as leverage, succeeded in guiding local officials to accept and sometimes even to embrace enthusiastically the assigned role for the countries of the region.

Nicaragua: Surface to Air Missiles

SAMs (Surface to Air Missiles) in Nicaragua were a legacy of the Cold War. As part of the Ronald Reagan administration's efforts to overthrow the Sandinista regime, the CIA created a Contra army and air force. Because Nicaragua lacked air defenses, CIA aircraft were free to supply Contra units from the air and also to raid the country with impunity. The first response of the Sandinista regime was to seek to acquire Soviet MIG fighter bombers as the best defense against incoming CIA aircraft stationed in Honduras. But the advice of Fidel Castro proved decisive, as he persuaded the Sandinistas to obtain helicopters instead of MIGs and to rely on missiles for air defense. The Soviets duly complied and provided Nicaragua with a fleet of transport helicopters and 2,051 SAM-7s.[3]

The Contra war ended in 1988 and the Sandinistas lost power in 1990 after their defeat in the election of that year. In the politically hectic years of the 1990s, the U.S. government forgot about both the SAM-7s and Nicaragua. Nobody seemed to care that in the mainland between the U.S. border and South America no other country possessed SAMs. Actually, in the entire Caribbean region only Socialist Cuba had large stocks of similar missiles, and the U.S. government always resented having to tolerate this huge defensive arsenal in that island. U.S. indifference to Nicaraguan SAMs changed to panic after the terrorist attacks of 9/11 in 2001. The U.S. government declared a war on terror and soon was making wild and irrational responses to the new terrorist threat, the most blatant being the unnecessary invasion of Iraq in 2002.

Among other paranoid fears, the George W. Bush administration fanned the worry that terrorists could use SAMs to shoot down commercial airliners and cause hundreds of deaths. Remote as was that possibility, nobody could guarantee that terrorists would never use those missiles. The argument that Nicaraguan SAMs might fall into the wrong hands was considerably weakened when the U.S. government provided assistance to establish secure storage facilities in that country.[4] What was almost never mentioned was that by depriving countries of these defensive missiles, the U.S. government acquired freedom of action to bomb and attack those countries by air with complete impunity. The U.S. government had never approved Nicaragua's acquisition of SAM-7s in the 1980s and used the terrorist threat after 2001 as leverage to correct that historical wrong. If Nicaragua destroyed its SAM-7s, the right of the U.S. government to dictate what weapons a particular country could have would also be publicly confirmed. As far as armament was concerned, the U.S. government was the "Big Brother" making sure that Caribbean countries and Mexico conformed to U.S. security requirements and never possessed any weapons capable of making meaningful resistance to any U.S. military intervention.

By 2002 a full media campaign was underway to persuade the Nicaraguan public that SAMs were unnecessary and actually useless for the country's defense. Nicaraguan army officers resisted the idea, and to overcome their opposition U.S. diplomats offered helicopters and other military hardware in exchange for the destructions of some missiles. But U.S. diplomats were too clever to sign formal agreements and merely promised in 2002 that if

Nicaragua destroyed the SAMs now, then the U.S. might increase its military assistance at some indeterminate future. Not surprisingly, vague promises were unsatisfactory for Nicaragua and the talks went nowhere.

Undaunted by this initial setback, ingenious U.S. officials decided to bring pressure on the Nicaraguan government from another side. In return for destroying the SAMs, "Honduras has been encouraged to reduce or eliminate its fleet of F-5s as a quid pro quo to Nicaragua."[5] Whatever chance this proposal had of securing Nicaraguan approval, the firm refusal of Honduras to participate doomed the exchange. The government of Honduras, in another example of its unreliability as an U.S. ally, insisted that this bilateral issue concerned only the United States and Nicaragua. Having a proxy country defend U.S. interests against a third country has been a favorite ploy of U.S. imperial diplomacy, but in the case of Honduras and Nicaragua, the scheme collapsed.[6]

An ultimatum during the October 2003 visit by Secretary of State Colin Powell to Managua and a threat to reduce military assistance by two million dollars provoked a panicky response from the Nicaraguan government in 2004. Between May and November of that year, the government eliminated one thousand SAM-7s, and the destruction was interrupted only when the National Assembly intervened in November 2004 and stipulated that the elimination of the remaining missiles required an approval vote of a two-thirds majority. The National Assembly was manifesting the concern of senior army officers who insisted that they could not guarantee the defense of Nicaragua without the SAMs. In particular Nicaragua without any air force was incapable of stopping the Honduran F-5Es from bombing the country at will.[7] The Sandinistas, in political exile since their electoral defeat of 1990, seized upon the missiles as a valuable campaign issue to rally nationalistic feelings for their party.

Early warnings that the Sandinistas were poised to make a political comeback by exploiting this issue should have counseled Washington to desist from the ambitious goals of eliminating all the missiles. Instead, U.S. officials intensified their efforts to try to destroy all the SAMs before the Sandinistas had any chance of returning to power under their candidate Daniel Ortega. In a first maneuver in March 2005, parliamentary allies of the U.S. Embassy succeeded in reducing the approval vote in the National Assembly from two-thirds of the deputies to only a simple majority. As late as February 2006 politicians still predicted that the missiles would be reduced to scrap metal before the SAMs became a campaign issue in the November 2006 presidential elections.[8]

These victory predictions notwithstanding, that same month of February 2006 the U.S. government found its demands crumbling. The army repeatedly had stated that it could not be left without these missiles and consequently was adamantly opposed to destroying all of the remaining 1,051 SAM-7s. Deputies seeking a compromise persuaded the army to accept 400 missiles as the lowest acceptable number. Exhausting as it had been for Nicaraguan politicians to extract this low number from the army, they found it even more difficult to secure U.S. approval. In particular, the Bush administration did not take kindly to conditions from the second poorest country in the Western Hemisphere. But even this last push to destroy nearly half of the remaining SAMs failed, and the Bush administration faced unfavorable prospects when the Sandinistas won the presidential election of November 2006.[9]

By pressing for the destruction of SAMs, the U.S. government gave the Sandinistas a wonderful campaign issue to gain control of the presidency. And when Honduras announced its acquisition of additional airplanes, supposedly for drug interdiction, the outcry in Nicaragua was massive and "causing the issue to spin out of control."[10] But undaunted U.S. officials did not let up their campaign for the destruction of the missiles even after the inauguration of Daniel Ortega on 10 January 2007. The persistence of U.S. diplomats was remarkable, and they explained to Ortega that SAMs "belong to another era and the U.S. government and Ortega's government have a rare opportunity to put much of that past behind them by reaching an agreement to turn over these missiles."[11] Previously the vague offers of military assistance had not softened the objections of the Nicaraguan army, and now in 2007 U.S. diplomats took another tack and appealed to the politician in President Ortega.

To generate support for eliminating the SAMs, the U.S. government dangled the offer of extensive medical assistance. In a well-organized publicity campaign, a study mission came to Nicaragua to examine its health system. Not surprisingly, its report described in graphic detail all the shortcomings in the country's hospitals and clinics. The U.S. government was willing to provide a generous aid package to address the country's health deficiencies on the simple condition of destroying the SAMs.[12] This quid pro quo of classical diplomacy was a particularly astute move, and for a time President Ortega seemed interested. But the failure to reply with a "no" actually reflected the traditional Hispanic courtesy. As 2007 gave way to 2008, in a last attempt the "Ambassador reiterated that the offer to provide Nicaragua with funding for health care and medical equipment in exchange for the destruction of Nicaragua's stockpile of [SAMs] remained 'on the table.'"[13]

The new Sandinista government quietly let lapse the offer, and the Bush administration failed to eliminate all the SAMs in Nicaraguan arsenals. In the struggle to impose its will on a small and poor Central American country, the U.S. government suffered a partial defeat. When democratically elected nationalist regimes were in power, the U.S. government faced major obstacles to impose its views on what constituted proper armament, as the next section on Venezuela confirms.

Venezuela: Arms Buildup

The task of eliminating Nicaragua's shoulder-fired Surface to Air Missiles was a Cold War legacy dating back to the Contra wars of the 1980s. In contrast, only in the initial years of the twenty-first century did the U.S. begin a campaign to limit the arsenals of Venezuela. The inauguration of Hugo Chávez as the new president of Venezuela on 2 February 1999 at least initially did not signify any sharp break in arms procurement. Chávez, as a former commander of paratroopers in the Venezuelan army, saw no reason to change the traditional reliance on American weapons for the bulk of the country's arsenal. However, the preference for the United States had not always signified exclusivity, and in a previous major exception, Venezuela purchased tanks from France in the 1970s. But when it came to strategic weapons, in 1983 Venezuela became the first country in Latin America to acquire F-16 jet fighters from the United States.[14]

Chávez was concentrating his efforts on transforming the economic and social structures of Venezuela and seemed in no rush to expand the country's arsenals. For their part, Venezuelan citizens considered corrupt deals as inevitable in arms purchases. The most pressing item facing the military was to install the latest technology on the aging fleet of F-16s. Because this was a costly and time-consuming process, both countries agreed to a staggered schedule stretching over many years. This arrangement operated smoothly during the Bill Clinton presidency, but soon after the inauguration of George W. Bush in January 2001 the new Republican administration interrupted the completion of the overhaul for all the F-16s. Sensing an opportunity to reenter the arms market in Latin America, Russia offered to sell Venezuela a fleet of MIG fighters for around 2 billion dollars in late 2001. It is not clear how seriously Chávez regarded this Russian offer, and most likely he considered it as a bargaining chip to press the U.S. government to complete the overhaul of his fleet of F-16s fighter jets. In any case, Russia's insistence on advance payment with almost no financing of the remaining balance was too big an expense for the Venezuelan budget to absorb. The Russian MIG proposal floundered and gradually disappeared.[15]

The Russians did not come home empty-handed, however, because their sturdy MI-17 helicopters captured the attention of the Venezuelan military. In September 2004 Venezuela signed a contract to purchase ten of these helicopters which, although designed primarily for transport, were fitted out as gunships when delivered in 2006. The helicopters were supposed to arrive in 2005, but training of Venezuelan pilots in Russia took much longer than anticipated. Even the U.S. Embassy in Caracas was not unduly worried about these helicopters: "The MI-17/s probably do not upset the regional balance of power. Venezuela's poorly run armed forces seem incapable thus far of integrating sophisticated equipment into combat."[16]

By the time Venezuela signed its first contract for Russian armament in September 2004, the situation had vastly changed since the 2001 Russian offer to sell MIGs. The most decisive event had been the discovery of U.S. complicity in the coup to overthrow Chávez in April 2002. In a probable exaggeration, the Venezuelan government believed that the U.S. government masterminded the entire coup attempt, and in response, Chávez became a permanent opponent of U.S. foreign policy. Yet a full break with the Bush administration over arms purchases did not occur until 2004 when the Venezuelan air force tried to buy Super Tucanos, a subsonic jet, from Brazil. Most importantly the Venezuelan military still insisted on the completion of the overhaul of the F-16s as the best solution for the air defense security of the country.[17] The situation seemed critical, because by 2004 "only six of the 21 remaining F-16s in the Venezuelan fleet are fully mission capable."[18]

To try to secure U.S. support, Venezuela carried out a two-fold strategy. Early in September 2004 Venezuela revived the 2001 Russian fighter proposal and through select leaks to the press magnified the purchase offer from two to as high as five billion dollars. Because U.S. officials knew that Venezuela could not come up with two billions much less five billions, the ploy seemed to be a crude bluff. But when a Venezuelan delegation headed by Vice President José Vicente Rangel returned from an October 2004 trip to Moscow, the Venezuelans reported that for a big ticket item of possibly as many as 40 MIG-29s, the Russians were willing to provide supposedly generous financing. A purchase of this magnitude was extremely worrisome to U.S. officials, and the Venezuelans were ready for a counterproposal to allay fears. All that the U.S. government had to do was lift its restrictions on the refitting of the F-16s and also resume Export-Import Bank financing for the overhaul of the jet fighters and for arms purchases.[19]

U.S. diplomats by not responding were essentially calling the bluff of Venezuela. Because the Russian financing terms were really not that generous, Venezuela delayed any decision on purchasing a new fleet of supersonic jet fighter-bombers. Later a different opportunity to acquire equipment came from Spain. The government of José Luis Rodríguez Zapatero was feverishly promoting Spanish exports to Latin America and offered extremely generous financing for the purchase of five patrol boats and a fleet of transport aircraft. The contract was signed in Caracas in January 2005 and immediately set off a storm of protests from the U.S. government. Spanish officials explained that the equipment was of a purely defensive nature and did not pose any threat. The overriding reason for making the ship sale was to save the financially troubled state-owned shipyard, Izar; without this huge contract the company faced bankruptcy.

Spanish officials repeatedly reminded the Bush administration about their extensive and almost unanimous Spanish support for almost all American foreign policies and complained about the obsession with the few areas of disagreement. U.S. diplomats bullied the Spanish government to such an extreme that an exasperated Defense Minister exclaimed "We are the eight-largest power in the world, but the U.S. government treats us like a fifth-rate power."[20] When threats no longer sufficed to force Spain into complete compliance with Washington's wishes, an indignant U.S. government, already outraged that Madrid had refused to obey the repeated warnings, simply prohibited the transfer of the technology needed to complete the airplanes. Madrid offered to manufacture a stripped-down version of the transport craft with basic technology, but the U.S. government did not want to compromise and insisted on teaching a lesson in obedience not just to Venezuela but also to Spain.

Spain was heading to an open clash with the Bush administration. The required technology was not available anywhere else in the world, and Spain could hope to reverse the decision only by taking risky reprisals, such as limiting the blanket authorization for U.S. airplanes flying over Spanish territory on the way to Iraq. On the verge of a costly confrontation, tensions suddenly eased when the Spanish airplane manufacturer stated that the Venezuelan contract was not essential to the firm. Instead for the Spanish shipyard the sale was indispensable for its survival, and at the same time Spain possessed all the technology needed for building the five patrol boats. Washington realized that no Spanish government could afford the political costs of driving Izar into bankruptcy, and to remove that pressure, U.S. diplomats became traveling salesmen and lined up purchasers for the patrol boats on the condition that Venezuela not receive them. The Spanish government refused and duly delivered the completed patrol boats to Venezuela. Madrid understood that the struggle was no longer just over the naval craft but about the Bush administration imposing its will.[21]

The half victory of the Bush administration with Spain was in contrast to complete success with Brazil. Since 2002 Venezuela had considered buying the subsonic Super Tucanos as a supplement to the supersonic F-16s. But because the Super Tucanos contained a large quantity of American components, including the jet engine, the U.S. government needed to approve the transaction. Brazil pleaded in vain with an inflexible Bush administration to allow the sale of Super Tucanos to Venezuela, but to no avail. The president of Brazil, Lula (Luiz Inacio Lula da Silva) warned that U.S. refusal to sell Super Tucanos was only serving to drive the Venezuelans into the arms of the Russians and Chinese. Brazil not only lost a lucrative contract but also reduced its influence over Chávez.[22] Years later President Lula still was angry "Brazil can't afford the type of embarrassment caused by not being able to sell Super Tucanos to Venezuela."[23]

After having halted the sales of airplanes from Brazil and Venezuela, the Bush administration concluded that part of the difficulty in pressuring foreign governments was the lack of a publicly articulated policy; in particular, the U.S. government did not want its opposition to Venezuelan arms sales to seem motivated only by commercial rivalry. During the Cold War the U.S. government would have designated Venezuela a Communist regime, but such a designation no longer carried the same moral outrage as before. The new buzz word was terrorism, and on the grounds that Venezuela was not cooperating fully with the global war on terror, the U.S. government imposed an arms embargo on 15 May 2006.[24]

The consequences of the U.S. arms embargo on Venezuela were not long in coming, and indeed Madrid officials predicted the outcome. "If Spain had been able to sell planes to Venezuela, the U.S. and Spain would have had some ability to control Venezuela's use of those planes by controlling parts and service. Instead, U.S. policy has pushed Venezuela into the arms of the Russians, who recently had not been major players in South America."[25] Venezuela did begin buying weapons from Russia, such as

100,000 AK-103 assault rifles, and most importantly an initial lot of 24 Sukhoi fighter jets. A grateful Chávez could not hide his pleasure when on a visit to Moscow he told President Vladimir Putin, "We would like to thank you for freeing us from a blockade."[26] As part of the AK-103 deal, Russia shipped machine tools and other supplies adequate to establish a rifle factory, and the equipment reached Venezuela in March 2008.

The first two Sukhoi fighter jets arrived in Caracas in December 2006 from Russia. A beaming Russian official commented on the weapons deliveries: "This is the result of four years of our painstaking, hard work—that our beautiful Sukhoi aircraft, our helicopters, our Kalashnikovs, with which all servicemen at all checkpoints are armed, as you have seen, have appeared in Venezuela. Russia has come here to stay for a long time."[27] On 5 December 2007 the Venezuelan air force announced the purchase of ten Ilyushin 76 cargo planes and also of two Ilyushin 78 with the capability of providing air refueling for three aircraft at a time. Nobody doubted the usefulness of these cargo planes to deliver humanitarian loads to areas of Venezuela ravaged by flooding or other natural disasters, but the cumulative impact worried the U.S. Embassy in Caracas.[28] "While many of these purchases reflect needed upgrades to Venezuela's aging defense forces, some weapons systems are in excess of Venezuela's needs."[29]

Decline in world prices for petroleum exports threatened to slow down Venezuela's purchases from Russia when the crisis over Georgia in the Caucasus in August 2008 brought an unexpected resurgence in sales. At a time when Russia was feeling international isolation because of its handling of the Georgia crisis, Hugo Chávez spontaneously stated "we know well what caused the conflict; we once again express our complete and firm support for your actions."[30] Chávez's comments were favorably received in Moscow, and a grateful Russian government expressed its appreciation by granting a more than one billion dollar loan on very easy terms so that Venezuela could continue its weapons purchases from Russia even in the face of falling petroleum prices. The amount of the credit rose to over two billion in 2009 to finance Venezuela's buying spree in Russia.

Still, supersonic jet fighters were very expensive, and the other branches of the military also needed new weapons such as T-72 tanks for the army. To reduce the costs of air defense, already in 2008 Venezuela was moving toward a greater reliance on Surface to Air Missiles (SAMs). One of the most important acquisitions was the S-300 SAM able to target 100 airplanes at once. The S-300 also had a range of nearly 300 kilometers, thus allowing Venezuela to shoot down any hostile aircraft attempting to take off from airstrips in neighboring islands, such as Curaçao (Map 1). But Venezuela did not acquire only long-range SAMs.[31] After exploring various options, the Venezuelan military also settled in January 2009 on the Igla-S, "considered one of the most lethal portable air defense systems ever made."[32] Already in 2005 the U.S. government was repeatedly stating to Russian officials its worries about arms sales to Venezuela. Moscow carefully examined U.S. concerns and concluded that because no international restrictions existed on selling weapons to Venezuela, "Russia recognized the U.S. as a competitor in the international arms trade, with the motivation of restricting Russia's market access ... Russia respected the U.S. right to determine U.S. policy on arms sales to Venezuela, but added, that is your decision, not ours; we have our own policy."[33] Nevertheless, because U.S. diplomats claimed that these missiles could fall into the hands of terrorists, Russia complied with the U.S. request to halt the shipment of the Igla-S until both governments could investigate any possible harmful consequences from the sale.

In July 2009 Russian and American experts met to discuss the sale of the Igla-S to Venezuela. Russia was more than satisfied with the safeguards Venezuela provided about the missiles and saw no reason to cancel the sale. When the U.S. and most governments turned against Russia during the short war in Georgia of August 2008, Venezuela had been among the minority of countries who passionately defended Russian actions in that conflict, and a grateful Russia was more than willing to sell weapons and anything else to Venezuela. In one last attempt to block the sale, U.S. diplomats used the argument that the weapons would fall into the hands of the FARC guerrillas in Colombia and urged Russia to sell only the vehicle-mounted version of the Igla-S missile. The Russians promptly replied that the Venezuelans specifically insisted on buying only the shoulder-fired models of this SAM. The U.S. government most reluctantly admitted failure in its efforts to pressure Russia on arms sales to Venezuela.[34]

As Spanish officials and President Lula of Brazil had predicted, the arms embargo of the Bush administration drove Venezuela to Russia and to a lesser extent to China for the purchase of weapons. By 2011 Russia claimed to have contracts for selling weapons to Venezuela totaling nearly eleven billion dollars.[35] The U.S. policy not only cost American manufactures many valuable contracts but also unleashed an arms race and a growing militarization in Venezuela. In spite of all its efforts, the U.S. government had not been able to prevent weapons from reaching Venezuela. For U.S. diplomats it had been a long, exhausting, and ultimately futile struggle stretching across several continents to keep the Venezuelan armed forces from deviating too far from their assigned role in the U.S. sphere of influence in the Caribbean.

Mexico: Fighter Jets

The failure to halt the flow of weapons to Venezuela was a major setback for the George W. Bush administration. Not since Nicaragua obtained Soviet weapons to defeat the Contras in the 1980s had a Caribbean country openly defied the U.S. government in weapons procurement. The Venezuelan breach, serious as it was, could also be a bad precedent that other countries might be tempted to follow. As U.S. diplomats canvassed the Caribbean for possible outbreaks of defiance, the search failed to uncover any other country interested in making significant weapons purchases. But the reassurance proved only temporarily comforting, because early in January 2007 a startled U.S. Embassy learned to its horror that the Mexican navy was interested in purchasing six Russian Sukhoi Su-27 supersonic jet fighters for the protection of its off-shore oil platforms. In the American mind, Mexico has been a source of labor, profits, and recreation, but was never supposed to be a place with a substantial military establishment. U.S. diplomats had a potentially major crisis in their hands because they could not bully large Mexico around like the smaller Caribbean countries. Mexican nationalistic sensitivities also could make negotiations cumbersome and volatile. In a first panicky response, the U.S. ambassador tried to deal with the issue as commercial competition and stated "that the U.S. had a similar product that we would like Mexico to consider and noted that there are more effective ways for Mexico to protect its oil platforms than Russian fighters."[36]

Mexico's military aviation was long overdue for upgrading. In 2007 the current fleet of the air force (under the Defense Ministry) consisted of 10 F-5E supersonic jet fighters purchased from the United States in 1981. Even back then the U.S. government blocked the purchase of more advanced jet fighters from Israel and, instead out of mainly commercial reasons, allowed Mexico to purchase only 12 of the lower priced F-5Es. This airplane was already nearly obsolete and rapidly reaching surplus status in 1981. When trial runs showed that the F-5Es were ineffective against drug aircraft, the U.S. government lost all sympathy for Mexico's desire to maintain even that dwindling fleet of jet fighters. Already in 1983 one jet plane crashed in flying exercises, and another plane crashed during the Independence Day celebrations of 1995. Mexico was left with a fleet of only 10 F-5Es, and by 2012, only seven of them were still operational.[37]

In spite of the sorry state of its jet fighters, Washington decided that Mexico did not need any new warplanes whether Russian or of any other origin. The U.S. State Department promptly announced that no U.S. warplanes were available for sale to Mexico and thus deprived American manufacturers of export opportunities. Washington defined the issue not as one of commercial competition over a contract but as a struggle to impose on the Mexican government the U.S. view on what was suitable for defense. In a blitz campaign among high officials of the recently inaugurated Felipe Calderón administration, U.S. diplomats relentlessly plugged the line that Mexico did not need those six Russian jet fighters. U.S. diplomats went so far as to express "our apprehension that this expensive purchase would complicate North American security cooperation."[38] It was simply preposterous to claim that the United States, with the largest and most powerful air force in the world, could face a threat from the puny number of six Russian jet fighters in Mexican hands. The United States had learned to live with a much larger number of Russian supersonic jet fighters in Cuba, but apparently six more in Mexican hands broke the limit of tolerance.

U.S. diplomats pressured their views forcefully on officials of the Calderón administration all too eager to please the powerful northern neighbor. U.S. diplomats also had a powerful ally in the Mexican Defense Ministry violently opposed to the expansion plans for naval aviation. The Defense Ministry was still smarting from the setback in 2007 when the Mexican Congress refused to authorize funds for the purchase of 12 probably second-hand F-16s to replace the obsolete F-5Es. The intense rivalry between the Navy and Defense Ministries meant that should the navy acquire the six low-cost Russian fighters, then those airplanes vastly outclassed the obsolete and barely operational F-5Es. U.S. diplomats adroitly lobbied officials of the Calderón administration until the Finance Ministry quietly removed from the budget those funds the Mexican navy had been accumulating for this purchase. As damage control for this embarrassing episode, official statements categorically denied that the navy had ever intended to acquire Russian jet fighters.[39]

As a consolation prize, the U.S. government sent teams to discuss air defense strategies for the offshore oil platforms in the Gulf of Mexico. U.S. officials hoped that training and visits would make Mexican officials forget the proposed purchase of jet fighters, but the issue of aerial defense did not fade away. After having eliminated the possibility of purchasing jet fighters, the only military alternative left to Mexico was to acquire Surface to Air Missiles (SAMs). After the long and tiring struggle to reduce Nicaragua's stock of these missiles, the last thing the U.S. government wanted was for Mexico to acquire an arsenal of SAMs. U.S. diplomats effectively downplayed the need for these missiles, but anxiety arose when the possibility loomed that Mexico might obtain advanced weapons from Israel. U.S. diplomats heard with considerable relief in early July 2008 that as far as relations with Israel were concerned "military cooperation and weaponry, such as an anti-missile system, are not important because Mexico does not feel threatened by its neighbors."[40]

U.S. diplomats had scored a major success when they convinced Mexican officials that the country needed neither jet fighters nor SAMs. But success did not mean neglecting careful observation, and U.S. diplomats attentively searched for any signs that Mexico might reverse its policy and try once again to acquire advanced weapons. A motive for worry came when Mexico was elected a temporary member of the U.N. Security Council. The Russian foreign minister decided to cultivate both a strong professional and personal relationship with his Mexican counterpart. Traditionally Russia converted those friendships into favorable opportunities to promote arms sales abroad. With great relief U.S. diplomats learned in February 2010 that in very extensive conversations between the Mexican and Russian foreign ministers "there was absolute zero talk of any arms sales."[41] In spite of this reassurance, U.S. diplomats remained vigilant trying to detect any indications of attempts by Mexico to acquire sophisticated armament. The U.S. government was unwavering in its determination to make sure that the Mexican armed forces possessed only those weapons suitable to fulfill U.S. goals in the region.

7

New Complications

> *But men of little prudence begin something which appears attractive without noticing the poison lurking beneath.*—Niccolò Machiavelli[1]

As the U.S. government tried to impose its military policy on the Caribbean during the first decade of the twenty-first century, the English-speaking Caribbean had been largely absent from U.S. concerns. And in Colombia having President Álvaro Uribe Vélez as the most loyal friend of the George W. Bush administration seemed to rule out any new problems in the oldest ally of the United States in the Western Hemisphere. But as this chapter shows, U.S. officials unexpectedly came to face new complications both in seemingly peaceful Jamaica and in turbulence-prone Colombia.

Colombia: False Positives

As the presidential term of George W. Bush was coming to a close, administration officials frequently devoted their last months in office to issuing congratulatory statements highlighting eight years of accomplishments. For the Bush administration, Colombia was high on the list of success stories. A widely distributed report explained that "The Colombian military and police, with U.S. government assistance, have made great strides and are now one of the most professional and best trained and equipped militaries and police in the hemisphere."[2] The report was quick to point out that the rosy situation emerged out of the mess the Bush administration inherited: "This was not always the case. At the end of the 1990s Colombia's army was mainly one of conscripts and it was on the ropes."

Certainly the Colombian army was in bad shape during the 1990s, and not just because of its repeated defeats in clashes with the Revolutionary Armed Forces of Colombia (*Fuerzas Armadas Revolucionarias de Colombia*) or FARC, the country's largest guerrilla group. Human rights abuses and extrajudicial killings had been a recurring problem since the partisan warfare of *La Violencia* in the 1960s. Even though members of the Liberal and Conservative parties no longer murdered each other, the army could not break from the bad habit of executing suspects, this time on the pretext that they were guerrillas.

An early precedent occurred on 7 June 1990 when the army claimed to have killed nine guerrillas in combat. It turned out that all the victims were civilians who belonged to a single family and one of them was 87 years old. To cover up this extrajudicial execution (which in all likelihood had been a case of mistaken identity), the soldiers hurriedly dressed up the cadavers in combat fatigues. But when a military judge arrived and reported that the bullet holes in the cadavers did not coincide with those in the fatigues, a civilian law-enforcement agency began to investigate the killings. Resistance from army officers blocked the inquiry, and when detectives tried to return to the scene of the crime, they had to abort the inspection trip when military helicopters opened fire on the investigative team. Not surprisingly, recurrent death threats persuaded the detectives to drop the case.[3]

Subsequent reporting strongly suggested that the 7 June 1990 massacre had not been an isolated incident but formed part of a recurring pattern of behavior inside the army. Already in 1994, "Body count mentalities persist, especially among Colombian army officers" who if they "cannot show track records of aggressive anti-guerrilla activity ... disadvantage themselves at promotion time."[4] Even more ominously, in 1997 a former colonel explained that "There is a body count syndrome in the Colombian army when it comes to pursing the guerrillas ... this mindset tends to fuel human rights abuses by otherwise well-meaning soldiers trying to get their quota to impress superiors."[5] The Colombian army became a graphic illustration of how society corrupts individuals and of how army elites manipulate an institution for their own benefit.

Without shocking and overwhelming evidence of this otherwise inconceivable practice, it seemed that the body count mentality would remain undetected as the driving force in the army's war against FARC. U.S. officials neatly separated combat performance and human rights into two different compartments. U.S. officials and the Colombian government replied to denunciations of human rights violations by considering them collateral damage and thus inevitable in any war or at worst as isolated cases of a few bad apples. The overriding goal of the public relations campaign was to preserve untarnished the image of the institution. As abuses multiplied in Colombia during the first decade of the twenty-first century, what was even more significant was the link between body counts and army promotions. The Vietnam War seemed to have discredited body counts of enemy deaths as a valid indicator of victory, but in reality the U.S. military had deeply internalized number crunching and developed a wide range of quantitative indicators to measure success in counterinsurgency warfare. U.S. diplomats soon adapted the mind frame of the U.S. military and peppered embassy reports with numbers supposedly reflecting the progress of the war against FARC.[6]

Human rights activists were not completely free of blame, because by focusing narrowly on the horrific aspects of army brutality, they failed to examine the dynamics of the war. Whether a tendency for violence is an innate characteristic of the Colombian people remains a hotly disputed point, yet it was obvious that few incentives were necessary to provoke violent and often sadistic cruelty. When U.S. officials imposed their metrics and quantitative procedures on the army, they were unleashing survival instincts. Trying to make the Colombian army resemble the U.S. army was not necessarily a bad idea, yet such a transformation required making major structural changes not just inside the army but more importantly throughout the entire society.

U.S. officials should have been aware of two major characteristics inside the Colombian army. One was the violence and ill-treatment enlisted men usually received during their mandatory military service. Enlisted men have generally come from lower class groups such as peasants, ethnic minorities, and slum dwellers. Officers, in the tradition of eighteenth-century armies such as Britain's, have defended iron discipline as the only way to make an army operate. Not surprisingly, middle-class families have been terrified at the possibility that their sons might have to serve in this exceedingly harsh army. Obtaining a military deferment became a coming-of-age ritual for boys, and if families failed to land an exemption, then the back-up plan was to secure an assignment to the few elite units treating soldiers as citizens and not as criminals. U.S. officials should have been aware of the brutal behavior inside the majority of the army units because the phenomenon was widespread in Latin America, including Colombia's bordering neighbors of Peru and Venezuela.

A second characteristic was the intense competition for promotion inside the officer corps. Colombia, as almost the rest of the Latin American countries, never seriously considered using universal military service to create a large standing army of conscripts. Mandatory military service became the curse of the unfortunate marginal groups unable to wield influence to obtain a deferment. For officers the small size of the military establishment left a very limited number of openings for promotion. As in usual practice in most armies of the world, after an officer has been passed over for promotion a certain number of times, then he must retire or resign. But the economic underdevelopment of Colombia means that officers have little chance of obtaining comparable employment in the civilian sector, in contrast to opportunities for former officers in the United States and Europe. In Colombia an early dismissal means that former officers lose their economic position and are in real danger of falling back into the ranks of the lower classes from where many probably originated. Corruption usually was not generally available as an opportunity for personal enrichment because drug trafficking, the largest source of illegal payments, fell mostly under the jurisdiction of the police.

Without easy access to drug money, officers in Colombia, driven by survival instincts, engaged in an almost Darwinian struggle to secure assignments and promotions. The need to report large numbers of dead guerrillas sparked two practices originating in the 1980s and continuing into the twenty-first century. The first was to count all accidental civilian victims as guerrillas. The officers were already going beyond any accepted principles when they counted dead civilians as guerrillas rather than dismissing those deaths as collateral damage. The second practice was to kill individuals suspected of being guerrilla sympathizers and then to count them as dead combatants. In both variations, whether accidental or deliberate killing of civilians, officers might dress up the cadavers in military fatigues to justify the claim that these were confirmed guerrilla deaths. And having a few corpses of real guerrillas scattered among the dead civilians helped to enhance the credibility of the "kills."

In the final stage beginning roughly after 2006 when the army offensives were losing steam, officers faced increasing pressures to meet their quotas for number of guerrillas killed. But with most army units retreating into a static defense, any inhabitants suspected of guerrilla sympathies had long since either been killed or had left the region. Grabbing and killing unwary civilians was much easier than tracking down elusive guerrillas. The practice became widespread, and officers carried out a form of social cleansing by sending out soldiers to kill individuals who might not be missed by the rest of society and then dressing them up in uniforms to claim that the cadavers were FARC guerrillas.

Human rights groups denounced the mysterious disappearances of many individuals, and starting in 2006 the press began to report suspicious incidents. A July 2006 report claimed that the Fourth Army Brigade stationed in Medellín had dressed up 30 civilian cadavers as guerrillas over a fifteen-month period, and a similar case appeared in Bogotá in September of that same year; in both instances official promises to investigate silenced the critics. Irrefutable proof surfaced in February 2007 that the army had falsely claimed 19 paramilitaries as guerrilla kills, but because the paramilitaries were illegal groups, public opinion still did not stir. Additional details appeared in January 2008 when a former sergeant explained how soldiers chipped in to buy weapons, usually a cheap pistol, to place next to the bodies of civilians; when properly staged, the photographs provided seemingly irrefutable evidence that the soldiers had killed guerrillas. In a profoundly disturbing revelation, the former sergeant recalled that officers rewarded soldiers with five days of paid leave for each dead body.[7]

Human rights groups had been reporting the crimes and collecting figures, but because the claims of such deliberate savagery seemed too extreme to be credible, the Álvaro Uribe Vélez administration routinely dismissed the accusations as fabrications intended to discredit the army. At the same time, civilian analysts were baffled by the high body counts the government was reporting for guerrilla deaths. In spite of the very large number of guerrillas supposedly killed, FARC retained its ability to maneuver and even to launch attacks. A theory of spontaneous regeneration was necessary to explain how rapidly FARC was able to replace its large losses when at the same time the expanding army suffered at most a tenth of the casualties attributed to the guerrillas. The real explanation for the contradictions in the figures only came with the revelation that army promotions were directly linked to the number of dead civilians. The shocking discovery took place at last when an army patrol ambushed and killed students from Soacha who were on an outdoor excursion near Bogotá in September 2008. The army duly dressed up and armed the bodies of the kids as guerrillas and then took the requisite photographs. At first the ploy seemed to be working because the army explained away the young age of the minors by using the frequent propaganda line that FARC was in such desperate straits that it was forcibly recruiting boys to fill its ranks.[8]

When the parents identified the bodies of their children, the adults angrily rejected the official version that their sons were battle-hardened FARC fighters. Newspapers picked up the story and demanded explanations for this strange occurrence. President Uribe Vélez, who angrily had rejected all previous accusations on the army as attacks from left-wing groups, finally relented and appointed a special army commission to investigate the Soacha massacre. In a few weeks the commission confirmed that "False Positives," as these murders came to be called, had been standard operating practice in the Colombian army since the late twentieth century. Virtually the entire high command was implicated in the scandal and also almost all the officers in combat units.[9] The army massacred thousands of innocent civilians as "False Positives," and the number likely exceeded ten thousand victims.[10]

President Uribe Vélez, who was deeply involved in trying to secure constitutional authorization to begin the campaign for his second reelection, realized that this scandal could derail his candidacy. To try to put the crisis behind him, at the end of October 2008 and just after receiving the findings from the special army commission, the President fired 27 officers and NCOs. At first sight the number was impressive because it included two division and three brigade commanders, but in reality he had barely scratched the surface. Next to go was the army commander, General Mario Montoya, who resigned under pressure on 4 November 2008. Minister of Defense Juan Manuel Santos wanted to press forward with the purge of criminal officers, but President Uribe Vélez felt that the investigation had to end before the unfolding scandal diminished his chances of a second reelection. The President, against the advice of Santos, appointed General Óscar Enrique González Peña, as the new army commander. The appointment was very significant because Gonzalez Peña was a protégé of the outgoing army commander and could be counted on to block the ongoing investigations into criminal actions.[11]

The "False Positives" story did not go away, and the media had a field day digging into the previously secretive life of the army. Newspapers, radio, and television gave extensive coverage to the story and complicated the efforts of General González Peña to carry out a cover-up. The General shifted his attention to halting the purge, and when Army Inspector General Carlos Suárez wanted to fire additional officers because of criminal behavior, the new army commander, in a break with past practice, insisted that they could be fired only if the accused already faced formal charges from the very slow moving criminal justice system. Previously officers had been routinely dismissed over practically insignificant mistakes, but now to set such a high bar guaranteed impunity to criminal officers who could continue their careers without restrictions for many years if not until their retirement.[12]

And getting criminal charges filed against guilty officers was proving harder by the day, as army personnel constantly intimidated potential witnesses. When threats did not suffice to keep individuals from providing incriminating testimony, death squads killed civilian witnesses and terrorized the rest into a trembling silence. In spite of the determined opposition of the army commander, army Inspector General Suárez continued to gather abundant evidence to increase the number of fired personnel to 51 and many more were in the pipeline. To halt the steady purge of criminal elements, army commander General González Peña concocted a maneuver to remove the upright army Inspector General by sending him as military attaché to Chile as a final step prior to an early retirement. Existing regulations did not permit the new Defense Minister Gabriel Silva Luján to reverse the removal, but he came up with his own counter maneuver and promoted General Suárez to the newly created position of Inspector General of the entire armed forces rather than just of the army.[13]

Despite the efforts of General González Peña and other criminal officers inside the army, the False Positives story did not go away, and finally the scandal wrecked the selfish scheme of President Uribe Vélez to seek a second reelection in 2010. Contrary to the claims in the new official version, the practice of False Positives did not disappear entirely from the army. New cases of counting innocent dead civilians as guerrillas continued to occur, although apparently less frequently and certainly more secretly than before.[14] And the reason that False Positives could not disappear was that the officer corps had internalized the U.S. principles of quantifying success: "Despite having implemented a balanced scorecard approach designed to de-emphasize body counts, the Colombian military leadership reportedly continue to react very negatively to lower numbers reported by subordinates."[15] More than any other episode narrated in this book, the False Positives scandal of the Colombian army supports the conclusion that the mechanical application of U.S. methods to the military of the Caribbean brought unforeseen results and sometimes also extremely harmful consequences. U.S. assistance and training had converted an incompetent and cumbersome institution into an efficient criminal enterprise.

Scandals in the Colombian Army

Additional proof that the Colombian army had a vested interest in the perpetuation of the guerrilla war came in early February 2014. The revelation by *Semana* magazine that army officials had been intercepting the digital messages of the official peace commissioners in Havana unleashed a new scandal. Code-named "Andromeda" this intelligence operation went into high gear in September 2012 precisely when President Juan Manuel Santos announced the start of peace negotiations with the Revolutionary Armed Forces of Colombia (*Fuerzas Armadas Revolucionarias de Colombia*) or FARC. At an innocuous looking Internet café in Bogotá called Buggly Hacker, military intelligence set up a sophisticated operation with a wide array of computers and software programs. On the ground floor while customers enjoyed the daily menu specials and took mini courses about the Internet and computers, on the top floor military intelligence officials conducted secret operations. The restaurant cover provided also a great opportunity to recruit naïve civilian hackers to carry out the wire intercepts. The eventual goal of Andromeda was to outsource the entire operation to private individuals for the purpose of hiding any military involvement. But throughout its existence, the civilian hackers sat right next to the uniformed personnel in the top floor of Buggly Hacker.[16]

The U.S. government was deeply involved, because as part of the military assistance to wage the war against the guerrillas, the CIA had provided the equipment and the training for these operations. As one military officer explained, the Americans "know absolutely everything that goes on at headquarters and they know what, who, and why is being intercepted."[17]

When negotiations began in Cuba in November 2012, this intelligence unit intercepted digital messages and telephone conversations both of the FARC delegates and the government peace commissioners. Army officers then distributed the information in ways most harmful to the peace process, but exactly which senior officers actually received and shared the message intercepts remains a topic of criminal investigation. The practical impact was obvious, however, and all sorts of obstacles kept hampering the peace talks. It seemed that when the negotiators were on the verge of a breakthrough, something always appeared at the worst possible moment to derail the conversations.[18]

During the 1980s the army had gone so far as to use force to sabotage peace talks taking place in Colombia under the Belisario Betancur administration, but under President Santos the relative inactivity of the guerrilla zones in Colombia and the location of the peace talks in Havana left digital intercepts as the only way for the army to block the negotiations.[19] However, in 2012 Operation Andromeda did not target just the peace negotiators, and in the tradition of political espionage in Latin America, Buggly Hacker actually devoted most of its efforts to digital surveillance of political parties, labor leaders, and opposition groups. Just the year before in November 2011

President Santos abolished the domestic intelligence agency (*Departamento Administrativo de Seguridad*) or DAS precisely because it had engaged in widespread eavesdropping of opposition political figures. And as Operation Andromeda confirmed, other agencies, in this case Military Intelligence, soon filled the espionage void left by the abolishment of DAS.[20]

Right after *Semana* magazine exposed the existence of Operation Andromeda on 3 February 2014, the army removed the ranking officer of the covert unit and the director of Military Intelligence. The army also shut down Buggly Hacker, after having transferred all the valuable information to undisclosed locations. The senior officers believed that they had contained the damage from the revelations, but in reality the magazine had opened a Pandora's Box.[21]

The Colombian public was only slowly starting to absorb the revelations when in a subsequent issue *Semana* magazine revealed the existence of massive corruption rings inside the army. Prior to 2014 whatever cases of corruption occurred in the army had remained closely guarded secrets because above all else the institution wanted to preserve its honorable image. At least since the 1980s it was popular knowledge that retired senior officers during their years of service had acquired substantial properties beyond what their salaries could justify. Whether the generals had received illicit payments or simply were extremely shrewd about taking advantages of opportunities to buy land with inside knowledge remained unclear. But because the wealth did not seem outlandish, a grateful public was hesitant to question the integrity of army officers who had fought to defend the country.[22]

The impact was shocking once *Semana* magazine revealed on 16 February 2014 that several corruption rings existed inside the army. The largest was a corruption network centered in army installations in Cali. In the Bogotá network, the central figure was Colonel Robinson González del Río, who reaped profit margins of as high as 50 percent through fraudulent contracts for replacement parts and fuel. As an all-around entrepreneur, he also sold weapons to criminal organizations, particularly in the headwaters of the Sinú and San Jorge Rivers (Map 8). He also devised a scheme for soldiers to smuggle weapons out of arsenals. Soldiers dismantled the arms and sneaked out the pieces aboard official vehicles. Later the parts were reassembled at workshops in the countryside before being handed over to the criminal groups. Colonel Robinson sold a wide assortment of weapons, ranging from large numbers of pistols and rifles to M-60 machine guns. Obviously to make his corruption enterprise function, he needed to recruit many collaborators ranging from privates to generals.[23]

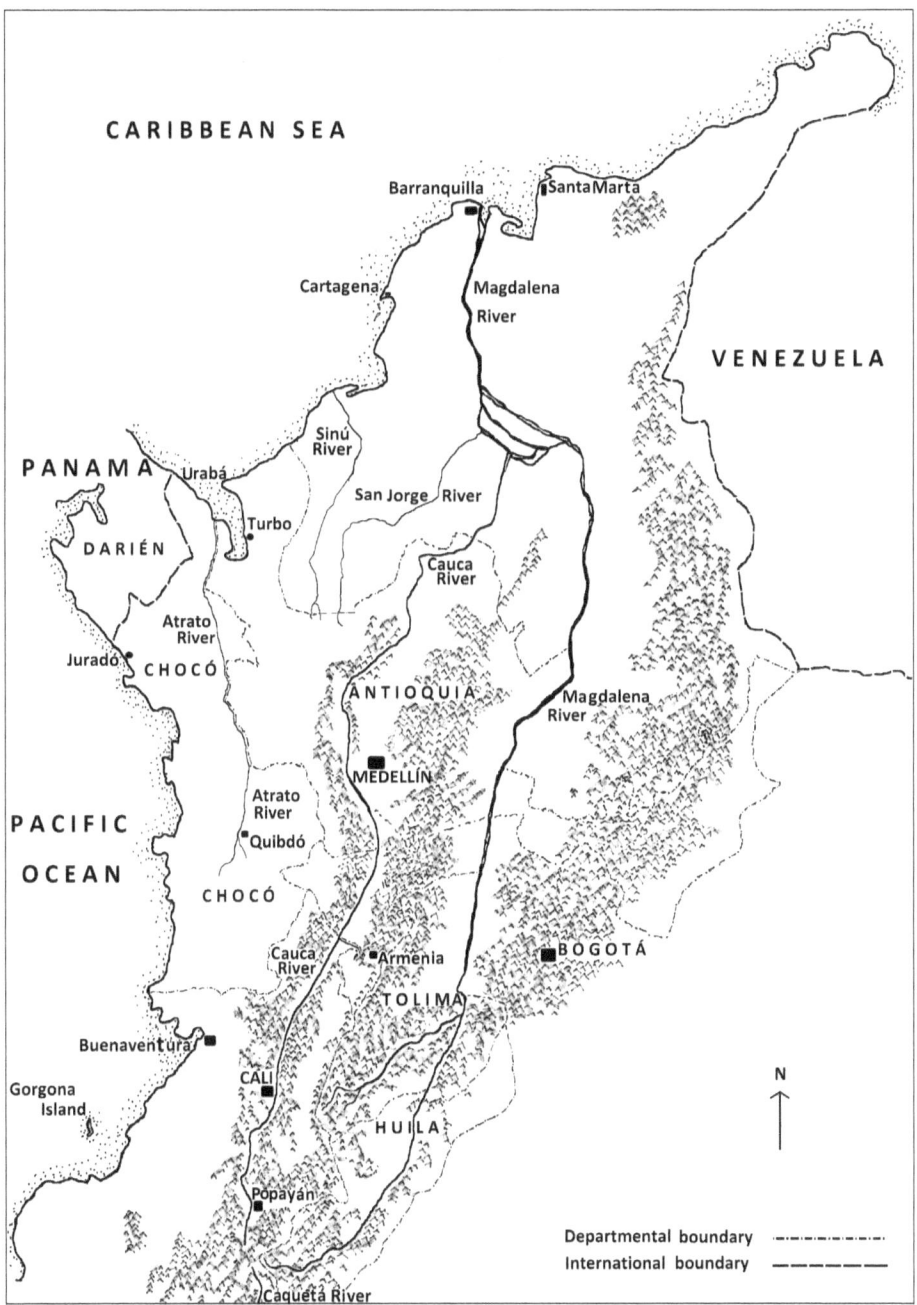

Map 8. Northwest Colombia.

The revelation that such a smooth running corruption network existed inside the army right in the country's capital was shocking to the public, but worse was to come. It turned out that Colonel Robinson was in prison on charges of participating in the False Positives massacres. At this point the new corruption scandal merges into the murder of at least 10,000 innocent civilians or False Positives. Partly to cover the corrupt practices, the network of Colonel Robinson channeled some of its ill-gotten gains to pay for the legal defense bills of the 401 officers accused of having murdered innocent civilians. Defense attorneys received airplane tickets, travel in military vehicles, and stays in army clubs, with all the expenses paid from the corruption money.[24]

Colonel Robinson was doing more than providing legal defense funds, because money also went to keep officials from talking. By accepting bribes from the corruption network, officers stayed quiet and declined to testify against their colleagues involved in the ongoing False Positive investigation. The army remained in denial about the False Positives and continued to regard the civilian investigation as an attack intended to destroy the military institution. The most graphic illustration of this attitude came from General Leonardo Barrero, later to be appointed commander of the armed forces, when he told Colonel Robinson in a recorded telephone conversation "Don't let those SOB district attorneys screw you, and to denounce them arm yourself as a mafia with the rest of the accused inmates"[25]

The situation of Colonel Robinson was very peculiar to say the least. As early as 2011 *Semana* had been denouncing that those officers imprisoned because of their participation in the False Positives scandal were enjoying extraordinary privileges in prison, but as usual the authorities did nothing to correct the anomalous situation. Now that the corruption scandal was out in the open, the media jumped to denounce some of the abuses that went beyond living in luxury inside prison suites. The accused officers wielded influence or paid bribes to judges to obtain transfers from the less comfortable civilian prisons to friendly military installations where they enjoyed many privileges including unlimited telephone service. Colonel Robinson actually spent little time behind prison walls; usually he was outside operating his network of corruption and investing his gains into private businesses. Army chauffeurs drove him and his family around in official vehicles, just as they did for many of the officers in prison because of the False Positives. The Colonel also used this free transportation to take a three-week vacation with his family.[26]

As more revelations came out, a beleaguered President Santos was faced with a tough dilemma as he prepared for his reelection in the August 2014 campaign. If he did nothing, he could lose the support of many outraged citizens, while if he came down too hard on the army, conservative groups might turn against him. As a first measured response, he fired five generals and other senior officers, including General Barrero already infamous for his "mafia" comment. But the President was reluctant to make a clean sweep in the senior ranks of the institution. Soon the media was reporting on the holdings of the fired generals, one of them had acquired 31 properties while he was in the army, others placed considerable real estate under their wives' names, while many stashed away large sums in off-shore banking accounts.[27]

Conservative groups soon began to attack Santos for being the President who fired the largest number of generals and senior officers, and the pressures mounted to halt the investigations into corruption and to forget about the False Positives. The specious argument was that in an army of 228,000 uniformed personnel in 2014, a few bad apples were inevitable. Because Santos showed no signs of stopping, unknown army officers began the selective leaking of e-mails by the President. It turned out that Operation Andromeda had hacked into the personal e-mail accounts of Santos and his family and had obtained at least one thousand e-mails. The e-mails did not harm the reputation of Santos, and some such as his failed attempts to buy paintings at a discount price from the world famous Colombian painter Fernando Botero merely confirmed that he did not have money to splurge as did the corrupt army officers.[28] Botero paintings were status symbols for Colombian drug lords who loved to decorate their residences with expensive artwork, and to have the President admit that he could not afford to buy these paintings was another indication of his personal honesty.

It came as no surprise to see the peace talks in Havana stall in March 2014, but both sides promised to return to the table for another round of negotiations in the future.[29] The army's real goal had been to sabotage the peace process, because if the war came to an end, the opportunities for personal enrichment drastically declined for army officers. Not just U.S. aid would end completely, but the Colombian government intended to slash the budget drastically once war ended, and the people would be most reluctant to support an institution that had so betrayed public trust in a mad scramble to accumulate material possessions.

The Mothers of Soacha, the women whose children the army had murdered and dressed up as guerrillas, were still clamoring for justice in 2014.[30] Because the public learned about the False Positives in 2008 more than five years before the scandals of eavesdropping and corruption, at first glance the latter might appear to be results of the False Positives. Actually the corruption scandal and the False Positives are two sides of the same problem, while the digital espionage is simply the latest of many tools used by the army to perpetuate the guerrilla war in Colombia.

The application of U.S. metrics of success to the insurgency created the False Positives in the last decade of the twentieth century. Because the long war against guerrillas had steadily eliminated moral scruples among officers, they had to take few additional steps to adopt the criminal practice of False Positives. But once killing innocent civilians to promote army careers became the acceptable norm, then officers without any moral scruples no longer could have any hesitation about stealing from the government. The complete moral degeneration of officers, something much worse than simple decay, made possible the simultaneous appearance of corruption networks inside the army and the generalization of the False Positives massacres. Theft and murder became two sides of the same coins, as officers rushed into a headlong dash to accumulate as much wealth and property while their army careers lasted. And the degeneration was so deeply ingrained, that simply dismissing some senior officers could not end these evil practices. The constant exposure to war and slaughter degrades the souls of humans, and Colombia cannot think of its moral regeneration until first the guerrilla war comes to an end. Peace is no longer just a material imperative but also a spiritual necessity.

Jamaica: Trouble in Paradise

After gaining independence from Britain in 1962, Jamaica, "despite its reputation as a happy-go-lucky island of sun, sand, and reggae,"[31] gradually sank into a state of lawlessness, corruption, and poverty. During colonial rule, a constabulary maintained law and order throughout the English-speaking Caribbean. The officers were mostly white, and to prevent collusion and corruption, the British enforced the rule that the rank-and-file members of this police force could not serve in their island of origin. This changed after independence when recruits for the new Jamaican Constabulary Force (JCF) came exclusively from that island. White officers and black Jamaican soldiers had also comprised the single army unit stationed in the island, and this too became exclusively Jamaican and took the title of Jamaican Defense Force (JDF).

Map 9. Central America.

A new phenomenon appeared in the 1970s when the two main political parties established close links with "garrison" neighborhoods in Kingston the capital (Map 9) and in other cities in the island. In exchange for organizing and arming these neighborhoods, the political parties received the votes of the residents. Armed street gangs ended up controlling these garrisons, and "beginning in the 1980s and continuing in the 1990s, the criminal gangs branched out and diversified into drug, gun, and human smuggling."[32] Jamaica became the largest producer and exporter of marihuana in the Caribbean, while its convenient position between the United States and Colombia soon called the attention of drug cartels. Puerto Rico and the Dominican Republic were less hospitable to the drug trade because of their location too far to the east and the presence of extensive law enforcement, while Revolutionary Cuba ideally situated was off-limits to drug traffickers. Jamaica remained as the last convenient country in the West Indies for moving cocaine from Colombia to the United States. Jamaica had the added advantage of being relatively close to the Colombian island of San Andrés (Map 9), and drug traffickers brought cocaine both by speed boats from that island and by airplane directly from the Colombian mainland.[33]

All this drug trade was in the hands of the "dons," the heads of the criminal gangs or "posses," and the real rulers of their respective garrison neighborhoods. The dons also gave money and jobs to individual residents and made donations to their communities; in a sense, the dons replaced a Jamaican government unable to provide services and employment for the people. The gangs, estimated to number 269 in the island by 2013, possessed the muscle and firepower to control the 49 garrison neighborhoods in Kingston. The dons even administered a crude but swift system of justice to settle disputes among residents and to maintain order in the garrison. A hierarchy existed, with the street dons responsible for several blocks in Kingston and community dons in charge of several garrisons. At the top of the hierarchy were the mega or super dons who acted as Godfather figures. To maintain their control, the dons ordered gang members to carry out many murders, so soon Jamaica came to have one of the highest homicide rates in the world. The most infamous of these gangs, the Shower Posse, got its name because of its practice of showering rivals with bullets. The dons also were political bosses who delivered votes to their preferred candidates in elections. The elected officials reciprocated the favor by ignoring the dons' illegal activities. As chiefs of criminal enterprises, the dons also operated the entire chain in the marihuana business from cultivation in the fields to delivery in the United States. And with the bulky marihuana shipments the dons also shipped cocaine from the drug lords in Colombia.[34]

Of the mega or super dons, the most important was Christopher "Dudus" Coke, who after his father's murder in 1992 replaced him as the head of the Shower Posse, the most powerful gang in Jamaica. The light-skinned Christopher expanded his father's inheritance until he operated the most extensive criminal enterprise in Jamaica. The Shower Posse had its base in the Kingston garrison of Tivoli Gardens, which Coke ran as a personal fiefdom, all the time remaining a close supporter of the Jamaican Labour Party.[35]

Throughout Latin America local political bosses have traditionally delivered votes to favored candidates and parties on election days. What made Jamaica special and probably unique was the combination of this electoral function with street gangs and drug trade in the hands of dons. The Jamaican Constabulary Force was no match for these powerful dons and, already in 2000, threats and the killing of policemen forced the government to deploy the army around police stations as a temporary measure for their protection. Under permanent threat, not surprisingly many police officials rather than fight a hopeless and deadly war preferred to accept bribes from the dons. During the twentieth century the war against the drug trade had been essentially a police function, because under the traditions of British law, soldiers could not arrest citizens. Even the reintroduction of flogging as a punishment did not seem sufficient deterrent against criminal activity, and the public clamored for the return of the death penalty as well. Clearly the police was unable to control drug trafficking, and the Jamaican government, worried that the violence might scare away the more than a million tourists who annually visited the island, increasingly turned to the army to try to limit the rapidly expanding marihuana and cocaine trade. Businesses for their part hired many private security guards, who already numbered over 20,000 at the start of the twenty-first century.[36]

Obviously the roughly 3000 active duty personnel of the army and the around 7000 policemen faced an impossible task, because trying to stop the drug trade while maintaining close ties with the dons was inherently contradictory. The three main criminal activities of the dons were the drug trade, money laundering, and extortion, and because the drug trade was the most lucrative of the three, not surprisingly the dons were determined to defend the source of their largest share of income. The JCF created special military-style units to eradicate marihuana fields, but the task was beyond its capabilities. Soon the Jamaican Defense Force took on crop eradication as a major function in the countryside. The U.S. government provided training, supplies, and equipment to the JDF. But as the army destroyed easily reached fields, the dons adapted and moved cultivation to more remote areas. The eradication of marihuana fields stalled because the U.S. government was slow to provide the helicopters indispensable to locate and destroy the hidden fields. Also the farmers very cleverly adopted the practice of growing marihuana plants alongside food crops, and this effective camouflage complicated even more the labors of both police and army units.[37]

While the JDF had been roaming around the countryside, the situation inside urban areas actually deteriorated. The gangs of rival dons were engaged in bloody warfare, and because "the populace in the neighborhood will often turn out to defend the don" the police felt powerless to intervene and "normally the area is simply sealed off to prevent the contagion from spreading, and personnel do not enter until the riot has blown itself out."[38] Although the U.S. government donated old models of M-16 rifles to the JCF, the police were still outgunned and were not trained in combat. In a fateful decision, the Jamaican government of the newly elected Bruce Golding sent the army to patrol the garrison neighborhoods inside urban areas, particularly in Kingston. These supposedly joint patrols were composed primarily of soldiers with only one or two policemen participating in case someone had to be arrested. In contrast to police whose training has the goal of protecting citizens, the primary mission of soldiers is to destroy the enemy. Consequently when the government sent army units into Kingston the risk of killing innocent civilians greatly increased.[39]

When the Jamaica Labour Party of Bruce Golding returned to office after 18 years of opposition, the public expected that unlike the weak and ineffective previous Prime Minister, Portia Simpson-Miller, at last the new government would stop the gang warfare and the constant murders in the garrisons. The situation had become so bad, that police commissioners brokered agreements between warring dons as the only way to reduce the gang violence.[40] But the new Prime Minister Golding won the elections precisely by making a pact with the devil, and his own electoral district was centered in Tivoli Gardens, the headquarters of none other than Christopher "Dudus" Coke, the most powerful don of Jamaica. How could Golding turn against the backer whose money and votes made him Prime Minister?

One of the most cherished assumptions justifying independence in 1962 was that black Jamaicans aware of local conditions could govern the island better than white British officials who educated in the Greek and Roman classics were seen as distant and uninformed. Whatever might have been the achievements of Jamaicans in electoral politics, it had become clear by the twenty-first century that only by turning again to white British officials could JDF operate efficiently. To combat the rampant corruption, Prime Minister Golding created a new anti-corruption branch inside the police and staffed it with British officers. As a precaution Canada and the United States also provided token police officers in an attempt to dilute the appearance of Britain returning to rule its former colony. Golding also obtained additional military assistance and training for the troops to give the army greater capacity to operate against drug traffickers on land and in the bordering seas.[41]

These efforts at reducing corruption and the drug trade could yield arrests and spectacular raids, but ultimately were futile, because the government was only cutting off a few tentacles from the self-regenerating criminal enterprises of the dons. As long as the dons retained their cozy relationship with the government, their impunity meant that they merely adapted and prospered under the new circumstances. Already in late 2008 "the conventional wisdom held by Jamaicans and foreigners living in Jamaica is that the country is sliding slowly but surely into the abyss."[42] And the police, by becoming more aggressive in their struggle against the street gangs, inevitably generated many accusations of brutality and even human rights abuses, while the gang members cleverly used women and children as human shields to escape capture and to stop the joint army police patrols.[43] Because innocent residents died in the shootouts between police and the gangs, the public began to sympathize with the dons, and "the JCF has not done a good job of making the broader public aware that it is conducting operations in horrific conditions in urban squatter developments where it is truly difficult to know who is the enemy."[44]

If policemen were having trouble distinguishing innocent residents from criminals, the soldiers had little chance of making correct identifications, and not surprisingly Prime Minister Golding did not want to risk many more civilian casualties by sending the army into the garrison neighborhoods. The Prime Minister was carrying out a very dangerous balancing act as he used the improved police and army forces to crack down on gang murders while he maintained his relationship with "Dudus" Coke. But the U.S. Justice Department forced his hand when early in 2010 it demanded the extradition of Coke to face charges of drug trafficking and weapons smuggling. Sensing the risks to his political future, the Prime Minister hired an influential law firm as a lobbyist to block the extradition request. When the attempts to persuade the U.S. government failed, Prime Minister Golding faced the impossible task of trying to convince Coke to surrender for extradition to the United States.[45]

When the Prime Minister finally ordered the arrest of Coke, the don struck back on 24 May 2010 and sent his gangs to attack three police stations in Kingston in an obvious attempt to intimidate the government. The situation became even more critical when other dons started to support Coke in his attacks, and for a moment it seemed that the government was facing a full-scale revolt in the capital. The gifts and services the dons traditionally provided had gained the support of the residents of the garrisons, and public opinion seemed to sympathize with the dons and not with the government.[46]

Because the police lacked the firepower, Prime Minister Golding had no other alternative than to send the JDF into Tivoli Gardens to find and arrest Coke. But the don was waiting for the army, and his gangs had been feverishly fortifying the neighborhood. Sandbags appeared on balconies and roofs, while overturned vehicles formed barricades to block all the entrances to the garrison neighborhood. Behind the barricades the gang members were not only well armed with automatic weapons but also had detonators for the many improvised explosive devices scattered throughout the area. As the soldiers approached, gang members opened fire and set off many explosions. The soldiers replied with overwhelming firepower, and soon large segments of the neighborhood were gutted and in flames.[47]

The worst was to come when the soldiers reached the easy conclusion that any youngster in the area was a gang member and should be promptly shot. The number of these extrajudicial killings could have reached two hundred, and at least 70 civilians died but only three uniformed agents. The operation lasted several days and only on May 26 did the shooting finally die down. Although human shields allowed Coke to slip away, this display of army power awed many of the other dons into turning themselves in to the authorities. For a few days it looked like the government at last had shattered the power of the dons and was well on its way to restoring order to the country. However, the impression was misleading, and gradually the other dons filled the power vacuum left by the departing Coke.[48]

A month later the police seized Coke when he was trying to escape dressed as a woman and wearing a pink wig as a disguise. But not even this picturesque capture and his extradition could make Jamaicans forget the blatant slaughter of so many civilians during the attack on Tivoli Gardens. The public held the Prime Minister responsible for the deaths, and he barely survived a no-confidence vote in parliament. He realized that his days in office were numbered, and in a desperate attempt to avoid dragging the Jamaican Labour Party down in defeat with him, he resigned from office on October 2011. He hoped that a younger candidate could save the party in the elections of 29 December 2011, but to his dismay, none other than Portia Simpson-Miller, the candidate Golding had defeated in 2007, won the election.[49]

To everyone's surprise, the elections of 29 December were the first ones in decades to have taken place in relative quiet, in contrast to previous electoral cycles characterized by violence and killings. Hopes were not high when the new Prime Minister Simpson-Miller, or "Sista P," as she was nicknamed, was inaugurated early in January 2012. Although she had before been the first female prime minister in Jamaica's history, if the posses had spread during her first term in office, she could not destroy them in her second term as prime minister. Violence intensified, and in 2014 shootings were occurring on an almost daily basis in the garrison neighborhoods. In a country where many boys dreamed of growing up to become dons, the appearance of new dons to fill the vacuum was a foregone conclusion.[50]

8

Central America Turns to the Left

The early twenty-first century has witnessed the reconstitution of U.S. hegemony over Central America.—Mark B. Rosenberg and Luis G. Solís[1]

Beginning with the CIA coup against Guatemala in 1954, U.S. policy has been hostile to leftist regimes in Central America. The degree of U.S. intervention varied across time in each of the countries, but the preference of the U.S. government for pliant client states in its sphere of influence remained a fundamental principle. When a round of elections starting in 2006 brought leftist candidates to office in Central America, U.S. diplomats went on high alert. The real fear in Washington was that the new governments in Central America might follow the path of Hugo Chávez in Venezuela or even worse of Socialist Cuba. The alarm bells went off when Sandinista leader Daniel Ortega, who had led the fight against the U.S.–financed Contra War of the 1980s, was elected president of Nicaragua in November 2006.[2] And when Guatemala and El Salvador also elected leftist presidential candidates, the fear of this contagion spreading throughout Central America provoked tremendous anxiety in the George W. Bush administration.

Nicaragua: The Return of the Sandinistas

As explained in chapter 6, the main consequence of the attempts by the George W. Bush administration to deprive Nicaragua (Map 10) of its Surface to Air Missiles (SAMs) was to unleash a wave of nationalist feelings. An unrepentant Bush administration tried again to pressure the Nicaraguan government on another highly volatile issue. Secret manipulation forced Nicaragua to contribute troops for the occupation of Iraq, but the ensuing nationalist uproar was so violent that the government had to recall the troops after just six months. Not surprisingly, the nationalist revival was largely responsible for sweeping the Sandinista political party back into

Map 10. Nicaragua.

office. The Sandinistas had been out of power since their electoral defeat in 1990, and the Bush administration was seriously worried that the regime would again be as radical and revolutionary as during the Cold War years of the tumultuous 1980s.

After the inauguration of Daniel Ortega on 10 January 2007, the U.S. government watched Nicaragua closely and with deep misgivings. The initial impressions were far from reassuring, and U.S. diplomats reported that the Sandinistas aimed at "the eventual imposition of an authoritarian, undemocratic regime friendly to countries who oppose us. The preliminary signs suggest that Ortega aims to consolidate his power and attempt to transform the Nicaraguan political model to ensure he remains at the helm for years to come of a leftist, centralized and largely non-democratic state."[3]

And the situation did not seem to get any better after sixty days when the governing structure of the new regime was starting to take shape:

Our sources tell us that all government decisions are being made by a very small cabal at the top of the pyramid—Ortega himself, his wife, and at times economic guru Bayardo Arce (although there have been some signs that his influence may be fading), former state security chief Lenin Cerna, and national security adviser Paul Oquist.... The inner circle consists largely of clandestine operators—paranoid cave dwellers who are afraid of the light and openness. Transparency is their worst enemy.[4]

For U.S. diplomats still mired in the Cold War, the Sandinista regime of 2007 was "fast reverting to its improvisation, secrecy, and centralization of the 1980s, forging an operating style more appropriate for a revolutionary junta than a modern, democratic state."[5] U.S. officials were particularly concerned about any changes to the army and the police. The new-found U.S. sympathy for the army was quite unexpected, because in the struggle to deprive Nicaragua of its SAMs, the army had fought hardest against U.S. demands. But suddenly U.S. officials regarded the Nicaraguan army as an invaluable ally in any future confrontation with the Sandinista regime. For his part, President Ortega was determined to guarantee the loyalty of the army and police, and his concerns increased when the National Assembly blocked his two nominees for the position of Defense Minister. Ortega feared that the senior army commanders were no longer devoted to the Sandinista cause. A memorandum to Ortega stated that "You know more than I that mentalities have changed since the times of the Sandinista Popular Army. In the senior ranks ... people have new and self-interested ideas."[6]

As a solution, President Ortega decided to control the army and the police without civilian intermediaries. In the case of the army, he eliminated most of the civilian positions within the Defense Ministry and ruled directly through senior commanders and in particular through its chief, General Omar Halleslevens. The army dated back to 1979 when the Sandinistas separated it from the police as replacements for the despised *Guardia Nacional*, the U.S.-created constabulary which had combined army and police functions. At the start of the twenty-first century the army enjoyed the confidence of the public and numbered 15,000 active duty personnel, just a thousand more than the hated *Guardia* in the last year of its existence. Both the army and the police maintained order and prevented gangs and other criminal groups from infiltrating into Nicaragua as happened in the rest of Central American countries. The drug trade was of limited proportions, because drug cartels preferred to channel their operations through the more receptive countries of the region.[7]

Once President Ortega made it clear that he was not planning any large expansion of the armed forces, the U.S. government should have stopped worrying about any military threat from Nicaragua. An even more comforting signal came when the army insisted on maintaining the close ties it had been cultivating with the U.S. military since 1990. The new Sandinista officials were "completely taken aback by the depth and the diversity of our assistance relationships with the Government of Nicaragua at all levels."[8] A crucial indicator of Nicaraguan interest came with the renewal of the Status of Forces Agreement (SOFA) in 2008; these agreements granted blanket immunity to U.S. military personnel against prosecution for any crimes committed inside Nicaragua. Sandinista officials were uncomfortable with this immunity, but opposition to the agreement declined once U.S. diplomats explained that Nicaragua had the final authority on whether to invite U.S. military personnel into the country for humanitarian or training missions. Senior officers were particularly insistent on renewing SOFA because the delays had already cost the loss of training exercises and the cancellation of important invitations.[9]

The renewal of SOFA clearly signaled that the leftist government of Nicaragua wanted to maintain close military ties with the United States. For his part, President Ortega was asking the U.S. government for more assistance in the struggle against the drug cartels. Meanwhile the Nicaraguan vice president was on a goodwill tour in the United States promoting the many investment opportunities the country offered to American businesses. In foreign policy statements the Sandinistas sided with Cuba, Venezuela, and Russia, but leftist Nicaragua showed no signs of wanting to break economic or much less military ties with the United States.[10] The shift to the left in Nicaragua, in spite of the dark predictions at the start of Ortega's first presidential term in 2007, had failed to create a threatening or dangerous situation in Central America for the United States. The task of maintaining internal order and the immense difficulties of trying to improve the material conditions of the Nicaraguan people imposed on the mature Sandinistas a policy of cooperation in bilateral relations with the United States. Washington would have preferred an unconditional ally but could live with the independent streak of Sandinista Nicaragua.

Guatemala: The First Leftist President of the Twenty-First Century

The inauguration of Álvaro Colom on 14 January 2008 brought a leftist to the presidency of Guatemala (Map 11) for the first time since the CIA overthrew Jacobo Arbenz in 1954. After the return to power of the Sandinistas in Nicaragua in January 2007, the last thing the George W. Bush administration wanted was another leftist regime in the U.S. backyard. But initial reports strongly suggested that the fears of a sharp turn to the left in Guatemala were vastly exaggerated. Of the individuals the new president nominated for his cabinet, "many of the nominees are well and favorably known to the Embassy, and we believe we can work well with them."[11] Many

Map 11. Guatemala.

additional actions confirmed that the new president was taking a moderate and centrist course. In the face of this unmistakable evidence, the outgoing Bush administration concluded that Colom posed no real threat and that preserving close ties with Guatemala was in the best interests of the United States.

The moderate approach of Colom, however, appeared extremely radical to wealthy groups more comfortable with right-wing civilian governments and military regimes. The Guatemalan upper class, fanatically immersed in the anti–Communist crusade of the twentieth century, could not accept that the situation had changed completely. "The private sector is deeply concerned by the spread of leftist populism in Latin America. With the FMLN poised for electoral victory, El Salvador would likely be the next to fall. The Colom government would feel empowered by the FMLN's victory and turn further left."[12] As it became clear that the incoming Barack Obama administration was not interested in participating in a campaign to destroy the Colom presidency, business leaders "anticipated having to fight the Cold War here again, and once again without U.S. support."[13]

As the Guatemalan upper class did everything possible to destroy the Colom administration, the absence of an embarrassing scandal hampered efforts to turn public opinion against the new president. An opportunity at last appeared when noted opposition leader Rodrigo Rosenberg appeared dead on 10 May 2009. A few days after his death, a video surfaced in which the deceased claimed that if he was found dead, it was because President Colom had ordered his assassination. The murder sparked violent demonstrations in Guatemala in May and June. The issue would not go away, filled news headlines for months, and practically paralyzed the country. To try to overcome the crisis, the U.S. Embassy brokered a deal between the opposition and the government. All parties agreed to accept as final the verdict by the United Nations Commission against Impunity, which already was operating in Guatemala.[14]

In a Latin America already full of extreme and bizarre events, the revelation that attorney Rosenberg had staged his own murder and had hired assassins to have himself killed still managed to shock the world. After having completed an exhaustive investigation, the United Nations Commission against Impunity announced these startling findings on 12 January 2010. As news of the revelations sank in, slowly President Colom gained a grudging acceptance from the upper class.[15] At last "private sector leaders expressed regret for the political turmoil that followed the publication of Rosenberg's allegations, and said they were willing to work with the Government of Guatemala on tax reform and other issues."[16]

By 2010 Guatemala was overrun by a crime wave of a magnitude the country had never experienced before. Drug traffickers had gained control over large areas of the country during previous years, and the arrival of the Zetas, a Mexican drug cartel, accelerated the descent into lawlessness in northern Guatemala. Actually a Guatemalan drug lord trying to defeat his rivals had invited the Mexicans into Cobán (Map 11), but then the Zetas, who were particularly savage, eliminated the other drug traffickers and took over the entire business. Most of the Zetas were former members of the Mexican Special Forces, and they completely outclassed other criminal gangs in combat skills and weaponry. Also the Zetas bribed Guatemalan officials to obtain bullets from the munitions factory and to steal weapons from army arsenals.[17]

The Zetas had easily neutralized the National Civilian Police, which "by 2004 had become penetrated by organized crime and otherwise corrupted."[18] The corruption in the police force became so rampant, that most of its high officials, including the chief, were caught red-handed while manually loading a shipment of 1100 kilograms of cocaine on 6 August 2009. Only the prompt arrival of army troops made possible the arrest of these corrupt police officers. At least a hundred policemen were implicated in this drug bust, and obviously the National Civilian Police lost any credibility about its ability to stop either drug traffickers or the crime wave engulfing Guatemala.[19] The reputation of the police was so bad that "individual citizens in rural communities are really frightened of an abusive, corrupt, and inefficient police force."[20]

Because Guatemala was involved in a potentially explosive dispute with Belize on the Caribbean side, the government of Colom could not afford to lose control of that strategic border region. "Narcotraffickers have gained the upper hand along Guatemala's eastern coast. Area police are corrupt, inefficient, and outmanned and outgunned; other local authorities intimidated."[21] A drastic solution was needed, and this came when President Colom reopened the military base at Puerto Barrios (Map 11) as the new headquarters for the Kaibiles, the Special Forces of the army, one of the units accused of participating in acts of genocide against Indians during the last two decades of the 1961–1996 guerrilla war. When Colom attended the inauguration ceremony of the base, he was sending the clear message that whatever human rights violations the Kaibiles might have committed in the past, the Special Forces were now indispensable in the new struggle against drug traffickers. Operating out of bases in Puerto Barrios and Poptún further to the north, the Special Forces restored control over the region bordering Belize. The Kaibiles also carried out joint patrols with the police in the neighborhoods of Puerto Barrios.[22]

By 2009 it had become clear that only the large-scale deployment of army units could disrupt the activities of the drug traffickers. The Zetas in Cobán controlled large regions of northern Guatemala, and to try to restrict their activities, President Colom established another Special Forces unit in that region. But the new mission required at least an additional 1000 Kaibiles and the army could spare only about 300. The army was already overstretched and lacked the numbers to assume new and pressing missions. For example, authorities were present in only four of the eight official crossing points on the border with Mexico, while nobody guarded any of the other 43 unofficial crossing points. The army was simply too small to meet all the demands, and in a historic reversal President Colom enlarged the army from 15,500 to 20,000 in November 2009. Because this increase did not include the separate unit of 3,000 reservists operating under police direction, the government substantially increased the number of soldiers available to face the crime wave sweeping the country.[23]

In reality the enlargement of the army had been a classic case of too little and too late. Crime became the most pressing issue in the Guatemalan elections of September 2011, and the voters trusted a former general, Otto Pérez Molina, to do a better job than the leftist parties. President Pérez Molina, his new Defense Minister, and his new Chief of the army were all former Kaibiles, and they promised to restore order by militarizing the country and applying an iron fist (*mano dura*) to the criminals.[24] The very mild leftist experiment of Guatemala was over, and right-wing conservative groups regained the complete control of the country that they had traditionally enjoyed since 1954. And Guatemala ratified its status as an unconditional ally of the United States.

After his inauguration on 14 January 2012, the new president ordered the military to wage a campaign against drug traffickers in the country and to control the crime wave. Previously, citizens had taken the law into their own hands, and 230 lynchings occurred just in 2011. Not wishing to exaggerate the image of militarization, President Pérez Molina concentrated on increasing the size of the National Civilian Police by 10,000 to a total of 35,000. Quietly he also increased the size of the army by 2,500 and emphasized a larger role for the Kaibiles in dealing with the heavily armed Zetas. But clearly Guatemala was going to need many more soldiers to win the war against the drug cartels and the gangs terrorizing the civilian population.[25]

El Salvador: The FMLN in Office

In the minds of Salvadorans, the National Liberation Front Farabundo Martí (*Frente Farabundo Martí de Liberación Nacional*) or FMLN was permanently associated with the bloody and long war in El Salvador from 1979 to 1992. Unlike the Sandinistas in Nicaragua, FMLN never gained power and did not have experience ruling the entire country. After the Peace Accords of 1992, FMLN reconstituted itself as a political party, but the smear campaigns of right-wing groups succeeded in keeping FMLN out of the presidency for over a decade. FMLN facilitated this slander campaign by nominating former guerrillas and hardline Marxists as candidates to the principal elected positions.

By 2005 growing dissatisfaction with the extreme free-market policies of the right-wing governments was persuading many Salvadorans to take a second look at the leftist opposition. Sensing an opportunity, FMLN nominated mainly moderates to run for the National Assembly, and most of these candidates won seats as deputies. Their surprising victory triggered intense debates inside the party, and in a decisive break with the past, FMLN nominated the moderate Mauricio Funes in November 2007 to be its next presidential candidate. The previous nominees to the presidency had all been former guerrilla commanders who reinforced the paranoia about a Communist takeover. In contrast, Funes was a well-known television journalist who had never fought in combat during the war. Because the presidential election was not scheduled until 15 March 2009, this early nomination gave Funes plenty of time to campaign extensively across the whole country and even to visit Salvadoran voters living in California. He repeated to any American officials caring to listen that he intended to maintain close ties with the United States including keeping the dollar as the official currency and remaining inside the CAFTA-DR free trade agreement.[26]

Funes organized a support group of businessmen under the name "Friends of Mauricio" who helped to confirm his moderate credentials. His electoral promises to make the campaign against poverty his highest priority gave him an early lead in the polls. But the right-wing groups responded with the tried and proven accusations of Communism, and soon television attack ads showed elderly women sobbing and crying that they were going to lose their houses if the Communist Funes won. An unexpected boost for FMLN came with the start of the global Great Recession in 2008 when the already bad economic conditions in the country became even worse. The rising crime wave fueled discontent, and the right-wing government of Tony Saca, seeing its candidate starting to trail, decided to jettison excess baggage in a last desperate attempt to prevent an FMLN victory.[27]

The election of Barack Obama to the presidency in November sent a clear message to El Salvador. The U.S. intention to get out of the Iraq War made pointless the continued Salvadoran presence in that Middle Eastern country. President Saca on 23 December announced that on 31 December the last Salvadoran troops would come home from Iraq, but the withdrawal could not stop Funes who won the presidential elections on 15 March 2009. In spite of soothing assurances from the winning candidate, some U.S. officials and the Salvadoran upper class still remained anxious about any excessive drift to the left.[28]

The lack of major ideological differences between the Funes and the Obama administrations made it easier for U.S. officials to accept the new Salvadoran president. Consequently, Funes did not have to face the difficulties Daniel Ortega experienced in Nicaragua during the last years of the George W. Bush administration. Both President Obama and Secretary Clinton telephoned to congratulate Funes after his electoral victory, and even before his inauguration on 1 June 2009, the Salvadoran president-elect briefly met with high-level U.S. officials including Vice President Joe Biden at a summit conference in Costa Rica. Secretary of State Hillary Clinton personally attended the inauguration of Funes on 1 June 2009, and in a symbolic move, she wore a bright red dress, the party color of FMLN. Although on his first day in office Funes established diplomatic relations with Cuba (El Salvador was the last Latin American country without an embassy in Havana), he remained genuinely friendly to the United States and was determined to maintain excellent relations with the Obama administration.[29]

At first the incoming leftist FMLN administration seemed to be making an even smoother transition into office than Álvaro Colom in Guatemala. Funes appointed the businessmen of the "Friends of Mauricio" to head the economic ministries and the other positions affecting the economy, and their policies were far from radical and resembled those of the Obama administration. These appointments were reassuring to business leaders who promised to work with the government and did not engage in that bitter campaign of hatred that so paralyzed the initial year of the Colom presidency in Guatemala. The leftist views of Funes remained largely symbolic, as when later in March 2010 he publicly apologized on behalf of the Salvadoran state for the assassination of Archbishop Óscar Romero in 1980.[30]

After his election Funes expressed his intention to appoint a senior officer to be Minister of Defense, thus putting to rest fears of any major change in the military. But most senior officers were so closely tied to the previous right-wing administrations, that Funes had difficulty finding an acceptable nominee. He finally had to reach into the ranks of retired officers to identify a congenial colonel, David Munguía Payés, for the job; only on May 31, the day before his inauguration, did Funes announce the appointment. Munguía Payés was U.S. educated and spoke English with near native proficiency after having spent his high school years in San Francisco. As a businessman he was a member of the "Friends of Mauricio" and also enjoyed a close friendship with Funes.[31]

In many ways Munguía Payés was the ideal candidate to be Minister of Defense, and the military establishment should have accepted him without hesitation when he took office on 1 June 2009. But his status as a civilian bothered the Salvadoran generals who insisted that only a general could hold that position. Munguía Payés correctly saw the objections of the generals as a power play, and he along with President Funes carefully plotted a counterblow to secure complete control over the military. During the month of June the Minister of Defense consulted individually with high-ranking officers and confirmed that while the generals were opposed to Funes, most colonels were eager to support the new government. In a surprise move, President Funes recalled the retired Munguía Payés to active duty and promoted him to general based on the recommendation of the promotion panel in 1998. It was common knowledge that a grave injustice had been committed against Munguía Payés when he had been denied promotion to general in 1998, and President Funes corrected that injustice on 30 June 2009.[32]

Now on active duty, General Munguía Payés promptly dismissed the entire class of generals and replaced them with officers he promoted from the rank of colonel. By this masterful move, the Minister of Defense and President Funes assured the loyalty of the officer corps and gained full control of the military establishment. In friendly meetings with U.S. officials, the Minister of Defense, speaking in fluent English, reaffirmed the government's commitment to strengthen the close ties with the U.S. military at all levels. He stressed the need for more assistance for the Salvadoran army, because it had been without new funding and had not increased in size since the end of the insurrection in 1992.[33]

Military and economic relations with the United States seemed to be running smoothly until a first snag appeared over the appointment of the new Minister of Public Security. The Funes administration was determined to obtain control also over law-enforcement agencies. As part of the balancing act of satisfying all sectors of his electoral coalition, Funes put FMLN militants in change of intelligence and police functions. The U.S. Embassy initially was worried about hardline FMLN militants occupying these sensitive positions. Upon closer examination, although some of these leaders still held to extreme Marxist positions, all had mellowed and no longer supported violent actions. However, the new Minister of Public Security Manuel Melgar was a special case, because he was implicated in "the 1985 Zona Rosa massacre which killed four off-duty unarmed marine security guards, two other American citizens, and several additional innocent bystanders."[34] As soon as the U.S. Embassy learned of the possibility that Melgar might be appointed, U.S. diplomats repeatedly expressed their objections to Funes administration officials, but to no avail. The U.S. Embassy was at a loss on how to handle Melgar and requested guidance on whether to leave his history in the past, work around him through other officials, or simply refuse to have anything to do with him. Pressing as the matter seemed to diplomats on the spot, the State Department did not share the same sense of urgency, and months passed without the Embassy receiving any guidance from Washington. By the time the State Department replied, Melgar was already firmly established in the position, and by then all the U.S. Embassy could do was to deny him entry visas into the United States.[35]

Because the National Civilian Police was under Minister Melgar, he may have indirectly influenced the decision to stop buying small arms from U.S. manufacturers. In reality one of the "Friends of Mauricio" owned an ammunition factory and also imported cheap weapons so that it was a foregone conclusion that future police contracts would go to this campaign supporter. And already even under the previous government, the National Civilian Police "had been moving away from U.S. weaponry in favor of equipment from Brazil, Argentina, and Israel because they could buy it more cheaply and get it faster."[36] As in most Latin American countries, in El Salvador the military have traditionally controlled the production and importation of weapons. As a partial solution to the crime wave, the National Civilian Police wanted to place orders for large quantities of automatic weapons from foreign countries. But the army refused to authorize these purchases because it felt that only a few tactical units in the police were adequately trained to handle these rapid-fire weapons.[37]

The deteriorating situation in El Salvador was disturbing similar to that in Guatemala where repeated efforts over nearly twenty years had failed to produce an efficient police force capable of controlling crime. The

National Civilian Police was not attracting recruits in adequate numbers, and the task of building up the police in El Salvador proved much more difficult than anyone had expected. To try to gain time, President Funes issued an emergency decree on 5 November 2009 authorizing 3,500 soldiers to reinforce the roughly 2,000 soldiers already serving under the National Civilian Police. These additional troops went to the five departments with the highest crime rates in the country, such as San Salvador and San Miguel (Map 3). In contrast to the previous operations with the police, this time the army retained full control over the troops, and only a handful of policemen accompanied the soldiers to make arrests in those cases beyond the jurisdiction of the military.[38]

The deployment was supposed to last for only 180 days, but as expected the government extended it on 7 May 2010 to last for a year, and in 2011 for still another year. The repeated extensions were ample proof that the National Civilian Police could not stop the growing number of homicides and the spreading wave of drug and gang violence. To help the police, Defense Minister General Munguía Payés was deploying the army in the struggle against crime, but its numbers were vastly insufficient. The Chapultepec Peace Accords of 1992 imposed sharp reductions that kept the army down to 11,000 soldiers. FMLN in negotiations in 1992 had demanded the abolishment of the army, but now in a stunning historical reversal, Funes, as the first FMLN president, enlarged the army until it numbered 17,000 in 2011. The expansion of the army was inevitable, because as even President Funes noted, the people trusted the army more than the police.[39]

The enlarged army and the many joint-army patrols were hampering the activities of the criminal gangs. The deployment of one battalion to guard the border with Guatemala also proved beneficial, but the greatest results came from the creation of a separate battalion to operate a security ring around the prisons. Previously, inmates not only received illegal items such as guns and drugs but also directed gang operations from the safety of prisons. A common trick had been for visitors to throw cell phones and other items over the prison walls and into the hands of waiting inmates. Incarceration for criminal leaders had become a welcome vacation from the rigors of life on the outside. Criminals easily bribed prison employees, but the additional ring of troops proved to be a real barrier as the soldiers constantly patrolled the walls on the outside and meticulously searched all visitors to the penitentiaries.[40]

The enlargement of the army could not have taken place without U.S. assistance, but this help came at a price, and U.S. officials proceeded to extract their pound of flesh in exchange. As mentioned earlier, El Salvador had withdrawn its last soldiers from Iraq on 31 December 2008, and because the United States was also accelerating its withdrawal, it seemed that El Salvador was at last off the hook as far as having to make more troop commitments to the Middle East. But Afghanistan was another matter, and almost from his inauguration U.S. officials had been badgering Funes to contribute troops as proof of being an unconditional ally of Washington. The request put the president in a dilemma, because the leftist base in FMLN violently opposed U.S. military interventions in the Middle East, but he knew that he could not say no. He tried to play for time and also equivocated to try to satisfy both his political base and the U.S. government:

> Regarding sending troops to Afghanistan[,] President Funes said it had been discussed and that they would approach it gradually, likely starting with a small site survey, then sending an observer or two, followed by posting staff officers in one of the safer areas, before ultimately deploying a unit.[41]

Such an almost invisible participation was inadequate for U.S. purposes, but American officials were sensitive to the political dilemmas Funes faced and were more than willing to work out a compromise solution. El Salvador agreed to send this time not combatants but trainers to Afghanistan. The tour of duty of the 22 instructors was supposedly short term but everyone expected its extension for as long as the U.S. government wanted. In its efforts to pressure the government of El Salvador, the Obama administration was undistinguishable from the previous Bush administration.[42] And in the longer historical perspective, the contrast was striking between the strenuous efforts the U.S. government had to exert to obtain a few dozen instructors in 2011 versus its vigorous rejection of the hundreds of thousands of Latin American volunteers who wanted to fight under their countries' flags against the Axis powers in World War II.

Funes realized that he had to pair his very modest Afghanistan deployment with something else to remain in the good graces of U.S. diplomats. The growing crime wave paradoxically gave him the opportunity to shoot two birds with one stone. The effectiveness of the army contrasted with the lagging performance of the National Civilian Police under the jurisdiction of the Ministry of Security. Funes decided to make the two civilians heading those two agencies the scapegoats for the growing crime wave and to replace them with army officers. The U.S. Embassy had been pressuring insistently for the removal of Manuel Melgar from the position of Minister of Security and had refused to sign an agreement providing assistance to El Salvador unless Melgar was first removed from office. He was the first one to resign on November 2011, and as his replacement Funes brought over General Munguía Payés from the Defense Ministry. U.S. officials were very pleased to work with the general in his new position as Minister of Security. For director of the National Civilian Police, Funes appointed an army officer who resigned his commission the day before taking office because the Constitution prohibited active duty officers from heading the police force.[43]

With army officers whether on active duty or retired in charge of both the Defense and Security ministries and the National Civilian Police, a coordinated offensive against the criminal gangs went into high gear. Militarization of the campaign against crime soon started to bring results in 2012. In March of that year, the Roman Catholic Church brokered informal conversations with leaders of the two main gangs in the country, Mara Salvatrucha–13 and Barrio 18. Civilian intermediaries were indispensable because the government could not participate in talks to avoid giving legitimacy to the gangs. Actually, the main negotiations were between the two gangs, because killings between the two groups rather than clashes with soldiers or the police accounted for the majority of the deaths. But the army, by bringing greater pressure upon the gangs, was in a sense intensifying the violence as the gangs fought over diminishing sources of illegal revenue and dwindling territory. In late March the two gangs agreed to a truce, and in exchange on 4 April 2012 the government lifted the army security ring around the prisons.[44]

Visiting prisons in Latin America has never been a pleasant experience, and placing the army security ring around the prisons had forced visitors to run a gauntlet of humiliation by the soldiers and abuses by the prison guards. The drastic measure had disrupted the lives of inmates who lost touch with the outside world and no longer enjoyed the luxury of communicating through cell phones. More than anything else the army security ring struck hard at the most vulnerable point of the gangsters who had seen their relatives and friends repeatedly abused in entering prisons. The gangs reciprocated the good-will gesture of the government by sharply reducing their violent clashes, and for the first time El Salvador experienced a drop in its horrendous homicide rates.[45]

A second round of negotiations took place in July 2012 under the auspices of Roman Catholic leaders and officials of the Organization of American States, including its Secretary General José Miguel Insulza. The leaders of Mara Salvatrucha–13 and Barrio 18 presented additional demands to make the truce permanent. Although the gangs wanted to reduce police presence in the neighborhoods, their main demand was to get the army out of the streets and back into its barracks. Army officers had never wanted to assume law-enforcement functions, and the government was also eager to reduce the military presence. But authorities knew that any precipitate withdrawal risked reviving the gang violence, so the truce remained informally in force into 2013. The government may also have authorized unofficial payments to gang leaders and certainly approved prison transfers to more comfortable facilities. This uneasy *modus vivendi* was too fragile to last in El Salvador and sporadic outbursts of violence continued to disrupt the relative calm.[46]

By 2013 public attention was starting to focus on the upcoming presidential elections in 2014. The Constitution did not allow reelection, so Funes could not be a candidate. FMLN wanted to remain in control of the presidency, but was torn by internal debates over whether to nominate a centrist to continue the policies of Funes or a hardline leftist candidate. The decision was crucial, because Funes had won in 2009 by a margin of only 60,000 votes, and any far left candidate was sure to face a hard time getting elected. FMLN decided to gamble on Vice President Salvador Sánchez Cerén, a former guerrilla commander, who became the party's standard bearer. He possessed strong anti–American credentials and previously he had opposed both CAFTA-DR and the use of the dollar as the official currency.[47]

The leftist candidate enjoyed the huge advantage of deep divisions inside conservative groups, and the right-wing parties actually fielded two candidates. Wisely Sánchez Cerén moved slightly to the center, and he very vigorously emphasized the need to maintain the closest economic ties with the United States. In a major reversal from his previous position, he now defended the need to keep the U.S. dollar as the official currency of El Salvador. Facing a divided opposition, not surprisingly he won the most votes in the election of 2 February 2014, but because he did not win a majority of 50 percent, the Constitution required a run-off election on Sunday 9 March 2014.

The polls indicated that he was likely to win by a wide margin, but the opposition candidate fought back ruthlessly with charges that the FMLN intended to follow the example of Hugo Chávez in Venezuela. In the end Sánchez Cerén squeaked by with a lead of 6,634 votes, considerably less than the 60,000 votes that brought victory to Funes in 2009. While the extremely narrow victory came as a surprise to Sánchez Cerén and the FMLN, citizen absenteeism was not to blame and actually voter turnout was four percent higher in the run-off than in the 2 February election. The opposition conservative candidate refused to accept defeat, and the electoral authorities proclaimed Sánchez Cerén the winner only on Thursday 13 March.[48]

Because the opposition still refused to accept the verdict, it was up to the army to cast the decisive vote. The many colonels Funes had promoted to general felt comfortable working with the FMLN, and once senior officers announced that the army unanimously and solidly supported the verdict of the electoral authorities, the campaign for president was over. The shift to the left from Funes to the new president had been a difficult sell to make to Salvadoreans, and it was clear that the FMLN had to achieve major results after the inauguration of Sánchez Cerén on 1 June if it wanted to remain in office for another term. Under President Funes El Salvador had dampened its enthusiasm for remaining an unconditional ally of the United States, and Sánchez Cerén would most likely follow the path of Sandinista Nicaragua of combining collaboration on economic and military topics while pursuing a more independent foreign policy.

The first order of business for the new president after he took office on 1 June 2014 was not foreign policy but the collapse in May 2014 of the truce between Mara Salvatrucha-13 and Barrio 18. Homicides in 2014 rose to the pre-truce level of 2011, and El Salvador was fast becoming one of the most violent countries in the world. As a first step, Sánchez Cerén appointed General Munguía Payés to be Defense Minister. Because the General had helped broker the truce between the gangs in 2012, it was felt that he could help restore order.[49] But as of September 2014 the crime wave and the war between the gangs was still raging unchecked in El Salvador.

9

Honduras: The Coup of June 2009

Honduras will not return to be what it used to be.—Truth and Reconciliation Commission[1]

Under most considerations, Honduras easily qualified to be a fourth unconditional ally of the United States. However, chapter 3 did not include Honduras in the select group of Colombia, El Salvador, and Guatemala, and the omission was far from accidental or capricious. Nobody doubted the intention of Honduras to belong to the club of unconditional allies, but its erratic behavior was too annoying. Decades of observation had persuaded U.S. diplomats that the country was unreliable. Ruling the third poorest country in the Western Hemisphere posed great challenges to Honduran leaders, and always looming over the horizon was the threat of becoming a failed state. Until the late 1990s the existence of an admittedly inefficient military had kept the country together, but as the armed forces retreated to occupy a marginal role in the life of Honduras, the burden of creating a new governing structure for the country became the most pressing task facing its political leaders as the twenty-first century began.

The Presidency of José Manuel Zelaya

In 1996 Honduras adopted U.S. suggestions to place its military under civilian control in a single defense ministry. A constitutional amendment in 1999 codified civilian control over the military and reduced the standing of the highest ranking officer from that of Commander of the Armed Forces to Chief of the General Staff. All officers came from the military academies and received adequate training. The officers often complemented their education with courses in U.S. installations. After the abolishment of conscription in the late 1990s the army relied on volunteers to fill the ranks, but most recruits rarely had more than a sixth-grade education. Their educational level was so low, that the army employed teachers to impart elementary education to many soldiers.²

A career in the army did not attract many recruits, and battalions declined to one-third of their strength by the end of the 1990s. U.S. military assistance in the past furnished a modest arsenal of small arms including 12.7 machine guns, but the rest of the equipment was old and worn out. In the intricate balance of power calculations of Central America, Honduras counted on its air force to deter potentially aggressive neighbors. In particular Honduras wanted to dissuade El Salvador from invading again as happened in 1969 with the Hundred Hours' War. To counter the numerical superiority of El Salvador in manpower, Honduras relied on its fleet of F-5E jet fighters, and originally started out with 12, the same number of F-5Es as Mexico. And in both countries the number of F-5Es diminished through attrition: in 2003 Honduras had operational at most eight of the jet fighters. The maintenance of these sophisticated airplanes required recruiting individuals with at least a high school degree. The U.S. base at Soto Cano, a driving distance an hour and a half from the capital of Tegucigalpa (Map 9), provided a great opportunity for air force personnel to learn by observation and instruction. Outside of Guantánamo in Cuba, Soto Cano Air Base with 565 American soldiers was the only U.S. military base in Latin America, and its existence partly helped fill the void left by the American withdrawal from installations in the former Canal Zone of Panama.³

The serviceable army of Honduras proved convenient to U.S. interests when the George W. Bush administration was looking for countries able to contribute troops to the "Coalition of the Willing" for duty in Iraq. President Ricardo Maduro saw the opportunity to ingratiate his country with the U.S. government and sent Honduran troops to Iraq in August 2003. The commitment pleased Washington and earned Honduras considerable goodwill "at the highest levels of the U.S. Government, raising Honduras's profile."⁴ The Honduran contingent served as part of the larger Spanish unit, and with bravado the Defense Minister stated "that the Government of Honduras has considered the ramifications of taking casualties in Iraq in support of the war against terror and that they were prepared for that eventuality."⁵

The U.S. Government did not seem to have any reason to complain about its apparent ally, but the outcome of the Iraq adventure soon showed why Honduras could never become an unconditional ally. As the presidential elections approached in November 2004, the candidate of Maduro's National Party was facing an increasingly hard campaign. In an attempt to improve the candidate's chances, Honduras informed the U.S. ambassador on 16 April that it was withdrawing its troops from Iraq. The simultaneous recall of Spanish troops gave a good excuse to justify the withdrawal of the Honduran contingent. The Iraq war had been very unpopular in Honduras, and at least 70 percent of the public always remained very opposed to any involvement. Unlike in 2003 when the Maduro administration claimed to be ready to accept casualties, in less than a year the attitude completely changed. The same administration could no longer accept any combat deaths, and Honduras rushed to extract its 368 soldiers from Iraq in May 2004.[6] And this sudden reversal in position was just one of many incidents proving the unreliability of Honduras as an ally.

The return of Honduran troops from Iraq was not enough to save Porfirio Lobo Sosa, the candidate of the National Party, in the elections of 27 November 2005. Experts had underestimated the appeal of the candidate of the Liberal Party, José Manuel Zelaya Rosales, who had run and lost in the previous presidential election. The real architect of the Zelaya victory was his vice-presidential running mate, Elvin Santos, a young businessman who attracted new entrepreneurs into the Zelaya campaign. With the help of Santos, Zelaya was able to create that multi-class coalition indispensable to electoral victory.[7]

In the Caribbean countries the U.S. government cast the most important vote, and without U.S. approval no candidate could expect to win. Zelaya had no difficulty on that score because U.S. diplomats rated him as pro–American and the relationship became so friendly that they often referred to him as "Mel," the nickname his loyal followers used to call him. Throughout his previous career and during his presidency he assiduously consulted with the U.S. Embassy in Tegucigalpa. Even on the day of his inauguration he flattered the ego of American imperial diplomats when outside his office he kept waiting President Vicente Fox of Mexico and other Central American presidents in order to see first the U.S. inaugural delegation. Because the head of the U.S. delegation was only the Attorney General, this was not just a breach of protocol but also a deliberate snub at the other presidents. For years imperial U.S. diplomats fondly savored recalling this episode.[8]

If any doubts still existed about his slavish compliance with U.S. preferences, overwhelming confirmation came when "in an early test, Zelaya has already shown that he is willing to take a politically painful decision on an issue of great concern to the U.S. government by switching his publicly announced choice for Foreign Minister."[9] The U.S. ambassador raised the objections in a morning telephone conversation with Zelaya, who explained that he had not been aware of the problem. That same afternoon Zelaya phoned back the ambassador to report that he had pulled the nomination and was appointing another individual who was acceptable to the U.S. government.

Zelaya knew very well the rules of holding office inside the informal empire of the United States in the Caribbean, and he was not about to defy the superpower as the tragic Jacobo Arbenz futilely tried to do in Guatemala in 1954 by refusing U.S. demands about dismissing officials. U.S. diplomats could fume about defiance by Mexico after 1920, Cuba after 1959, and Venezuela after 1999, but in the rest of the Caribbean countries presidents had internalized well the lesson that it was dangerous to defy the U.S. government on the matter of appointments to high office whether civilian or military. Part of the duties of U.S. diplomats was to compile information on all officials holding high government positions. This has been a normal task for diplomats of all countries, but in the Caribbean region, the job took on added significance. Submission of the biographical information to Washington sent the implied message that no obvious reason existed to block the appointment. Because usually the Caribbean country at least initially had better information on candidates than the U.S. Embassy, its rulers generally avoided any problems by excluding unacceptable candidates. But sometimes some nominations slipped through the cracks, as was the case for the foreign minister in Zelaya's Honduras. But the U.S. government usually was not a hard overseer and did not always want to micromanage the Caribbean countries. Even when only inadequate biographical information was available on a nomination for high position, the U.S. Embassy preferred to leave the benefit of the doubt to the local president. If objections later arose, the excessively frequent reshuffling of cabinet ministers so typical of Caribbean countries gave plenty of opportunities to dispose of the unwanted individual.[10]

The last question U.S. diplomats had to answer was whether Zelaya was a leftist or a socialist, and more precisely if his views coincided with those of Hugo Chávez in Venezuela and Fidel Castro in Cuba. The U.S. Embassy received unequivocal evidence that President Zelaya was a moderate and centrist, a conclusion in sharp contrast to the wild and unfounded accusations from right-wing members of the Honduran upper class. His new ideas consisted of creating "citizen power" by increasing the participation of Hondurans in the running of governmental affairs through greater transparency. Zelaya urged decentralization as a way to foster citizen participation, and he promised to distribute funds and to delegate authority to the 298 municipalities in the country. These ideas seemed hard to implement but they were harmless enough and did not raise any particular concern.[11] The U.S. Embassy had no objections to Zelaya completing his presidential term, although it admitted that "his personality, administration, and likely government plans could make working with him an interesting challenge."[12] In a litmus test, Zelaya enthusiastically implemented CAFTA-DR, the free trade agreement of the United States with the Central American countries and the Dominican Republic. Zelaya personally intervened to make sure that the treaty, ratified by Honduras in March 2005, went into effect that same year.[13]

Very presciently U.S. diplomats detected that personality rather than policy was going to be the real issue in dealing with Zelaya. This early revelation was remarkable, because outwardly he had a very pleasant and amiable character and was a very enticing public speaker. As upper-class Hondurans worried about the large number of supposed leftists in the cabinet and other high official positions, U.S. diplomats did not share those concerns. Conservative opponents often cited Patricia Rodas, the daughter of a prominent Liberal politician, as a radical extremist pushing the President toward the left. It turned out that she was one of many Zelaya loyalists who kept the U.S. Embassy minutely informed about the maneuverings inside the government. Pro-American officials such as Rodas repeatedly reassured U.S. officials that Zelaya had no intention of moving to the left, but then what direction was he taking? A mystified U.S. Embassy reported in May 2006 that "one hundred days into his term, President Mel Zelaya remains somewhat of an enigma."[14]

Until January 2008 Zelaya still had wide representation in his cabinet from different sectors of the Liberal Party, but already in his first year he had allowed Vice President Elvin Santos to be marginalized from his administration. Santos, a rising star in the Liberal Party, was responsible more than anyone else for the electoral victory of Zelaya in 2005, and the Vice President was very likely to be the candidate for the November 2009 presidential elections. Relations between the two men were very close and friendly, yet in actual practice "he was effectively shut out. Santos was forced to hire his own staff, locate his own office space outside the presidency, and even set up his own computer system."[15] A bitter public and private campaign was under way to discredit Santos, as political rivals resented his rapid rise and were doing everything possible to sabotage his nomination for the 2008 presidential elections. Santos was so dispirited that already in February 2006, barely a month after the inauguration, he wanted to resign. His close friendship with Zelaya and the advice of the U.S. ambassador persuaded the strongly pro–American Santos to remain as vice president. However, the failure of Zelaya to consult Santos or to use him in any meaningful role was the single most important political mistake of his first year in office.[16]

The support of Santos was crucial for Zelaya, because although the Liberal Party had a majority in the unicameral legislature, many deputies represented other factions in the party and did not automatically support the President. As long as representatives of those other factions were in his cabinet or in other high positions, Zelaya could count on their votes to approve legislative proposals. Santos was the ideal person to broker deals with the legislators, but because Zelaya did not want to rely on him, the President increasingly turned to the military.[17] Since the late 1990s the military had stayed out of politics, and the senior officers wanted to keep their distance from civilian matters. In the history of Honduras this non-involvement was unusual, because until the mid-1990s traditionally civilian presidents had appointed many officers to important civilian posts. The shortage of qualified civilian candidates already made President Ricardo Maduro appoint some army officers to civilian positions, and Zelaya turned even more frequently than his predecessor to the military to fill administrative positions, but without altering the essentially civilian character of his presidency.

A major step came on 26 November 2007 when Zelaya appointed General Romeo Vásquez Velásquez to a second three-year term as Chief of the General Staff. The announcement came as a bombshell and sparked a flurry of rumors that Zelaya intended to carry out a "self-coup" (*auto golpe*) to remain in office. General Vázquez Velásquez was strongly pro–American and had been appointed for the first three-year term by President Maduro. Zelaya upon his inauguration had kept the general and allowed him to finish his term, and everyone had expected a new general to be the replacement for the next three-year term.[18]

> The reaction within the Honduran Armed Forces was very negative, as the extension goes against all accepted procedures and protocol. By accepting this extension to six years, Vásquez effectively cuts out four generations of military officers (those who graduated with him, those who should be taking over the position in January, and those who should be ready for promotion in 2008 and 2009) from the possibility of promotion to Chief of Honduran Defense. This is extremely detrimental to the morale of the Honduran Armed Forces.[19]

The decision to renew the general's term of office for three more years seemed perfectly natural to Zelaya who had become his close friend. The

President also carefully cultivated relationships with the top generals in the three branches of the armed forces, and he was confident that the personal friendship was strong and deep enough to trump any institutional loyalties or legal obligations. Counting on complete military loyalty but enjoying little support in the legislature or even his own Liberal Party, he felt confident about appointing a cabinet composed of officials totally subservient to his views on 6 January 2008. "The cabinet changes punish those Zelaya distrusts and bring in people he thinks he can control or those who are extremely loyal ... in short, Zelaya is increasingly isolating himself and having a harder time finding anyone to trust."[20]

The Path to the Coup

Rising gasoline prices had been fueling discontent among the population of Honduras even before the election of José Manuel Zelaya in November 2005. He pledged during his campaign to reduce the high cost of gasoline, but it was not clear what the Honduran government acting alone could do to control fuel prices. Barely one hundred days into his term U.S. diplomats somewhat regretfully came to accept the reality that the only way Honduras could lower fuel prices was to strike a deal with Venezuela. Precisely Hugo Chávez had created Petrocaribe as the organization to channel Venezuelan petroleum at below market prices and on easy credit terms to energy-poor countries of the region. Already in June 2006 Zelaya was ready to sign a deal with Petrocaribe, but before taking this possibly risky step, he first wanted U.S. approval.[21]

For the Bush administration, the decision was not easy either. With the U.S. government pouring billions of dollars into the deserts of Iraq and Afghanistan, almost no funds remained to deal with the pressing problems of the Western Hemisphere. The ideal situation for the U.S. would be to offer substantial economic assistance to Honduras as a way to compensate for the high gasoline prices. But the U.S. Embassy did not enjoy this luxury and instead took a very pragmatic position about obtaining petroleum from Venezuela in 2006:

> Action Request: Post suggests that during Zelaya's June 5 meeting, POTUS emphasize U.S. government displeasure at the government of Honduras's decision to make such an arrangement with the government of Venezuela. A carefully modulated message that does not demand walking away from such a deal, but nevertheless pressures Zelaya, could be effective in encour-

aging his support for other important U.S. government interests in Honduras.²²

Either President George W. Bush failed to communicate these nuances or the Honduran leader missed the meaning, and Zelaya as a precaution shelved the Petrocaribe proposal for over a year. But by 2008 the Honduran public clamored for action on the high gasoline prices. Zelaya went on a speaking tour across Honduras seeking to pressure legislators to approve the Petrocaribe treaty. He was most effective in the rural areas where the majority of the population lived. His skilled manipulation of public opinion secured legislative approval in January 2008 for the treaty making Honduras a member of Petrocaribe.²³

The decline in gasoline prices boosted the prestige of Zelaya. However, his political opponents feared that Chávez would demand something in return for membership in Petrocaribe. The Venezuelan president soon came collecting the payment by asking Honduras to join ALBA (the Bolivarian Alternative for America), a free trade agreement exclusively for countries of Latin America and the Caribbean. Zelaya duly complied, and before a huge crowd in a grand public ceremony in Tegucigalpa, he signed the agreement making Honduras a member of ALBA. Several presidents, including Chávez, had come to attend the signing ceremony, but the ratification of the treaty was far from certain: "while current indications are that the Honduran congress will not ratify ALBA, the political environment is too fluid to make a firm prediction."²⁴

Zelaya considered membership in ALBA to be one of his legacies for the country, and he was determined to push the agreement through congress, but with few deputies willing to support him, he faced an impossible task. He seemed incapable of translating his enormous popularity in the countryside into votes in congress, until the presidential ambitions of the chief of the congress, Roberto Micheletti, provided a unique opportunity. Relations between the two men had been tense since January 2006 when Zelaya had tried to block his election as chair of the unicameral legislature. Both men belonged to the Liberal Party, and Micheletti controlled the legislative branch while Zelaya controlled the executive branch. The rivalry between the two Liberal Party leaders was a major reason why Zelaya faced a hostile legislature, but with the presidential election of 2009 fast approaching, Micheletti decided to mend his fences with the President. In exchange for passing ALBA in the legislature, he asked for Zelaya's backing in the party primaries of 30 November 2008. Zelaya agreed to the deal, and on 9 October 2008 the legislature approved ALBA after adding some meaningless amendments stating that the agreement did not cover military or political matters. The passage of the bill was a resounding triumph for Zelaya, and business leaders were shocked to see how swiftly the agreement was unanimously passed; those deputies who were opposed simply did not show up to cast their vote.[25]

The entry of Honduras into ALBA marked the high point of success for Zelaya who could confidently claim that along with other measures such as Petrocaribe and the implementation of the free trade agreement of CAFTA-DR, he had done more for the country than most previous presidents. To fulfill his part of the deal with Micheletti, Zelaya made many public statements supporting the candidacy of Micheletti to be the standard bearer for the Liberal Party. The President went so far as to take a campaign tour with the candidate. But try as hard as Zelaya might, he simply could not transfer his popularity to the drab Micheletti, and the winner that emerged in the primaries of 30 November 2008 was the spurned Vice President Elvin Santos. This victory of Santos had momentous consequences for later events in Honduras. Zelaya, in spite of the personal differences with Micheletti, felt closer to the latter than to businessman Santos from the most conservative wing of the Liberal Party.[26] Micheletti saw slip the last chance to become president by means of elections because he simply was not an appealing candidate. And even if he could somehow reinvent himself for the subsequent presidential election in 2013, by then he would be either too old or possibly not alive.

Micheletti burned with passion to become president, yet how could he attain his ambitious goal? According to the constitution, a sitting vice president could not run for president, so Santos, shortly after his primary victory, resigned his office in December 2008. In effect, his resignation left the country without a vice president, and according to the Honduran constitution, the next in line to succeed was the chair of the legislature, none other than Micheletti.[27] In a major oversight, Zelaya failed to consider the implications of having next in line as his constitutional replacement a political rival who was determined to do anything to become president of Honduras. At the same time, the failure of observers and U.S. diplomats to grasp the determination of Michelletti to reach the presidency led to incorrect predictions about the likely political outcome.

As the decisive year of 2009 began, Zelaya, rather than enjoying his last full year of his presidential term, was conflicted by two worries. First of all "on substance, Zelaya has been a mediocre president at best; with few tangible policy successes, while doing little to confront Honduras's long term problems (e.g. poverty, corruption, and crime)."[28] The crime wave, and in particular the spread of gang violence in Honduras, heavily damaged the image of his administration. Second, he was unable to give up political power. "In a primal way, Zelaya is finding it very difficult to accept his lame duck status. Zelaya's incapacity to let go and his deep need to remain the chief political protagonist on the Honduras state, may be the strongest driving force propelling Honduras towards a crisis." For the U.S. Embassy, this was not necessarily a bad thing and actually provided many opportunities: "The fact that Zelaya is driven by personality needs instead of ideology means ... also that he can be manipulated by us. By playing to his ego and to his respect for the United States, we have, over the past ten months, been able to move him to our side on some key issues confronting the country."

As a solution to his deep personal cravings, Zelaya mapped out a strategy to remain a relevant political player even after his presidential term expired on 27 January 2010. First, using executive authority, early in 2009 he announced a whopping 60 percent jump in the minimum wage, a measure that increased his popularity among the urban employees and enhanced the prestige he already enjoyed in rural areas. His political opponents criticized raising the minimum wage as an obvious political ploy. His rivals, both inside his Liberal Party and in other parties, were even more worried about his decision to give the military better benefits and pay raises. Firmly and decisively, Zelaya was bringing the military back into the political life of the country.[29]

Believing that he had laid a sufficient political foundation, on 24 March 2009 Zelaya interrupted the regular television programming to announce his decision to hold a "consultation" on Sunday 28 June 2009. Citizens would vote on whether to support or reject his proposal to include in the presidential elections of 29 November a fourth ballot box for voters to decide whether to convoke or not a constitutional convention in 2010. This announcement came as a bombshell in Honduras, because only the unicameral legislature had the authority to authorize bringing issues to voters. Without this legislative authorization, the agencies in charge of voting could not conduct any elections. To bypass this hurdle, Zelaya ordered the National Institute of Statistics to organize the voting for this "consultation." Legal challenges promptly tried to stop the government from holding the vote, and predictably in late May the Supreme Court declared the proposal to be illegal and unconstitutional.[30]

The fourth ballot box gained tremendous public attention for the President who relished the limelight. "His talent for tactical surprise, improvisation and genius for generating crisis serves his interest in keeping opponents off balance and remaining the chief protagonist on the Honduran political scene."[31] After the decision of the Supreme Court, it was time for Zelaya to pocket his political gains and to abandon a proposal that had not been even his original idea. Yet as soon as he received advanced notification of the Supreme Court verdict, rather that moving on to another issue to remain the center of attention, he began to seek a way around the judicial decision. He had understood that his agreement to support Micheletti in the primary election of the Liberal Party on 30 November 2008 included endorsing the fourth ballot box, but the President felt betrayed when after the primary loss Micheletti reneged in the commitment. Politicians remembered the promise differently and claimed that they had agreed to have the legislature authorize a referendum on whether to convoke a constitutional convention only after the new president took office and never to have a "consultation" with a fourth ballot box.[32] It may well be impossible to tell who was actually telling the truth, but the sense of betrayal of Zelaya was clearly exaggerated if not unfounded. A crucial quality in negotiations is the ability to realize when to trust a politician's pledge, and this difficulty reflects the underlying assumption that commitments in politics are always negotiable and reversible. For Zelaya to complain about betrayal was childish behavior more appropriate in a schoolyard.

Before the Supreme Court formally announced that the "consultation" was invalid, Zelaya once again interrupted television on 29 May to state to the nation that he agreed with the verdict and was instead going to hold just a poll (*encuesta*) on 28 June and not a "consultation" (*consulta*). Specious as was the legal reasoning of this sleight of hand trick, complications already arose about the logistics of carrying out the poll. Originally Zelaya ordered the National Institute of Statistics to run the voting process. In response, the Supreme Court gave that institute a cease-and-desist order from all activities related to voting including publicity. The dilemma for the Director of the National Institute of Statistics was real: he was subject to arrest if he defied the Supreme Court, but if he followed the judicial ruling the president could fire him. In what initially seemed a Solomonic solution, the Director tried to satisfy his two masters by replying that the Institute had not been able to do anything to implement the president's order because of lack of resources and consequently was logistically incapable of conducting the election. The argument of lack of resources was persuasive even to President Zelaya, who in his announcement of 29 May ordered the military to provide all the logistical support necessary so that the National Institute of Statistics could carry out the poll on Sunday 28 June.[33]

Having armed soldiers practically run the voting stations did not seem a great lesson in electoral practices and also raised the larger question of whether the military would agree to conduct elections even if just for purposes of polling. On one level, all the private polling showed that the citizens overwhelmingly supported convoking a constitutional convention, so an official poll to discover the state of public opinion was completely redundant. All experts agreed that the 1983 constitution needed to be heavily revised or replaced, and politicians were more than willing to have a constitutional convention but only after the new president took office in January 2010.[34] But when Zelaya decided to bring in the military to man the polls, neither participants nor observers realized that the controversy had entered a new and much more critical phase.

The U.S. Embassy correctly stated that "Zelaya is edging the country towards a major political crisis,"[35] but its failure to recognize the gravity and urgency of the controversy contributed to the unfolding crisis. Many reasons accounted for the diplomatic and intelligence failure. U.S. diplomats had grown accustomed to erratic and contradictory behavior on the part of Honduras, but because Honduran officials traditionally expressed their loyalty to the United States, U.S. diplomats became dulled to the warning signs. Career motivations were present, because U.S. diplomats do not want a coup or revolution to break out in their watch. Even if the condemnation of diplomats does not match the witch hunts of "Who lost China or Cuba" during the congressional hearings of the Cold War, a diplomat's career can suffer irreparable harm in internal evaluations. In the Caribbean countries, the obligation of diplomats was to maintain U.S. supremacy quietly and cheaply, without leaving a paper trail, and without distracting senior officials from their concentration on matters in other parts of the world. Not surprisingly on 1 April the initial reaction of the U.S. Embassy to the Fourth Ballot Box proposal was slightly upbeat: Zelaya "covets U.S. approbation and does not want to break with us. We will leverage this, stay in close touch and seek to influence him and the other players to ensure a legal, constitutional and consensual way forward."[36]

The translation of the last message into plain English was that "we are old hands at this game of Caribbean intrigue, don't worry, we have everything under control." But what the U.S. ambassador did not say in his 1 April message was that already on March 30 he had told Zelaya that the

Fourth Ballot Box proposal was a bad idea. The ambassador made the revelation in his message of 1 May: "In a two-hour private discussion with President Zelaya on March 30 and shorter exchange on April 27, the ambassador told Zelaya that he personally believed that the Fourth Urn proposal was a bad idea and a distraction."[37] This cable was remarkable in many ways, starting with the ambassador breaking the tacit rule that direct instructions to client states should not be written down. Also, the ambassador *twice* gave the instruction and still Zelaya did not drop the Fourth Ballot Box idea; such an open defiance normally qualified a ruler or high official for status as anti–American and would trigger reprisals. But the ambassador was clearly at a loss as what to do and even provided a fig leaf by stating that he had given only his personal advice to Zelaya, thus leaving plenty of room for the State Department to reverse the ambassador's statement. Clearly he was too proud to request guidance and was sending a muted cry for help from Washington. The message was too subtle for State Department officials to perceive, and in typical bureaucratic fashion the ambassador received no answer and was left to flounder as best as he could.[38]

The U.S. Embassy finally decided to ask openly for help from the State Department only because of an unusual request from Zelaya himself. In a long conversation with the President on June 7, the ambassador laid out a list of reasons almost like a legal brief in a court case for dropping the Fourth Ballot Box proposal. In a new and sympathetic argument, "the ambassador said that we wanted President Zelaya to end his term of office on a high note and avert a political crisis that could be tragic for Honduras, himself personally and his family."[39] Zelaya replied by asking the U.S. ambassador to participate in a series of talks or negotiations not with opposition parties but with the members of the Liberal Party, his own majority party controlling the unicameral legislature, the Supreme Court, and the independent Attorney General's Office. A long and honorable tradition existed of diplomats participating in negotiations among rival factions or groups about to go to war inside a country, and often diplomats working quietly behind the scenes had been instrumental in reaching agreements and maintaining internal peace. But to have diplomats help a president negotiate with his own political party was an unheard of proposal that totally blurred the lines between foreign and domestic politics.

The request for the U.S. ambassador to participate in these talks with members of Zelaya's own party could correctly be attributed to the massive and preponderant presence the United States enjoyed in almost every aspect of Honduran life. But U.S. diplomats were too wise to want to flaunt their power, and such a request could come only from an increasingly irrational Zelaya. The U.S. ambassador, at a loss as to how to respond to this unusual request, properly replied that he first had to consult with Washington. Obviously the real underlying reason, as the Truth and Reconciliation Commission later concluded, was that the entire Honduran political system was broken and unworkable. Because Honduras was heading toward a major crisis, the ambassador did what he should have done at least a month before and referred the matter to Washington as an "Action Request." In diplomatic practice an "Action Request" from an embassy means that the State Department has to provide a response in reasonable time. This mechanism allows senior officials to cope with the flood of cables reaching Washington and in effect delegates to the embassies the task of prioritizing the incoming cables. Of course, senior officials have the authority to respond to any cables by asking questions, requesting information, and sending out instructions or warnings. But unless the Secretary of State and his senior advisors are unusually dynamic and dedicated, the State Department normally is not proactive and rarely goes beyond reacting to embassy requests, as was the practice while Hillary Clinton served as Secretary of State.[40]

The U.S. Embassy did support participation in the dialogue among Liberal Party leaders, because "Zelaya may be seeking a face saving way out of this situation; he will probably want some guarantees against prosecution after he leaves office, both for himself and his family."[41] The novelty of the unusual request from Zelaya rather than any interest in Honduras resulted in a rather fast response from Washington, perhaps too fast because senior officials did not have time to immerse themselves in an issue they were just becoming aware of. Predictably the State Department endorsed the recommendation of having the ambassador participate in the talks and quite naturally insisted on avoiding publicity and keeping the involvement as discreet as possible. But the State Department could not go beyond repeating the usual platitudes on respecting the rule of law, democratic institutions, and constitutional procedures. And on the crucial topic of the Fourth Ballot Box, the State Department equivocated: "We would oppose any move to hold the poll in defiance of the constitutional order and authorities."[42]

The poll that Zelaya was planning to carry out on 28 June was already in defiance of the legal order, so there the State Department was not saying anything new. Was the U.S. ambassador supposed to tell Zelaya that it was the official position of the U.S. government that he needed to drop the nonsense of the Fourth Ballot Box immediately? Or was the U.S. Embassy supposed to wait and at the appropriate moment support those institutions that tried to block the poll because of its illegality? The U.S. Embassy was left adrift as to what step next to take because of the failure of the State Department to ratify or repeal the personal warnings the U.S. ambassador had already given to Zelaya. Without a ringing endorsement of his warnings to the Honduran president, the ambassador had been wise to label them as personal advice only.

As U.S. diplomats beginning in June 17 plunged into discussions with Liberal Party leaders, all participants believed that the Fourth Ballot Box poll would take place as scheduled on Sunday 28 June. U.S. diplomats expected the real crisis to erupt between July and September when the Honduran congress had to respond to the poll results by either rejecting them or agreeing to include the referendum on convoking a constitutional convention in the presidential elections of 29 November 2009. As the U.S. Embassy prepared for the crisis supposed to start after July, U.S. diplomats and many other observers failed to notice the signs that the political elite was not willing to wait so long for a showdown with Zelaya.[43]

The Coup of 28 June 2009

In a series of rulings and sentences dating back to 27 May 2009, both the judiciary and the independent Attorney General's Office consistently stated that changing the name from "consultation" to a "poll" did not alter the inherent illegality of the president's actions. These bodies also ordered the executive branch to desist from all actions to implement the Fourth Ballot Box proposal. Because the president had ordered the military to help the Institute of Statistics carry out the poll, General Romeo Vásquez Velásquez, the Chief of the General Staff, requested an opinion from army attorneys, and they too declared the court's decisions to be legal and constitutional.[44] José Manuel Zelaya remained unfazed, and twice in what turned out to be a prophecy, he publicly "taunted his opponents to try and arrest him."[45]

President Zelaya believed that he could carry out the poll on 28 June because of his close friendship with General Vásquez Velásquez and other generals such as air force commander Luis Javier Prince Suazo. In reality the military resented having been dragged back into partisan politics and wanted to remain a non-political body. "Vásquez described the military's situation as intolerable, with being told by their commander-in-chief to carry out an order while being told by a court that order was illegal."[46] In an effort born out of frustration, the military threatened to take direct action unless the politicians promptly solved the fourth ballot box crisis. The generals began floating these threats in the middle of June, and as soon as the U.S. ambassador got wind of them, he promptly summoned General Vásquez Velásquez to state categorically that the U.S. government was completely opposed to any military coup. The general explained that the purpose of these warnings was "to prod politicians into reaching an agreement,"[47] and with these reassurances the U.S. ambassador believed that the danger of a military coup had passed.

In an unexpected manner, the logistics of organizing the poll came to unleash the final confrontation. Zelaya, as a typical mediocre Latin American politician, did not know the mechanics of implementing any major operation. Had he from the start turned to private institutions and steered clear of the military and of governmental agencies, the judiciary would have been on very shaky ground to prohibit a poll relying on private funding and citizen volunteers. By mid-June a large group of devoted followers stood ready to conduct the poll, but the insistence of Zelaya on ordering the military to participate ultimately doomed him.[48]

It was clear that the military preferred to sit out this messy event, and when the general refused to obey the presidential order to distribute the ballots, an exacerbated Zelaya fired Chief of the General Staff Vásquez Velásquez on live television at 2200 hours of 24 June. The civilian defense minister and the three service chiefs, including the air force commander Prince who had been the general most sympathetic to Zelaya, also resigned. Supposedly the military was left leaderless, but the president failed to appoint any substitutes (the Vice Chief of the General Staff was away on a study trip in Brazil) and all the commanders remained at their offices handling routine matters. The gravity of the situation was obvious to all, and the U.S. Embassy scrambled to do damage control by urging politicians and generals not to take precipitate action. The U.S. ambassador finally told Zelaya that he needed to cancel the poll to avoid the approaching train wreck. But the President, who throughout his career had been so solicitous of U.S. wishes, now for the first time refused to comply.[49]

The U.S. Embassy did not have too much time to dwell on this defiance because a new crisis suddenly erupted in the Honduran congress. Robert Micheletti, the head of the unicameral legislature, immediately saw in Zelaya's firing of the Chief of the General Staff as the great opportunity to achieve his life-long ambition to become president of Honduras. Micheletti gathered support among all political parties until he was sure that he had the votes on the floor to obtain the removal of Zelaya from office. But he knew better than anyone else the rules of the game in Honduras, and before taking this action, he called on the U.S. ambassador on the evening of 25 June to obtain U.S. approval. The U.S. ambassador was horrified with this proposal, because in 2005 the Honduran congress had abolished its authority to impeach and remove presidents from office. Only the judicial branch could remove a president and then only after the independent Attorney General's Office had successfully proved criminal charges in court against a president. A flurry of phone calls from the U.S. Embassy to Honduran politicians soon persuaded the deputies to desist from deposing Zelaya, and later that night Micheletti sounded the retreat and referred the matter back to a legislative committee.[50]

The U.S. Embassy breathed a sigh of relief after having stopped this illegal and unconstitutional maneuver to remove Zelaya but had failed to detect the sinister role of the Supreme Court Chief Justice who had been instrumental in telling legislators that congressional action to remove the president was unconstitutional. The sly judge failed to reveal to U.S. diplomats that one of his arguments to persuade legislators was that a separate process was already underway both to remove Zelaya from office and to abort the poll just a few days away.[51]

U.S. diplomats proceeded on the naive assumption that there was still time to patch together a last-minute solution to avert the crisis. The U.S. Embassy accepted the request of Zelaya to set up a conference phone call between him and the fired Chief of the General Staff in order to negotiate a solution. On 25 June the President offered to reinstate the Chief of the General Staff in exchange for the military guarding the polling stations. This arrangement might have averted the showdown earlier, but because that same 25 June the Supreme Court handed down a ruling declaring illegal the firing of the Chief of the General Staff, not surprisingly the latter felt that his reinstatement with no preconditions was the only possible solution. The U.S. ambassador realized that the situation had changed with the Supreme Court ruling, and that evening he urged Zelaya to reinstate the fired commanders but the President continued to delay.[52]

Zelaya made a big mistake when he felt that he could leave the military leaderless while he opened up still another area of confrontation. With the poll just a few days away, the logistics of organizing the operation were becoming practically impossible. Still assuming that the military would support him, he authorized the entry of the 15,000 ballot boxes and the printed ballots into the air force base near Tegucigalpa. In the morning of 25 June officials of the independent Attorney General's Office and electoral judges went to the air force base to impound those materials and sealed them in warehouse FA-4 in that base. Civilian opponents seemed to have effectively checkmated Zelaya, because without those voting materials the poll could not take place.[53]

When news of the impounding of these materials reached the presidential palace, and outraged Zelaya called on his followers at 1330 hours on 25 June to march on the air force base. A caravan of automobiles and other vehicles set out for the military base, and along the way some marchers acquired weapons and became drunk. The air force commander gave his soldiers orders not to fire, and the enraged crowd burst through the barriers and spread out inside the military compound. General Prince was worried that with so many inflammable materials and explosives in the warehouses, the civilians might accidentally start a fire or set off explosions. In spite of these arguments, when Zelaya arrived he refused to order his followers to leave the base. The general, fearing a shootout with civilian casualties, authorized the marchers to take out of warehouse FA-4 the voting materials and haul them away in four trucks.[54]

Gradually the crowd dispersed and Zelaya believed that he had won a smashing victory over his political opponents. In reality, he had humiliated the last sympathetic general in the military, and the raid on the air force base galvanized his opponents to take the final decisive action. The evening of that same 25 June, the Attorney General filed suit with the Supreme Court requesting an arrest warrant against President Zelaya and an authorization to prevent his escape from the country. The opposition had felt insulted by the previous taunting of Zelaya to come and arrest him, and officials were working feverishly to secure his arrest and to bring him to trial. The next day after rush proceedings the Supreme Court authorized the military to arrest the president but kept the decision secret. Even days later the Chief Justice was denying to U.S. officials the existence of this arrest warrant. That same 26 June Zelaya concentrated his attention on beginning the distribution across the country of the 15,000 ballot boxes. He ignored the warnings of sympathetic politicians that his opponents were coming to arrest him and that as a precaution he needed to go into hiding.[55]

Zelaya long before had lost touch with reality and innocently remained living in the presidential palace. His capture did not take place on 27 June because the arrest warrant caused considerable consternation when it reached the Defense Ministry. As was the custom of Chief of the General Staff Vásquez Velásquez, he first consulted with the army's official attorneys, but even after they reported that under the circumstances the Supreme Court had every right to order the arrest for trial of the president, he still hesitated to act. The Chief Justice then telephoned him to find out why the order was not being executed, and the General agreed to come for a meeting in the Supreme Court building. After the conversation Vásquez Velásquez concluded that he had no alternative but to enforce the order for the arrest of Zelaya even in defiance of the express wishes of U.S. diplomats.[56]

A Zelaya totally oblivious to the fierce struggles raging around him was finally woken up by the sounds of gunfire near the presidential palace around 0500 hours on 28 June 2009. With the full support of army Chief of Staff Miguel Ángel García Padgett, General Vásquez Velásquez sent an overwhelming force to awe the presidential guard, which after putting up token resistance, allowed the President to be arrested. The soldiers forced Zelaya into a vehicle, drove him off to the air force base, and packed him aboard the presidential airplane. At 0630 hours the airplane took off for a flight to Costa Rica, because the generals had decided that the mere presence of Zelaya could incite disturbances and riots in Honduras. By the time the U.S. Embassy called around 0700 requesting that Zelaya not be deported, his plane was already on the way and landed in Costa Rica shortly after 0800 hours.[57]

The generals afterwards consistently claimed that their decision to exile Zelaya was spontaneous. His absence left the field wide open for Roberto Micheletti to achieve his life's ambition to become president of Honduras. After a series of parliamentary procedures, on 28 June the legislature duly elected Micheletti to be the provisional president of Honduras. Later that same afternoon he was sworn in, and he promised to hold the elections previously scheduled for 29 November 2009. It was understood that he would not be a candidate and that he would hand power over to his elected successor on 27 January 2010.[58] In spite of these conditions, nothing could hide the reality that Honduras had experienced the first successful coup of Latin America in the twenty-first century. Inevitably, repression and human rights abuses soon colored the Micheletti interim presidency as it vainly tried to stifle dissent.[59]

Remaining in office for little over six months, Micheletti received perhaps the most intensive diplomatic pressure seen in Latin America in the twenty-first century. In a strange alliance, both the U.S. government and Hugo Chávez of Venezuela joined forces to try to undo the coup. European countries and the U.S. government stopped almost all economic assistance, and Chávez halted the flow of subsidized Venezuelan petroleum. The sanctions fell heaviest on the Honduran population, but the slowdown in economic activity also reduced the profits of local businesses. In spite of the economic hardship and diplomatic isolation, it proved impossible to convince the Honduran ruling classes to accept the return of Zelaya, and Micheletti was able to complete his brief term as provisional president.[60]

Much more successful was the international pressure to make sure that free and open elections were held as scheduled on 29 November 2009. The contenders were very familiar figures with Porfirio Lobo Sosa, whom Zelaya had defeated in November 2004, facing off against the original favorite, Elvin Santos. Although Santos the businessman was from the conservative wing of the Liberal Party, the public still considered him too closely associated to the deposed Zelaya. Consequently, the electoral victory went to Lobo of the National Party. The experiment of Zelaya with the fourth ballot box had cost his party the presidential election of November 2009.[61]

Even after the crisis was over, the U.S. government continued to deny travel visas to all officials of the Micheletti regime. In particular, American diplomats wanted to single out for punishment those individuals who had defied Embassy instructions. The U.S. government could not do much about the judges in the judiciary, but whenever the opportunity appeared, inflicted reprisals on other officials.[62]

The best example came during the presidency of Porfirio "Pepe" Lobo. The U.S. government considered him to be strongly pro–American and was extremely glad to have him replace the usurper Micheletti. U.S. officials promised to resume economic assistance and to normalize relations with Honduras and seemed unusually eager to restore an electoral system. Even before the inauguration of Lobo on 27 January 2010 the U.S. ambassador had been demanding a complete change in the military high command. The primary motivation was not so much to make a clean sweep of officials connected to the coup as to punish those who had defied the U.S. government. The most visible target on this blacklist of undesirables was Chief of the General Staff Vásquez Velásquez, whom the U.S. government wanted removed promptly from office. This created a problem for President Lobo, who with considerable justification did not want to pick a fight with the military so soon after his inauguration. Besides, the second three-year term of General Vásquez Velásquez was set to expire in November 2010, and the President did not see any reason to rush to shorten the term of the outgoing

General. The U.S. ambassador on 14 February 2010 explained the facts of life to Lobo: "it was also critically important for General Vásquez to step down. The Ambassador made it clear that the U.S. was not in a position to reengage with the Honduran military and restore military assistance until new leadership was in place."[63]

Lobo agreed to remove General Vásquez Velásquez, but when the President tried to appoint the army Chief of Staff General García Padgett as his replacement, the ambassador also vetoed this appointment because that General was too closely connected with the coup against Zelaya. Important as it was to punish coup participants, it was much more vital to restore full U.S. control over Honduras. When Lobo pleaded for more time to shuffle his military commanders, the ambassador gave a stern warning that his time was up and "suggested that Lobo act sooner rather than later."[64] As a face-saving device, the President appointed General Vásquez Velásquez to head the state-owned national telephone company, the largest business in the country. The President also obeyed U.S. instructions on the appointments for the rest of the high military officials including the civilian minister of defense.[65]

The crisis provoked by the reckless moves of Zelaya had passed, but Honduras required close watching as one of the most troublesome allies of the U.S. The Truth and Reconciliation Commission believed that Honduras could not return to being what it was before the coup, but the U.S. government had certainly succeeded in reestablishing control over the armed forces and restoring its informal empire over Honduras.

10

Mexico: The Armed Forces Embrace the United States

> *No other country in the world has a greater impact on the drug situation in the United States than does Mexico. The influence of Mexico on the U.S. drug trade is truly unmatched.*—FBI and DEA agents[1]

U.S. officials saw a great opportunity to bring the armed forces of Mexico into the orbit of U.S. influence even before the 2 July 2006 presidential election. Whether the favorite, Andrés Manuel López Obrador, or his closest contender, Felipe Calderón of the PAN (*Partido de Acción Nacional*) won, the U.S. government realized that the new president had to turn to the military to fight the drug cartels. And the Mexican armed forces could be effective only with abundant U.S. assistance. These conclusions were not just an accurate prediction of future performance but also reflected the extensive American collaboration with the Mexican navy. Partly to maintain its independence from the Defense Ministry and partly to obtain much needed equipment and training, already during the Ernesto Zedillo presidency the Mexican navy enthusiastically developed a strong relationship with both the U.S. Navy and the U.S. Coast Guard. The navy's dependence on U.S. surplus warships to form the bulk of its fleet made any other attitude suicidal.[2] Rejecting the defiant attitude of the army toward the U.S. government, the Mexican navy cooperated with U.S. officers to become an effective and reliable force. The well-trained Marines soon came to play crucial roles in various operations. The navy also made itself indispensable to Mexican presidents who showed no intention of sacrificing its institutional independence.

The Start of the Felipe Calderón Presidency

The upset victory of PAN candidate Felipe Calderón over the favorite candidate, the leftist Andrés Manuel López Obrador, was so narrow, that immediately many observers called for a recount of the votes. But because under Mexican electoral law both candidates must agree to a recount, the predictable refusal of Calderón cast doubts on his legitimacy as the rightful president of Mexico. López Obrador has not ceased to claim that the election was stolen from him, when in reality his eccentricities and the highly effective attack ads of the Calderón campaign were responsible for tipping a slight majority of undecided voters at the last minute.

The cloud of fraud followed Calderón during his entire presidency, and repeatedly he strove to gain legitimacy. Barely a week after his inauguration on 1 December 2006 he decided to prove to the people that he was the rightful ruler of Mexico. He realized that the passivity of the outgoing Vicente Fox provided a great opportunity to highlight the contrast in presidential leadership. In a staged media event to boost his public opinion ratings, President Calderón declared on 8 December the start of a war on the drug cartels and the deployment of troops and federal police to Michoacán (Map 5). As troop-filled trucks rumbled out of Mexico City and headed for that state, the new administration believed that it was sending a message of strong action to the public. And to stay in the news, on December 18 the government was already making victory claims about its operations in Michoacán.[3]

Skeptics were soon questioning the supposed victory, and later it was learned that the drug traffickers had received ample warning about the deployments and easily avoided the arriving soldiers and policemen. Joaquín Guzmán Loera alias El Chapo Guzmán, the head of *La Federacion*, the drug consortium covering the entire Pacific region of Mexico, was more than happy to let the Calderón administration claim victory. He knew that the government had to show successes to justify its legitimacy to the Mexican people.[4]

In reality a much higher priority for El Chapo Guzmán was to end the war he had begun in 2002 against the Gulf Cartel. The armed bands of the Zetas had proved to be the decisive weapon of the Gulf Cartel in blunting the offensive of *La Federación*. Despite the support of the AFI (*Agencia Federal de Investigación*), the investigative police of the Attorney General's Office, *La Federación* had been unable to smash the Gulf Cartel, and by 2006 the bloody struggle between the two rival cartels reached a stalemate. The war was costing both sides too many casualties and was hurting the drug business. El Chapo Guzmán decided to play the role of peacemaker and proposed a peace conference between both cartels. The leaders of the Gulf Cartel accepted, but on the condition that the meetings take place in their territory because it was *La Federación* that had started the war.

10. Mexico: The Armed Forces Embrace the United States

El Chapo Guzmán swallowed his pride and came with his close collaborators to a peace conference at Valle Hermoso in Tamaulipas State not far from Matamoros on the U.S. border in June 2007 (Map 7). El Chapo Guzmán was taking a major personal risk in coming to the these meetings later continuing in other places including the exclusive Polanco neighborhood in Mexico City. He judged correctly that the mutual interest in maximizing profits made the Gulf Cartel prefer peace to war. In what came to be known as the Valle Hermoso accords, both cartels divided up Mexico into their respective spheres of control. Because the Gulf Cartel resented the privileged access *La Federación* had to high police and government officials, as part of the Valle Hermoso accords the drug cartels agreed to share the expenses of bribing the authorities. Unidentified government officials inserted the condition that the cartels try to export the majority of the drugs to the United States and reduce the amounts going to the local corner stores (*narcotiendas*). The drug traffickers sealed the agreements with wild orgy-like parties brimming with premium tequila, rowdy music, and flashy prostitutes.[5]

The Valle Hermoso accords were an immediate success, and Mexico breathed easier as the number of drug-related killings declined substantially. For a while in 2006 and 2007 it seemed that the Calderón administration, without really having done much, was actually reducing drug-related violence (Table 3). Skeptics believed that the arrangement was contrived and could not possibly last long, but in 2007 the first hints of trouble came not from a revival of the drug war but from the last effort of the Mexican guerrillas.[6]

As chapter 4 explained, of the two main guerrilla movements in Mexico in the 1990s, the Zapatistas fizzled out completely, and the People's Revolutionary Army (*Ejército Popular Revolucionario*) or EPR splintered into rival factions and seemed to have disappeared by 2000. The years of the Fox presidency passed without any significant guerrilla activity, but the army was not fooled and continued to maintain sizable units in Guerrero and Chiapas, the states hosting the original guerrilla actions. On 24 May 2007 two leaders of EPR disappeared under mysterious circumstances, and soon the group was accusing authorities of having arrested, tortured, and killed them. The government denied the accusations, and EPR responded by blowing up PEMEX gas pipelines in the states of Guanajuato and Querétaro on 5 and 10 July 2007. The explosions paralyzed manufacturing plants in that region and caused tremendous damage to the economy. On 10 September six of the 12 explosives charges EPR placed in gas pipelines in Veracruz and Tlaxcala exploded and caused major disruptions to the local economies heavily dependent on natural gas.[7]

The attacks on the PEMEX pipelines constituted the first time a guerrilla group targeted the infrastructure of Mexico. Because the state petroleum company kept secret the location of these underground pipelines, it was obvious that sympathizers within the state petroleum monopoly had leaked the information to EPR. Mexico braced for a wave of attacks on its infrastructure, and observers feared that the authorities were helpless to protect the vast pipeline network. The Colombian government tried to score propaganda points by repeatedly claiming that FARC, the Colombian guerrilla group, had trained EPR to set the explosives, but both Mexican and U.S. officials found no evidence to support the accusations. The next move of EPR did not come until 24 April 2008 when the guerrilla group requested conversations with officials to locate the two leaders missing since May 2006. The government agreed to set up a commission to try to discover what happened to the disappeared leaders but not to engage in peace talks. From the start the investigating commission felt constrained and in April 2009 dissolved itself. By then the EPR had splintered into bickering factions, and the guerrilla threat facing Mexico faded away.[8]

Historical perspective shows that the bombings of EPR in 2007 were the Swam Song of the guerrilla movement. The military was less sure and not only maintained large garrisons in the southern states of Mexico but also concentrated its intelligence efforts in tracking down the remnants of the rapidly disintegrating EPR. Once again this diversion of resources prevented the military from focusing on the activities of the drug cartels. By directing its scant intelligence resources into tracking a dying guerrilla movement, the military failed to detect the new trends in the drug business.[9]

A first revelation shocked the army. On 28 December 2007 a DEA surveillance aircraft filmed the unloading of a large cocaine cargo in a secret airstrip to the south of Mexico City. Army soldiers and federal police unloaded the drugs and also delayed the arrival of anti-narcotics units coming to intercept the shipment. By the time honest officials arrived, the entire shipment of drugs had vanished. Incidents like this one were routine across Mexico and resembled the unloading of drugs at Llanos de la Víbora in November 1991. But while the army had been in denial about the 1991 episode it considered a fabrication, in 2007 the video showing in excruciating detail the drug operation could not be so easily dismissed. The troops were under General Ricardo Escorcia, the zone commander, and although he was immediately removed from that post, bringing charges against him proved very difficult and not until May 2012 was he finally arrested and extradited to the United States.[10]

General Escorcia had been throughout his army career a crucial ally for *La Federación*, and his removal created strains in the relationship with the Gulf Cartel. But the anticipated war between the two cartels did not take place, and instead what began in early 2008 was their partial disintegration because of internal struggles. Shortly after the removal of General Escorcia, AFI arrested Alfredo Beltrán Leyva on 20 January 2008. El Chapo Guzmán was worried about the sudden rise to power of the Beltrán Leyva brothers and feared that they would remove him as head of *La Federación*, so before they could act he struck first and revealed to the police the location of Alfredo. The brothers, who had been the contacts with the Zetas, had become very close to the armed bands of the Gulf Cartel, and after the arrest of Alfredo, the other brothers united with the Zetas to fight against El Chapo Guzmán. This alliance between the Beltrán Leyva brothers and the Zetas shattered *La Federación* into smaller crime syndicates, such as the Sinaloa and Juárez cartels and *La Familia Michoacana*.[11]

This fragmentation of *La Federación* seemed to give the Gulf Cartel a great opportunity, but actually it was also beginning to suffer its own internal process of disintegration. The Zetas gradually started operating independently of its former master, the Gulf Cartel, and eventually became a separate organization. The forces of dissolution turned out to be present both in *La Federación* and in the Gulf Cartel, and as the drug business fragmented into smaller organizations, the danger increased of extensive warfare among drug cartels to control the most lucrative routes. The ones first caught in the crossfire were the corrupt officials in the state, local, and police forces, who had to decide from whom to receive their bribes. And by accepting bribes from one drug syndicate, the police officers became targets for rival groups.[12]

Along the way the Zetas became violently savage and perpetrated horrible atrocities such as mutilating and dismembering the bodies of their enemies and sometimes even feeding captives to wild beasts. As the number of horrific deaths climbed in these clashes among drug rivals, the Calderón administration faced a difficult decision. It could try to engage in a frontal battle against all the drug cartels or it could try to reestablish arrangements among the drug lords to reduce the levels of violence in Mexico. Because the periods of relative calm in Mexico coincided in the past with implicit arrangements between the government and drug lords, the Calderón administration decided to explore this possibility before engaging in a full-scale war. In September 2008 the Minister of the Interior, Juan Camilo Mouriño, the government official closest to President Calderón, sent retired general Mario Arturo Acosta Chaparro on secret missions to negotiate a truce among the drug cartels.[13]

Inevitably the former general met with the most important drug lord in Mexico, the already legendary El Chapo Guzmán. The leader of the Sinaloa Cartel explained that the government was fanning the flames of the struggle among drug groups. In his view, the violence turned vicious when state and federal officials in each major city of the country sold themselves to a rival drug group. If the government could not control state officials and sometimes even federal officials, there was no point for the cartels to negotiate with the Calderón administration. Besides, the Sinaloa Cartel had already won over to its sides many key officials for the struggle to destroy the Beltrán Leyva brothers.[14]

The slim possibilities of reaching an agreement to restore peace among the cartels vanished when the Minister of the Interior died in a mysterious plane crash in northern Mexico City on 4 November 2008. Not only had the attempt to seek a truce failed but the government later made every effort to erase all traces about the failed mission of Acosta Chaparro. The retired general, the last living witness in this tangled web of intrigue, was himself brutally gunned down on 20 April 2012 after an earlier failed attempt on his life.[15]

As cartel violence raged, everyone realized that the 447,922 municipal, state, and federal police officers were unable to stem the tide of drug-related violence. Not only were the police generally untrained, underpaid, and ill-equipped, but the vast majority was under strong suspicion of being corrupt. Organizational fragmentation prevented any coherent response to the crime wave, because the police belonged to 2,040 different agencies. A first proposal to merge the state and municipal police forces into a single body in each state failed in the Mexican congress. The legislative refusal reflected the fierce determination of mayors and governors to retain the police under their control as sources of patronage and political support.[16]

The federal police was too small to fill the gaps left by the state and local police. The Calderón administration increased funding for the federal police agencies and tried to enlarge the number of their personnel. Unfortunately the job of policemen lacked social status and was considered too dangerous and poorly paid to attract adequate numbers of recruits. The Ministry of Public Security supposedly contained the most prestigious police units, but in an embarrassing failure, a call for 8,000 college-educated recruits brought only half that number. The federal police forces desperately needed more investigators to solve crimes, and as a partial solution, the Attorney General's Office transferred all its eradication, fumigation, and interdiction functions to the Defense Ministry in the first months of 2007. Installations and all equipment such as airplanes and helicopters passed to the military thus releasing detectives to work on crime investigations.[17] Because the task of increasing the police agencies was going to be a very slow and tortuous process, the military had to step in to gain time until an efficient police force emerged in Mexico. But was the military ready to take on the drug cartels?

The Military Offensive Against the Drug Cartels, 2008–2010

With Mexico ablaze in drug-related violence, President Felipe Calderón made the desperate gamble of hurling the military against the drug cartels. Because he had already declared war on drugs in December 2006, a new declaration was not needed. But the decision in late 2008 was very different from that in December 2006 when the purpose of the modest military deployment in Michoacán had been to stage a public relations event for the sake of better ratings in public opinion polls.

The armed forces seemed at first glance to be formidable but did reveal serious limitations. The army with 200,000 soldiers (Table 2) was the largest branch and stationed troops in all the Mexican states. Yet the location of units tended to focus on the provincial capitals and in remote or strategic regions in the individual states, thus leaving vast stretches of territory and many towns covered only by municipal or state police. To fill these gaps, the army was improving its practices for patrolling large areas of the countryside. At the beginning Humvees seemed like the best vehicles for moving troops in remote areas, but experience showed that four-wheel drive pickups were a better alternative to haul supplies and to carry soldiers. The army reduced the number of Humvees and replaced them with one thousand four-wheel drive pick-up trucks assembled by General Motors in Mexico. The army's military workshops upgraded the pickup trucks for combat by welding roll bars, armor plating, and reinforced floors.[18]

Unlike the police, the army never had any difficulty recruiting enlisted personnel and officers. Candidates from all Mexican states applied to the

Chapultepec Military Academy, located in Mexico City, and because of its huge population, natives of the capital predominated in the officer ranks. Among the enlisted men, the situation was nearly the opposite. Although soldiers came from all Mexican states, the army had taken the deliberate decision to concentrate recruitment on the most remote rural areas in Mexico. For these peasants living in abject misery usually in southern Mexico, becoming a soldier represented a step up and an improvement in their living conditions. The army rejected the many pleas from the public and from social commentators to draft the "*Ni-Nis*" (*ni trabajan ni estudian*), the hundreds of thousands of idle youngsters roaming around urban centers. Spoiled brats of the wealthy and more frequently unemployed drifters formed the bulk of the *Ni-Nis*. The army considered them to be too degenerate to transform into obedient soldiers and preferred the less educated but more pliable peasants uncontaminated by the vices of urban life. With the available funds the army's preference made budgetary sense but left the field open for criminal syndicates to tempt the *Ni-Nis* with the prospects of easy money. In another consequence, the rural background best fitted soldiers for combat in the countryside but left them poorly prepared to conduct operations among the civilian population in the cities.[19]

The army also feared that forcing unwilling recruits into the ranks was counterproductive and only served to exacerbate the already acute problem of desertions (Table 5). In sharp contrast to the few desertions in the prestigious navy, the low pay, poor living conditions, and limited opportunities for training or promotion left almost no incentives to stay in the army. The harsh life of any army was particularly brutal in Mexico where mistreatment of recruits and iron discipline has been the norm. Not just the enlisted men but even many junior officers adopted a sullen, silent, attitude and learned to follow orders blindly. Not surprisingly, enlisted men and even junior officers sometimes felt that desertion was their only alternative, and these deserters often joined the private forces of the drug cartels. In long overdue measures, the Calderón administration insisted on improving the living conditions of the soldiers. Besides providing subsidized mortgages, in a dramatic step the government raised salaries by 46 percent in 2007 and by 40 percent in 2009 at a time when Mexico was experiencing very low inflation. Combat troops received even higher percentage increases each year and also housing allowances for families. Also because the increased risk of dying in combat signified a tremendous blow to the deceased's family, as compensation the government decided to provide a full military burial with honor guard and survivor's pension for any dependents.[20]

Table 5. Mexico: Desertions from Army, 1997–2012

Year	Number	Percentage of Army
1997	11,122	6.09
1998	18,861	10.34
1999	19,849	10.88
2000	22,205	12.17
2001	15,870	8.57
2002	15,503	8.24
2003	14,744	7.71
2004	18,267	9.55
2005	20,224	10.58
2006	16,405	8.33
2007	16,641	8.45
2008	9,112	4.50
2009	6,879	3.39
2010	4,398	2.13
2011	3,361	1.60
2012	8,644	4.12

Sources: George W. Grayson, *The Impact of President Felipe Calderón's War on Drugs on the Armed Forces* (Carlisle, PA: U.S. Army War College, 2013), p. 54; Iñigo Guevara Moyano, *Adapting, Transforming, and Modernizing under Fire: The Mexican Military 2006–2011* (Carlisle, PA: U.S. Army War College, 2011), p. 17; "Más de 55 mil deserciones en las Fuerzas Armadas," 9 June 2013, contralinea.info.

As Table 5 shows, the number of desertions sharply declined, and this increased retention of personnel contributed to greater cohesion inside the army units. The improved compensation constituted the most important contribution of the Calderón administration to Mexico's armed forces. The better pay made the military an acceptable career for young people otherwise lacking any other civilian employment. Although reservoirs of impoverished peasants living in remote areas will continue to exist in Mexico in the near future, the army soon will have to learn how to recruit enlisted personnel of urban origins. Good compensation during a lifetime serves as a great incentive for attracting recruits.[21]

Another factor also helping to mitigate desertions was the recruitment of women into the army. In a *macho* institution generals were most reluctant to accept women, and the first females entered the ranks only as nurses in 1938. Because enlisted men grumbled at having to do secretarial work and other office duties, the army allowed uniformed women to fill many of these clerical positions whenever reliance on female civilians proved impractical. Joining the army as enlisted personnel was such a radical decision for females that they carefully thought out the consequences unlike many impulsive males who after joining were left with desertion as the only solution to escape an unbearable situation. With the exception of army nurses, the possibility of rising to the rank of officer practically did not exist for females until the Military Medical School in 1973 and the Military Dental School in 1976 admitted women as students. The health professions (nursing, medicine, and dentistry) grouped almost all the female officers and in a sense segregated these officers from the rest of the army. A similar pattern was present in the U.S. military even as recently as 2012 when 39 percent of female officers were in health professions with the U.S. Navy having the highest figure of 46 percent.[22]

As Table 6 shows, the number of women in the army continued to rise steadily, reaching the high of 5.6 percent in 2012. The percentage share of women in the Mexican army will certainly continue to rise higher but probably not by substantially more. In the United States in spite of both rapid increases and slow decline in the total size of the U.S. military, the percentage of women in the armed forces has stayed steady at around 14 percent since 2000. That percentage is certainly beyond the reach of the Mexican *macho* society. It will be very surprising if the Mexican army ever reaches a 10 percent figure for female participation.[23]

Table 6. Mexico: Women in Army, 2006–2012

Year	Number of Women	Percentage of Army
2006	6,309	3.21
2007	6,831	3.47
2008	7,980	3.94
2009	8,714	4.23
2010	10,234	4.97
2011	10,301	5.00
2012	11,736	5.55

Source: George W. Grayson, *The Impact of President Felipe Calderón's War on Drugs on the Armed Forces* (Fort Carlisle, PA: U.S. Army War College, 2013), p. 44; "Mujeres reclutadas en el ejército," 28 August 2012, Alcalorpolitico.com.

Special characteristics absent in Mexico have fueled the enlistment of women in the United States into the military. In the United States only 53 percent of women in the military are white, and blacks at 31 percent make up the largest non-white group of females in the U.S. military; mixed racial background and other races comprise the rest of the non-white females in the U.S. military. No distinct ethnic or racial group of women in Mexico has found refuge in the army to a comparable degree. In another contrast, 48 percent of women in the U.S. military marry active-duty personnel, and the percentage is certainly higher if couples who met in the military and married after one or both left active duty is counted. While women do not state finding a husband as a reason for enlisting, for half of these women the U.S. military has become a matrimonial agency, a role largely absent in the more traditional Mexican society.[24]

In Mexico by the twenty-first century limiting the entry of females into officer rank only to medical personnel was no longer acceptable when many women were pursuing successful careers in business and in politics. A major breakthrough came in 2007 when the Chapultepec Military Academy admitted female cadets for the first time in its history. Chapultepec Military Academy provides the overwhelming majority of officers in the Mexican army and is the indispensable starting point for any military career. Its significance is much greater than West Point's in the larger U.S. army. In recent Independence Day parades uniformed women have figured prominently as standard bearers for marching units. However, the steady increase in female participation in officer ranks has not been accompanied by promotions and appointment of women to major command positions. The army claims that senior male officers first have to leave the institution before high-level positions can become available for female officers, but this claim sounds like an excuse. For example in 2007 only five women held the rank of generals, but their number dropped to 1 in 2012 when four retired, yet at the same time the total number of generals in the army rose from 537 in 2007 to 541 in 2012. Clearly the entry of female officers into senior positions has a long and hard road ahead.[25]

Another surprising phenomenon in the Mexican armed forces has been the near absence of sexual harassment, an endemic problem in the U.S. military. The smaller percentage of females in the Mexican army and underreporting because of fear of reprisals may account for a large part of the difference, while the apparent segregation of women into health and administrative positions certainly is a contributing factor in protecting women. However, the near silence about harassment also suggests that a more complex set of cultural dynamics has been at work. As Human Rights groups have documented, enlisted men have shown no hesitation about stripping, groping, and raping Central American, Indian, and any other females whenever the opportunity appeared, a behavior that makes hard to explain the apparent lack of sexual harassment inside the ranks.[26]

Supporting the army in the campaign against the drug cartels were the navy ships and the Marines. The destroyers, frigates, and gunboats of the Mexican navy were mostly surplus vessels from the U.S. navy, but because most of these ships were fitted with helicopter pads, these floating platforms were essential to discover and interdict drug shipments far from coast. Closer to shore the patrol boats, most of them built in local shipyards, intercepted drugs coming by sea. Because drug traffickers responded by purchasing fast speed boats able to outrun the slow patrol boats, the Mexican navy turned to foreign shipyards able to produce fast craft with cruising speeds of 45 and 50 knots.[27]

Another very valuable component of the navy was its Marine Corps. In an unwise decision, the Vicente Fox administration slashed the Marines by half to less than 5,000 by transferring unwilling Marines into federal police forces, a transfer that had no positive effect on the struggle against organized crime. The Felipe Calderón administration reversed the decision and rebuilt the Marines shortly after taking office. The Marines rose in number to 11,000 and worked closely with the Unit of Naval Intelligence created in 2008. Although small in size, this unit became the most effective intelligence service in the entire Mexican military and enjoyed the support and trust of U.S. agencies.[28]

In reality intelligence on drug traffickers remained a glaring weakness of the armed forces. The army could not repeat with rich drug traffickers the success it had in gathering information on poor guerrillas. The federal government had never found a comparable replacement to the Federal Directorate of Security abolished in 1985, while the municipal and state police gathered inadequate information and did not regularly share it with the armed forces. Under U.S. influence the air force began operating drones in 2009, but this step already reflected the decision of President Calderón to make the armed forces a full ally of the U.S. government in the drug war. U.S. agencies, such as DEA with its vast network of informants in Mexico, inevitably became the primary source of information for the Mexican army.

Early in April 2008 President Calderón reached the conclusion that the Mexican armed forces on their own were incapable of winning a war against the drug cartels. Since taking office U.S. diplomats had been pressuring him to order the armed forces to establish close working relations with the U.S. military, something the army had been most reluctant to do. President Ernesto Zedillo in the late 1990s had tried to get the Mexican army to work with the United States on specific cases, but those initiatives had largely withered by 2008. At the start of that year, President Calderón instructed his generals to work closely with the United States on all issues. These orders were superfluous for the Mexican navy already engaged in close contact with its U.S. counterparts. But for the Mexican army, establishing close personal relations at all levels with U.S. military officers meant a momentous break with the past history of distrust, distancing, and resentment.[29]

Results from the embrace of the United States were soon evident. The number of Mexican officers trained at U.S. institutions had dropped sharply since the Zedillo years and the downward trend was reversed in 2009 (Table 1). Annual staff talks between the U.S. and Mexican armies were for the first time organized. Reciprocal visits by high ranking officials of both countries became frequent, and in a major example, Secretary of Defense Robert Gates made an official visit to Mexico, the second one after the inaugural visit by William J. Perry in 1995. The U.S. military also provided considerable logistical support to the Mexican armed forces. And even greater amount of assistance came under the Mérida Initiative, a U.S. program designed to provide military and economic aid to countries of the Caribbean locked in the struggle against drugs.[30]

The alliance with the United States was bringing concrete benefits to Mexico, even though the U.S. military, bogged down at that time in the quagmires of Iraq and Afghanistan, could not devote the time, personnel, and resources that Mexico desperately needed for the war on drugs. Later events showed that this historic opening of the Mexican army was a unique opportunity that the U.S. government largely missed because of its commitments half way around the world. It is sad to have to reflect on how much the U.S. government could have accomplished in mending the relationship with the Mexican army with just a small fraction of the vast sums poured into the deserts of Iraq and Afghanistan. But those realizations had to wait some years, because the Mexican armed forces first had to fight a war against the drug cartels.

In the mind of the Mexican public the military offensive against drug traffickers began in December 2006 with the operation against Michoacán. But as explained at the start of the previous section, that large-scale deployment of troops was mostly a public relations stunt. The earliest indication about the possibility of an offensive came on 23 January 2008 with the presidential decree authorizing troops to assist state and local police forces throughout the country. The first implementation of this authority came when the government sent 2000 soldiers and 500 federal police to help the embattled authorities at Ciudad Juárez (Map 6) in March 2008. That city was racked in violence and killings because of the outbreak of the struggle between El Chapo Guzmán and the Beltrán Levya brothers for control of that key crossing point in the drug trade to the United States.[31]

The deployment to Ciudad Juárez in March 2008 did not need to signal the start of the military offensive, because the government was still involved in secret negotiations to establish a truce among the drug cartels. But the failure of efforts to restore a semblance of harmony among the drug cartels transformed the arrival of troops in Ciudad Juárez in March 2008 into the start of the offensive. A more dramatic step came on 18 November 2008 when soldiers, Marines, and federal police replaced the local police in half of the neighborhoods of Tijuana. The rationale for this extreme action was the need to purge and train the local police officers but only half of the force at a time, with the troops filling in as replacements.[32]

What was different about this large offensive was not just its vast dimension covering many states in Mexico but also its extension into urban areas. In the approximate division of labor existing until 2008, the soldiers and Marines concentrated on the countryside and coastlines, while the police, whether local, state, or federal, concentrated on urban areas. The division had never been absolute, and as a major exception, the Attorney General's Office had operated a vigorous eradication and interdiction program in the countryside until 2007 when full responsibility passed to the military (Table 7).

Table 7. Mexico: Crop Eradication and Drug Seizures, 1990–2013

Year	Crop Eradication (hectares)		Drug Seizures (metric tons)	
	Marihuana	Poppy	Marihuana	Cocaine
1990	8,778	8,660	594.7	49.8
1991	12,702	9,342	254.9	50.2
1992	16,802	11,222	404.5	38.8
1993	16,645	13,015	494.6	46.1
1994	14,207.1	10,958.6	529.9	22.1
1995	21,573.3	15,389.2	780.2	22.2
1996	22,768.6	14,670.9	1,016.9	23.8
1997	23,576.1	17,732.2	1,038.5	35.0
1998	23,928.3	17,449.1	1,062.5	22.6
1999	33,351.3	15,746.5	1,472.0	34.6
2000	31,061.4	15,717.9	2,050.8	23.2
2001	28,735.1	19,116.5	1.839.4	30.0
2002	30,774.9	19,157.9	1,633.3	12.6

10. Mexico: The Armed Forces Embrace the United States

	Crop Eradication (hectares)		Drug Seizures (metric tons)	
Year	Marihuana	Poppy	Marihuana	Cocaine
2003	36,585.3	20,034.0	2,247.8	21.2
2004	30,852.5	15,925.6	2,208.4	26.8
2005	30,856.9	21,609.2	1,795.7	30.8
2006	30,161.5	16,889.9	1,902.1	21.3
2007	23,315.7	11,410.5	2,213.4	48.0
2008	18,660.2	13,189.3	1,684.0	19.6
2009	16,703.5	14,810.8	2,094.7	21.5
2010	18,581.4	15,484.3	2,313.4	9.9
2011	13,430.3	16,389.4	1,798.9	11.3
2012	9,164.7	15,786.2	1,310.7	3.4
2013	5,364.2*	14,662.2*	971.9*	6.4*

*Preliminary figures.

Source: Presidencia de la República, *Segundo Informe de Gobierno, 2013–2014* (Mexico: Talleres de la Nación, 2014), p. 51; Instituto Nacional de Estadística y Geografía, *Anuario Estadístico de los Estados Unidos Mexicanos 2002* (Mexico: Instituto Nacional de Estadística y Geografía, 2003), Table 8.27.

Thus not only did the soldiers and Marines have nearly full authority over the countryside and coastlines, but the military after 2008 began to displace police forces inside many cities. Previously soldiers wearing masks on their faces had manned check points at the entrances to the cities as part of the network of posts along highways and junctions. The army expanded its role when it sent detachments to patrol city streets and also established posts at key places inside urban areas. This heightened army presence was supposed to intimidate drug traffickers, but to everyone's surprise, they frequently stood and fought until army reinforcements arrived. In response drug traffickers set up ambushes against army patrols and launched a series of attacks on troop barracks. Although the superior firepower of the soldiers generally prevailed, the days in the 1970s when the troops commanded fear and respect were long gone. The drug traffickers considered themselves almost the equals of the army and were not hesitant about engaging in firefights.[33]

One of many clashes illustrates well the nature of the combat in urban areas. On 10 February 2009 at Villa Ahumada, a town south of Ciudad Juárez (Map 6), gunmen of the Sinaloa Cartel broke into the homes of nine municipal policemen, killing six and taking three prisoners. As the members of the Sinaloa Cartel raced out of the town aboard many vehicles, airplanes tracked the convoy while helicopters and pickup trucks brought soldiers. In the combat the troops killed 14 gunmen and rescued the three remaining policemen. The victory had been more apparent than real, because later investigations revealed that the policemen had been killed or kidnapped because they were on the payroll of the rival Juárez Cartel.[34]

Other violent clashes continued to occur across Mexico, and as time went on and the drug traffickers acquired automatic weapons and grenade launchers, the confrontations became increasingly brutal and destructive. Only the army possessed the armament and the training to face the heavily armed drug traffickers who had effectively outclassed and outgunned the police forces carrying worn-out weapons, usually only revolvers. It was frequent in operations against major drug lords to have the army on standby ready to come to support outgunned police forces in case of very strong resistance, as happened in February 2009 at Reynosa, Tamaulipas in a raid against one of the leaders of the Gulf Cartel.[35]

Was the heavy military presence in major cities hurting the drug cartels? After a year of deployment in Ciudad Juárez, the results were at best meager, but rather than withdraw, the government decided to up the ante and sent an additional 5000 soldiers and 2000 federal policemen to that embattled border city in March 2009. The army divided the city into six sections, each under the control of a colonel who reported to the commander of the military zone. On 13 March the military also took command of the municipal policemen and deployed them only in the relatively safe commercial district of Ciudad Juárez. The army sent its units into the crime-plagued neighborhoods in the city suburbs. Even if the massive effort in Ciudad Juárez succeeded in restoring order to that troubled city, the Mexican government lacked the resources to sustain this deployment. The costs were too high to replicate this approach in other cities, and the government was too impatient to wait for favorable results.[36]

At the very least, the army deployment was supposed to gain time for the police forces to expand and improve their capabilities. The Mexican congress and the Calderón administration had been working on a comprehensive program for the police, but what passed congress on 30 April 2009 left much to be desired. The legislators largely ignored the incompetence of the state and municipal police agencies and concentrated on the federal police. By this time the evidence of the collusion of AFI with El Chapo Guzmán had become public knowledge, and the congress abolished the corrupt agency and replaced it with a new Federal Ministerial Police to serve as the investigative branch of the Attorney General's Office. This police force of 4,764 members received authorization to increase its strength by another 1,500 agents. The police force of the Ministry of National Security received authorization to increase to over 30,000 members. Most of the changes turned out to be cosmetic, and the increases in number of agents fell short of the needs.[37] Because the Mexican congress did not want to alter the local and police forces, then the federal police needed to increase at least to 100,000 members to pose a real threat to the drug cartels.

And the long awaited and disappointing police reorganization deprived the military of the justification for its involvement in the drug war. If the army was supposed to gain time until the police was ready to assume its tasks, the law of April 2009 confirmed that the police was not going to be capable of fulfilling its obligations for many years if not decades. The military lacked the resources to sustain its deployment for extended periods, and even if adequately funded, the number of soldiers was too small to replace the local and municipal police throughout the country. And the longer the army stayed in the campaign against the drug cartels, the greater the corruption. The revelation that an army major was receiving monthly payments of $100,000 dollars for cooperating with the drug cartels was particularly shocking. Although nobody doubted the reports of widespread corruption among civilian and military officials, specific figures on the actual payments were very scarce.[38] The case of this major also raised an intriguing question: if a relatively low-ranking officer received $100,000 a month, then how much were the generals and high officials receiving from the drug cartels?

Structural defects crippled the Mexican army in its war against the drug cartels. The rural origin of most enlisted men made them ill-prepared to operate in urban areas, while soldiers trained to track down, destroy, and kill the enemy easily could turn into armed peasant killers. As Human Rights activists reported, collateral damage, such as abuses and innocent deaths, was frequent inside the cities. The army continued to make its most efficient contributions in the countryside, whether in manually eradicating fields of marihuana and poppy or in intercepting shipments of cocaine. In the war against the drug cartels, the army was an instrument too blunt to reach the vital arteries of the drug trade. Already by early 2009 many generals, seeing their beloved institution falling prey to corruption, were eager to return to the barracks. As more and more officers were exposed to the temptations of corruption, the likelihood of striking any decisive blows against the drug cartels became increasingly remote. The army needed to find the least embarrassing exit from an unwinnable war.[39]

Even in operations in the countryside the lack of intelligence on the drug cartels severely limited the impact of the army's interdiction campaigns. It became a commonplace to state that no matter how many shipments of cocaine the army intercepted, U.S. consumers continued to receive their steady supply of illegal drugs. Also the drug cartels were sensitive to shifts in consumer demand, and after 2005 also set up laboratories to supply the newly acquired taste of U.S. consumers for methamphetamine. While it does not seem that this new product will replace cocaine as the drug of choice in the United States, the drug cartels have been most eager to push methamphetamine because this product allows them to achieve vertical integration in the business. Although some ingredients are not locally available and must be imported, the drug cartels find it much easier to import the needed chemicals from many parts of the world rather than have to deal with the fiercely independent suppliers of cocaine from Colombia.[40]

Inadequate intelligence on drug operations has been the main stumbling block in raids against the drug cartels. Repeatedly the army has been moving in the blind and hurling crushing blows at non-existent targets. The limitations on the Mexican army and police were so glaring, that the U.S. government, even while bogged down in the Iraq and Afghanistan operations, agreed in 2008 to provide funds, equipment, and training to Mexico under what has been called the Mérida Initiative. From 2008 to the middle of 2013 the U.S. government provided nearly $1.6 billion dollars of assistance to Mexico, principally in the form of equipment, materials, training, and helicopters to the Mexican armed forces and the police. Fears of embezzlement and graft persuaded the U.S. Congress to stipulate that Mexico could not receive any money.[41]

The Merida Initiative has definitely had a beneficial impact in Mexico, but the program has been crippled by fundamental flaws. In a typical Latin American trait, Mexican officials were fascinated by the opportunity of acquiring brand-new shiny equipment and weapons. High-tech gadgets, whether ranging from communications gear to digital detections devices, and helicopters, were the highest priorities for Mexican officials who believed that with this new technology and rapid deployment by air they could at last win the war on drugs. Most of this new equipment came from or through the U.S. military, and Mexican officials preferred this channel in contrast to dealing with European or Asian providers, because U.S. officials did not offer bribes to seal deals and thus eliminated the corruption typical of arms purchases in Latin America. But Mexican officials forgot or did not know about prior experiences with acquisitions from the United States. The U.S. military has apparently the most cumbersome bureaucratic procedures in the world, and soon Mexican officials, accustomed to massive paperwork in their country, were complaining about the delays because of permits and many other requirements. So by the time the equipment finally reached Mexico, considerable time had elapsed after the launch of the Mérida Initiative in 2008.[42]

Once the equipment began to arrive, Mexican officials realized that the personnel could not handle the new technology. Training should have been the first priority in using Mérida Initiative funds, an error that both U.S. diplomats and Mexican officials came to admit only years later. And this training and education could be effective only if the proper institutions existed. This was particularly true for the police forces, and they needed to be completely rebuilt, a mammoth undertaking for which the funds were vastly insufficient. In effect the fatal flaw of the Mérida Initiative was underfunding. At a time when the drug cartels were receiving anywhere from 18 to 39 billion dollars annually from the sale of drugs in the United States, to expect $1.6 billion dollars spread over five years to win the war on drugs in Mexico was completely unrealistic. The flood of money pouring into the coffers of drug cartels gave them the means to bribe and corrupt as many officials and officers as needed to keep the drugs flowing north.[43]

Nowhere else were the funding limitations most evident than in military intelligence. Gadgets were useful but what the intelligence agencies most lacked was adequate funding. Money was essential to create networks of informants inside drug cartels and required large sums and not the small change adequate for petty criminals. But money was also indispensable to pay high salaries and generous pensions to military and police personnel. Raising pay for enlisted personnel helped reduce desertions as explained earlier and also bolstered personal loyalty to the institution. By making officers less likely to succumb to the temptation of bribes, the military gained the best defense against infiltrations by drug traffickers.

Without adequate intelligence, the army was operating in the dark. And because the drug traffickers were increasing their armament, each confrontation with soldiers was becoming bloodier and costlier in lives. The army became very cautious about engaging in shootouts and was reluctant to arrest well-guarded drug lords.[44] Officers who were corrupt not only hid information about drug traffickers but also strongly supported the cautious officers advocating a slow response. As was already common with police operations, infiltrated agents in the army also tipped off drug traffickers about attempts to arrest them or to seize drugs.

The army was so slow to respond to requests for action that the government increasingly turned to the Marines to carry out difficult missions.

The tipping point came in December 2009 when U.S. intelligence sources found drug lord Arturo Beltrán Leyva in Cuernavaca (Map 5) in a residence about an hour's drive south of Mexico City. U.S. diplomats shared the location with the army, but it failed to act on the information, whether from caution, incompetence, or corruption. In desperation the U.S. Embassy contacted the Unit of Naval Intelligence, and soon Mexican Marines who had received extensive training in the United States were on their way to detain drug lord Beltrán Leyva. In a first attack, the Marines killed several of his bodyguards and arrested 23 of his followers. He did get away, but with U.S. intelligence guiding them, the Marines pursued him relentlessly. On 17 December 2009 the Marines surrounded Arturo in an apartment building, and after a long firefight, killed him. During the operation one navy cadet died, and in a shocking reprisal, drug traffickers tried to avenge the death of Beltrán Leyva by slaughtering family members of the dead Marine when they were mourning over his cadaver.[45]

Savage reprisals could not save the Beltrán Leyva brothers, and subsequent arrests by Mexican police and permanent pressure from the Sinaloa Cartel hastened the destruction of their organization. President Calderón however, was angry at the failure of the army to carry out the operations, and in a public sign of disapproval, he removed the army from command of Ciudad Juárez and handed control over to the federal police in the first week of 2010. Because the Sinaloa Cartel had already finished destroying the Juárez Cartel of the Beltrán Leyva brothers, shooting deaths declined in that city, and the return of relative tranquility allowed the army to claim that handing over control to the police represented a victory over the drug traffickers.[46]

Withdrawal after defeat is never pretty, and the transfer of control to the federal police in Ciudad Juárez was as elegant an exit as the army could hope for. The offensive to destroy the drug cartels by hurling the army at all likely targets and sending soldiers into cities had clearly failed. The next section gives a close-up look at how drug violence has racked Mexico's second most important city, and the subsequent section explains how the Calderón administration tried to shape a new campaign against the drug cartels in its last years in office.

The Battle for Monterrey

Intense levels of violence appeared in many regions of Mexico and lasted for years. The drug traffickers fought among themselves and also engaged in bloody clashes with the military and police forces. The armed confrontations were most frequent in the cities near the U.S. border and in areas of marihuana and poppy cultivation. The violence sometimes spilled over to areas, such as Monterrey (Map 7), a city at first glance not suitable for the drug trade. During the initial years of the twenty-first century, most drug violence had spared Monterrey, which remained an oasis of tranquility much like Mexico City.

The vibrant economy of Monterrey, in particular its large manufacturing and financial sector, made the city the second wealthiest in the country after the capital. And while Mexico City traditionally has been suspicious of the United States, Monterrey on the contrary has enthusiastically cultivated close business and cultural ties with Americans across the border. The industrious inhabitants were richer than the rest of Mexican citizens, and this wealth made Monterrey attractive to drug traffickers, in particular to senior cartel members. The prosperous business sector in the city routinely handled large cash flows, so at the beginning drug traffickers could slip their tainted money through banks and invest in local businesses without arousing suspicions. As the money flowing from the U.S. continued to rise and topped over 20 billion dollars annually, the drug traffickers became more sophisticated and learned to camouflage their piles of cash in legal transactions. In effect, Monterrey became the capital of money laundering in Mexico, but because this activity took place out of sight, neither the government nor the public showed any concern and kept quiet about this massive stimulus to the local economy.[47]

By the time the four-year long battle for control of Nuevo Laredo ended in 2007, many drug lords had moved their residences to Monterrey. Because neighborhoods of both the upper class and the upper middle class existed in the metropolitan area, the drug lords could buy mansions and expensive automobiles without sticking out as usually happened in provincial cities and even in Mexico City with its apparently endless slums. At least twenty of the wealthiest Mexican families lived in Monterrey, and drug lords by claiming to be "businessmen" or ranchers could purchase luxury homes in exclusive neighborhoods and blend in without too much difficulty. Although admission into the exclusive private clubs was completely out of the question for drug lords, having their children attend elite private schools and living in the same neighborhoods as the wealthiest Mexicans were major steps toward gaining social acceptance.[48]

Ironically, the wealthy neighborhoods of Monterrey became the most profitable market in Mexico for street sales of drugs. Since the 1990s sales in local corner stores (*narcotiendas*) had become an important secondary source of profits for drug cartels throughout Mexico, but in Monterrey this retail business became unusually lucrative because the spoiled children of the upper class paid top dollars for quality drugs. Monterrey became important for drug lords not just for money laundering but also for this lucrative street trade in wealthy neighborhoods, while the many amenities the city offered attracted senior members of the Gulf Cartel. However, their first order of business was to finish driving out the Sinaloa Cartel from this region. The Sinaloa Cartel had put many officers of the state and municipal police on its payroll, and the Gulf Cartel sent its armed bands, the Zetas, to murder police officers suspected of links with the rival cartel. This unexpected spike in the number of murdered police officers shocked the country, and in response President Felipe Calderón sent army units and federal police agents to try to stop the murders in Monterrey in 2007.[49]

The army set up check points on the roads leading to the city, while both the state and municipal police carried out extensive patrolling inside Monterrey. At the same time, the drug lords realized that what the government really wanted was to keep the murders out of public view, and the drug-related killings no longer took place in broad daylight but out of sight. And because most of the murder victims were small-time corner drug dealers, the public soon lost interest, and city residents prematurely breathed a sigh of relief. By the start of 2008 the worst seemed to be over, and most of the federal police reinforcements had already returned to their bases in other parts of Mexico. The only worrisome element was the appearance of gangs kidnapping Mexican owners of small and medium-sized businesses. But because the kidnappers did not seem to have any clear connection to the drug cartels, this early warning of a future increase in violence went unheeded.[50]

In reality, the Gulf Cartel through its armed gangs, the Zetas, was making a determined effort to gain full control of all police institutions in the Monterrey metropolitan area. The control the Gulf Cartel won over Nuevo Laredo had turned out to be largely meaningless for smuggling drugs into the United States because that border city was merely the end point of a transit route passing through Monterrey. This last city and not the border town of Nuevo Laredo was the real choke point in the drug trade. As the Gulf Cartel put on its payroll at least 40 percent of the personnel in both the state and municipal police forces, only the military remained outside the control of the drug lords. In an innovative tactic to drive out or at least discredit the army, the Gulf Cartel organized demonstrations in one neighborhood against the military in January and February 2009. Protesters denounced the army for committing human rights abuses against the civilian population and demanded that the government withdraw the military from law-enforcement duties. Most of the protesters came from a drug-controlled neighborhood and the rest of the demonstrators received a payment to attend, a practice the drug barons copied from politicians who paid followers to attend rallies.[51]

The ploy of demonstrations failed, and the Gulf Cartel redoubled its efforts to control the police. The authorities tried to use polygraph tests to detect corrupt police officers, but the measure was ineffective because the drug lords threatened the examiners. Rotating high police officials every three months only disrupted the command structure. So many police personnel died in the line of duty, that finding a sufficient number of recruits proved impossible. Even the more prestigious state police was two thousand officers short of meeting its authorized strength. "The drug cartels do not fear the state and municipal police; instead clean police officers are afraid of the cartels."[52]

Clearly the Monterrey area was spinning out of control, and the only institution left capable of restraining the drug lords was the military. Starting on 1 June 2009, the Mexican army began arresting many high-level police officers accused of corruption. An important piece of evidence had been the payoff lists the police had found in raids on the homes of drug lords. After four weeks of nearly continuous arrests, the army recovered a major payoff list which in turn fueled another round of detentions. The spectacle of over one hundred senior police officers arrested on charges of corruption shocked the public, and prominent citizens and business groups called for the military to take over all the security functions in the city. The military was careful to dampen those requests, because it knew that so much of the intelligence on the corrupt police officers actually had been gathered by honest police detectives. As was usually the case across Mexico, the army lacked the means to develop its own network of agents and relied on intelligence from the state police.[53]

The troops were needed because the Gulf Cartel had acquired grenades and automatic weapons vastly outclassing the usually puny armament of the police. Arrests of any drug trafficker were turning into violent and bloody affairs, and increasingly the police needed the backup of the soldiers either to capture or kill the drug traffickers. Many Zetas were former soldiers, wore army or police uniforms, and operated in military formations. Shootouts became common during 2009, and the most shocking incident took place in June 2009 during the arrest of a high-level police officer. The corrupt police officer commanded 100 heavily armed policemen, and a near battle almost erupted when they faced 150 soldiers and federal policemen. To prevent a similar confrontation in the future, the army confiscated all heavy weapons in the hands of state and municipal police.[54]

Business leaders had been pleading for the militarization of Monterrey and wanted the army to assume responsibility for all street patrols. The business community partially obtained their wish when army and state police took over all patrol duties in a northwest suburb of Monterrey in November 2009. But drug-related violence continued to escalate and reached the highest level on 4 December, which newspapers called the bloodiest day in the recent history of the city. That day proved particularly embarrassing to the army and showed that its will to prosecute the war on drug traffickers was waning. When police asked for backup from the army to apprehend a drug lord, army officers ignored the request. In desperation the police turned to the small detachment of Marines in Monterrey, and navy officials promptly sent the Marines to effect the capture. A half-hour shootout ensued before the Zetas were either killed or taken prisoner.[55]

But this was not the end of the clash, because when the Marines requested help from soldiers to escort the prisoners, another detachment of Zetas attacked the army convoy. The Zetas seemed to have confused the army convoy with that of the Marines and were determined to rescue the prisoners. A running battle ensued as vehicles in both convoys burst into flames and nearby cars exploded during the ferocious exchange of rocket propelled grenades and heavy machine gun fire. Photographs of the burning vehicles splashed the front pages of local newspapers, and the images resembled the destroyed convoys at the end of the Iraq War. The whole operation might have been a diversionary attack, because shortly after the shooting ended, the Zetas broke into a Monterrey prison and released 23 prisoners.[56]

All these clashes on 4 December 2009 seemed to foreshadow even more intense confrontations between the armed forces and the drug dealers, but actually they closed one cycle of violence. Although the army engaged in a public relations campaign to try to repair the damage from its failure to respond promptly to the initial request, in reality the generals wanted to extricate their units as fast as possible from an impossible campaign. At first it seemed that the violence was subsiding early in 2010, and the military grasped these signs of success as a clear indication that it was time to make an orderly withdrawal not just from Monterrey but from many other parts of Mexico. The opportunity came in February 2010 when unmistakable signs confirmed that a new struggle was beginning in Monterrey. The Zetas, the armed branch of the Gulf Cartel, became a separate cartel and began a long and deadly war against their former masters.[57] Because the murders that predominated were those among the two rival cartels, the rest of the population came to enjoy a relative security. By 2012 the Zetas were gaining the upper hand in their campaign against the Gulf Cartel, and the government, as an interested bystander, could do little more than watch this struggle play out. And the failures of the government were not limited to Monterrey but, as the next section explains, took place in many other areas in the war against the drug cartels in Mexico.

The Hunt for El Chapo Guzmán

As the Felipe Calderón presidency approached its mid-term, the failure of the nation-wide military offensive to deliver decisive results strongly suggested the need to modify the strategy against the drug cartels. Unable to sustain the high costs of the unproductive military offensive, Mexican officials reduced the scope of operations and concentrated on a more modest strategy consisting of the two goals: (1) arresting or killing drug lords; (2) restoring order to a few key cities, mainly on the border with the United States.

Of the two components, the most visible was the goal to eliminate major drug lords. Called the "Kingpin Approach," its goal was to cripple drug cartels by eliminating their top leadership on the assumption that the other elements in the cartels would then disintegrate and wither away. The increasing number of drug-related arrests (Table 3) and of extraditions to the United States (Table 4) reflected the increased involvement of soldiers and Marines in law-enforcement duties. Most if not the overwhelming majority of kingpin arrests were the result of cooperation between the armed forces of Mexico and U.S. agencies such as DEA. Naval intelligence and specialized army intelligence units helped follow the movements of drug lords.[58]

The kingpin approach filled the news with spectacular operations, and not surprisingly, the Calderón administration played up each successful capture as a great victory in dismantling the drug cartels. And of these stories, none captured more public attention than the pursuit of El Chapo Guzmán (Joaquín Guzmán Loera). Considerable circumstantial evidence suggests that the Vicente Fox administration showed little or no interest in recapturing El Chapo Guzmán after his easy escape from prison in January 2001. But whatever arrangements may have existed between the Fox administration and the drug lord, a similar modus vivendi failed to appear under Calderón, and already early in 2007 the government was trying to capture El Chapo Guzmán.

The hideouts of this drug baron had been in the Western Sierra Madre Mountain Range, in the area bordering the states of Chihuahua, Durango, and Sinaloa. DEA informants learned that El Chapo Guzmán was planning to attend a splendid coming of age (*quinceañera*) celebration for a daughter of a close associate in a mountain village in Durango on 19 May 2007. The army sent 30 soldiers to effect the capture, a surprisingly small number when 50 well-armed bodyguards protected the drug lord. When the "security rings" of lookouts near the village duly reported the slow approach of the army convoy up the dusty and winding roads, El Chapo Guzmán decided not to spoil his friend's celebration and rode away on all-terrain-vehicles with his bodyguards into the nearby mountains. By the time the army detachment arrived, the drug lord was nowhere to be found.[59]

Another opportunity to eliminate El Chapo Guzmán came shortly after with his fourth wedding celebration. With the support of the drug lord, an 18-year old female won the title of Queen of the Coffee and Guayaba Fair, and he wanted a splashy wedding on 3 July 2007 to celebrate his marriage to the beauty queen. Because she was a U.S. citizen, she could freely cross the border as later she did to give birth to their children on American soil. Hundreds of guests gathered at the location of the wedding in the mountains of Durango, and El Chapo Guzmán came three days before in the company of at least 70 bodyguards. Such a big gathering could not be kept secret, and soon local authorities and DEA informants were sending a stream of reports about the presence of the drug lord. The army this time mounted a massive expedition with many soldiers riding all-terrain-vehicles and a helicopter hovering above. But all the preparations were in vain, and when the expedition arrived the day after the wedding, once again a forewarned El Chapo Guzmán had made a leisurely escape into the mountains with his new bride and his large entourage.[60]

Mexican officers tried to learn from these failed attempts and carefully elaborated a three-stage strategy to capture or kill El Chapo Guzmán. In the first stage of the operation, the army saturated the Western Sierra Madre Mountain Range with units not so much hoping to capture him but to gather intelligence on his movements. Exploring the arduous terrain of

10. Mexico: The Armed Forces Embrace the United States 191

Durango proved exceedingly time consuming and only in October 2009 could the army confirm the location of as many as 15 sites the drug lord used as hideouts. The troops also learned that at least 300 guards formed his private security force, and this large number complicated the second stage, which called for surrounding his most recent hiding place. The army could not keep secret the deployment of a large number of troops for the second stage of the strategy. Informants and security rings allowed El Chapo Guzmán to stay at least one step ahead of his pursuers, who never reached the third stage of actually capturing or killing him in the mountains.[61]

Another opportunity to eliminate the drug baron came in February 2012 when with only a couple of guards and a prostitute, he was vacationing in a mansion in Los Cabos, Baja California (Map 6). This time the federal police carried out the operation, but once again the drug lord was tipped off and was able to make a hurried escape presumably back to his mountain hideouts in Durango. The very public incapacity of the Calderón administration to capture El Chapo Guzmán was proving a political liability, and as a way to help a loyal ally in trouble, the U.S. government proposed to repeat the Osama Bin Laden operation in Mexico. Relying on U.S. intelligence sources, Navy SEALs planned to disembark from helicopters and kill the drug lord, much in the same way as the SEALs did in Pakistan against the terrorist. President Calderón was enthusiastic about the proposal, but both the army and navy strongly rejected an operation that left them at best as idle and useless spectators.[62]

The kingpin approach was the first of two components of the revised strategy against drug traffickers in Mexico. The second part called for concentrating efforts on a few cities, principally along the U.S. border. The Calderón administration believed it had been trying to do too much with too little, but starting in late 2009 a new approach "targets a few joint projects in a few cities rather than doing a little of everything."[63] Consideration originally was also given to including in the new approach not just border cities, and for a few months Culiacán (Map 6), the capital of Sinaloa, was a strong contender for a determined effort away from the border. But in another indication that the Mexican government was overcommitted and its resources stretched to the limit, the suggestion of incorporating a city of the interior into the new strategy gradually faded.

The U.S. government became the driving force in the efforts to rescue the Mexican border cities. Bilateral commissions teamed a wide range of U.S. and Mexican officials from the border cities of Juárez-El Paso and Tijuana-San Diego. Cooperation on the border became very strong between the two governments, and even the Mexican army assigned two officers to serve as permanent liaison with the El Paso Intelligence Center. As part of the plan, the military lowered their profile in the border cities and delegated to the reconstituted police forces the task of patrolling the streets. With a myriad of U.S. and Mexican officials supervising the performance of the police forces in these two border cities, criminal activity went down and a semblance of order and relative calm seemed to have returned. In reality the success in reducing violence in both Tijuana and Ciudad Juárez by 2011 responded mainly to the victory of the Sinaloa Cartel over rival drug dealers. As the Sinaloa Cartel consolidated its control over those two border towns, it was in its own best interest to maintain order as a condition for the smooth operation of the tunnels and routes bringing drugs into the United States.[64]

The modest success in the two border cities could not hide the complete failure of the war on drug cartels of the Calderón administration. During his six years in office, the number of homicides increased and exceeded by more than a third the 74,586 victims killed during the term of office of his predecessor Vicente Fox.[65] Many observers had predicted the failure of the war against the drug cartels. Critics unfairly reduced the entire policy of President Calderón to trying to eliminate only the drug lords, when in reality the administration's strategy if anything had been too broad and overly ambitious except during its last years.

The kingpin approach was based on the assumption that eliminating drug lords by arrest, extradition, or execution meant destroying the drug syndicates. Because the drug cartels operated under a highly personalized form of command, when authorities eliminated this key person, no replacement was ready to fill the gap. This personalized structure contrasted with the hierarchical structure in the FARC, the guerrillas in Colombia. The kingpin approached failed in the Colombian government's war against the guerrillas because trained and eager subordinates were ready to replace a fallen leader. Instead among the criminal organizations in Mexico, violent power struggles arose among rival contenders and more frequently the constituent parts of the drug cartels took on a life of their own. Because only a few bands such as the numerous Zetas could become their own drug cartel, the majority of the leaderless gangs went on a rampage of criminal activities, in particular kidnappings and extortions to raise money. In executing these actions these gangs perpetrated murders and frequently degenerated into sadistic activities. Many of the killers went around offering their services to likely clients; accounts spread that when divorces turned messy and

10. Mexico: The Armed Forces Embrace the United States 193

ugly, these killers offered to eliminate the quarrelsome spouse for the right price.[66]

This crime wave sweeping Mexico and tormenting the civilian population made voters turn against PAN, the political party of President Calderón in the 2012 elections, and the PAN candidate finished at an embarrassingly distant third place. To the surprise of many observers, the PRI returned to the presidency after an absence of twelve years. Many commentators had dismissed the PRI as a political dinosaur and did not expect this resurrection, but the explanation lay in the disastrous policies of the Calderón administration. Along with the crime wave and the high number of deaths came mounting unemployment and a deteriorating economic situation. The siren calls of democracy that PAN constantly had broadcast turned out to be nothing but illusions, and Mexicans returned home to the PRI by electing Enrique Peña Nieto as president of Mexico.

11

Mexico: The Presidency of Enrique Peña Nieto

> The people ... are doing the work that the government is incapable of performing. —Estanislao Beltrán, chief of the self-defense forces[1]

Before his inauguration on 1 December 2012 Enrique Peña Nieto announced major changes in the war on drugs. The argument of the Conservative political party PAN that the politicization of the Interior Ministry had been responsible for the violence in Mexico had been totally disproved by the 12 years of experimentation with a separate Ministry of National Security. President Peña Nieto promptly abolished this ill-begotten Ministry and transferred all its police forces to the Interior Ministry. Only the Attorney General's Office retained its detective units as its traditional investigative force.

The centralization of federal police forces under the Interior Ministry also gave the Mexican government the opportunity to monitor contacts with the United States. Because many subordinates were accustomed to making separate arrangements with U.S. officials, in late May 2013 the president ordered that all non-military contacts with U.S. law-enforcement officials had to be channeled through the Interior Ministry. This change in procedure did not mean a distancing from the close cooperation with the United States existing under the two previous presidencies and rather aimed to improve the efficiency of the Mexican government.[2]

Because President Peña Nieto had openly refused to cooperate in the campaign of hostility against Venezuela and had shown support first for Hugo Chávez and then for his successor Nicolás Maduro, the concern arose that the new President was taking an anti–American position. But the new administration was not as interested in pursuing an independent foreign policy as in concentrating on its more than abundant internal security problems. U.S. military cooperation with Mexico was the decisive litmus test to measure relations with Washington. Just ten days after taking office, Peña

11. Mexico: The Presidency of Enrique Peña Nieto

Nieto had the Defense Ministry publicly ratify its commitment to host the Conference of the Armies of the Americas, a two-year cycle of technical meetings culminating in a large conference in the host country. As will be recalled from chapter 7, Chávez had pulled Venezuela out of this organization precisely to reduce the ties between his military and the United States. Instead Mexico splendidly hosted the Conference in October 2013 and duly passed on the two-year chairmanship to U.S. unconditional ally Colombia.[3]

Peña Nieto did more than simply uphold pre-existing military commitments. Knowing well the preference for Mexican generals to fester and sulk in isolation from their U.S. counterparts, he insisted on maintaining both formal and informal contacts with senior American commanders. At his orders, both the Defense and the Navy ministers went to Washington, D.C., in July 2013 to expand bilateral military cooperation with the United States. For the Mexican navy the visit merely ratified the many strong links already existing with American uniformed personnel, but for the army the trip was a major sacrifice. Subsequent reciprocal visits by both American and Mexican senior commanders made routine the practice of cooperation at the highest military levels.[4] The return of the PRI to power in December 2012 in many ways resumed the level of cooperation with the U.S. military that the previous PRI president, Ernesto Zedillo, had first started in the late 1990s.

Pursuing the Drug Lords

The determination to continue strong military ties with the United States did not stop the Enrique Peña Nieto administration from trying to emphasize civilian law enforcement. The President ordered the enlarged Interior Ministry to concentrate its efforts on reducing the widespread criminal activity terrorizing many Mexican citizens and placed the highest priority on stopping the wave of kidnappings. In spite of persistent rumors that the new administration intended to cut deals with the drug cartels, the campaign against the drug lords continued without interruption. However, because initially the administration shifted away from the "kingpin strategy" of going after drug lords, fugitives such as El Chapo Guzmán (Joaquín Guzmán Loera), the leader of the Sinaloa Cartel, seemed to be getting a free pass. The Peña Nieto administration never formally announced the resumption of the "kingpin strategy" but as reports poured in revealing the approximate locations of several drug lords, quietly the authorities prepared to seize the opportunity to strike some highly visible blows against drug lords.[5]

U.S. law enforcement agencies were using new technology to intercept the cell phones of the drug cartels. DEA took the lead in these inquiries and at times was supported by ICE (Immigration and Customs Enforcement). Since the last years of the Felipe Calderón administration the usual procedure was for U.S. agencies to develop the leads and then hand them over to the Mexican navy, presumably because civilian agencies and even the Defense Ministry were too riddled with drug informants to keep any operation secret. A first blow came on 15 July 2013 when the navy seized the chief of the Zetas and in August the same happened to the chief of the Gulf Cartel. Before cynical commentators could claim that the Peña Nieto administration was neglecting the Sinaloa Cartel, the largest, richest, and most powerful in Mexico, the authorities began a series of arrests of major figures of that organization, some very close to the fugitive El Chapo Guzmán.[6]

Because all these captures in no way halted the flow of drugs into the United States, the limitations of the kingpin strategy were obvious, but at least the arrest of top figures showed that the government was not collaborating with the drug cartels. The business was so lucrative, that new leaders promptly appeared to keep the drug trade functioning, for example the arrested chief of the Gulf Cartel had replaced the previous leader who had been captured a year before. Significant as were the arrests of 2013, the only leader with a legendary international reputation was El Chapo Guzmán. Because a large private security force constantly guarded him in his many remote hideouts in the mountains of Durango, the capture of the wealthiest drug lord realistically was considered impossible without some lucky break.[7]

In what El Chapo Guzmán considered a masterful move, he had chosen as his fourth wife a teenager who was a U.S. citizen. With her American passport she freely crossed the border and also returned to give birth to twin daughters on U.S. soil. Although she had no pending charges, border authorities considered building a circumstantial case against her to secure her arrest, but desisted upon deeper reflection because U.S. officials believed that she just might provide a trail leading back to her husband. The assumption was well founded, because for the first time in his life, El Chapo Guzmán had fallen madly in love with his young wife and their twin daughters, perhaps partly because she was so different from his other females. He had previously associated with sturdy peasant women who were content with tortillas and beans, but his fourth wife expected to live the good life in a city and was not about to follow El Chapo Guzmán along the dusty trails of Durango.[8]

As long as he stayed moving around in the countryside, he was safe and could even risk occasional visits to the cities. But as leftist guerrillas found out the hard way during the twentieth century, the cities were traps for individuals fleeing from the police. When in November 2013 border authorities arrested a mid-level operative of the Sinaloa Cartel when he was trying to enter the United States at Nogales, in his cell phones they found the telephone numbers of other Cartel members. At first the lead did not seem very promising, and authorities on both sides of the border located for arrest only low-level individuals of that criminal organization. The trail eventually led to the arrest on 17 February 2014 of a messenger who confessed that El Chapo Guzmán was living in several safe houses in Culiacán (Map 6). Steel armor doors protected the entrances to the houses, which were connected by a network of underground tunnels to escape routes linked to the city sewer mains. Navy personnel organized raids of those houses but had underestimated the thickness of the steel doors, and by the time the attackers broke through the doors, El Chapo Guzmán had reached the tunnel entrances hidden under the shower of the bathroom and escaped his pursuers.[9]

Everything argued for a quick retreat to the safety of the mountains of Durango, but his love for his young wife and their daughters clouded his mind. He knew that he could no longer stay in Culiacán, and as a partial solution he left with his family for the seaside resort town of Mazatlán. Had he possessed a comparable network of safe houses with escape tunnels, he might have been safe, but in a fatal decision, he decided to stay with his family in a condominium building. Mazatlán was a favorite destination for American retirees who comprised the majority of the tenants in the relatively modest building. As an additional precaution, El Chapo Guzmán had himself disguised as a sick patient and rolled on a wheel chair up to his apartment. Did he realize that being in an elevated building meant that as soon as uniformed personnel took control of the ground floor, no escape was possible?[10]

Very early in the morning on Saturday 22 February 2014 at 0500 hours, uniformed navy personnel knocked down the door of the apartment and soon confronted the wife who claimed to be alone with her daughters. Some accounts assert that the navy was merely following all leads to arrest any member of the Sinaloa Cartel. Because most leaders of the Sinaloa Cartel had already ditched their cell phones, the intercepts of conversations had almost disappeared, and very likely the navy was knocking on the doors of all the last detected places of conversations. The official account is that the navy came looking specifically for El Chapo Guzmán in that apartment, and as soon as the Marines recognized the woman as the wife of the drug lord, they knew that they were close to something big. After searching every corner of the apartment the naval personnel located a heavily armed El Chapo Guzmán locked up in the bathroom. Not wishing to harm his family, he agreed to surrender and soon was on the way to the maximum security prison in Mexico City.[11]

The capture of El Chapo Guzmán was a major success for the Peña Nieto administration and completely shattered the accusations that the President had planned to make deals with the drug cartels. However, the arrest of the head of the Sinaloa Cartel raised the possibility that he might denounce all the political figures that in the past had received his money. Extradition to the United States to face only drug charges rather than political corruption seemed a very tempting solution to this hot potato. Revelations about extensive corruption, even if under previous administrations, could not help in any way the reputation of the Mexican government. However, nobody expected the capture of El Chapo Guzmán and the other leaders of cartels to stop the drug trade. Whether under large cartels or smaller nimbler organizations, Mexico would continue to be a reliable supplier of a wide variety of illegal drugs to the United States.[12] And even if the cartels ceased trafficking in drugs, as happened in Michoacán, the criminal organizations could still become a destructive force inside Mexico.

The Challenge of Michoacán

The administration of Enrique Peña Nieto inherited a problem dating back to the presidency of Vicente Fox. In 2006 a new criminal organization, *La Familia Michoacana*, emerged as a rival to the Sinaloa Cartel and the Zetas. Receiving Colombian cocaine through the port of Lázaro Cárdenas (Map 5) and creating a large distribution network in the United States, this new cartel rapidly became a major supplier in the international drug trade. Its leader was Nazario Moreno González "El Chayo," and under his direction *La Familia* also was a pioneer in the export of large amounts of methamphetamines into the United States.[13]

From the very beginnings *La Familia* exhibited some unusual traits that made it unique among the Mexican drug cartels. Its leader El Chayo was a fanatical evangelical who brought an intense Christian Messianic ideology to the organization. Members of *La Familia*, in a style reminiscent of Shining Path in Peru in the 1980s, tried to impose a strict morality on the territories under their control. The new cartel punished thieves and kidnappers and took violent reprisals against any form of extortion. In particular rapists could expect prompt execution, because El Chayo made the protection of women one of the justifications for the existence of his organization. To the gratitude of many mothers, *La Familia* also policed the behavior of teenagers, in particular those who brought back from the United States bad habits such as graffiti spraying and outrageous street clothes. That *La Familia* was able to establish this strict control over the southern municipalities of Michoacán testified to the weak government presence in the area.[14]

During its early years *La Familia* maintained an uneasy collaboration with the Zetas to the east, but a falling out occurred and both became bitter enemies. As newcomers to the drug business, *La Familia* lacked the experience of the veteran drug traffickers in the Sinaloa Cartel, and El Chayo made many rookie mistakes, especially in the vast distribution network he organized in the United States. Law enforcement agencies saw a great opportunity to pounce on the new cartel, and in coordinated raids and arrests on 22 October 2009, police dismantled the distribution network of *La Familia* in the United States. As mentioned earlier, as part of the campaign to destroy *La Familia*, the Felipe Calderón administration in one of its first acts dispatched large numbers of troops to Michoacán in December 2006, and this deployment was the first of seven lasting until July 2009. The army deployments did keep the pressure on the cartel but failed to strike any decisive blows because local inhabitants were afraid to denounce *La Familia* members to the authorities. In effect, as had happened in many other similar army operations across Mexico, the lack of intelligence and especially the absence of informants left the soldiers striking blows in the dark.[15]

Despite the success of *La Familia* in eluding the army persecution, El Chayo realized that his organization was in danger of collapse. One faction inside *La Familia* decided to join the Zetas, while the majority under El Chayo sought a tactical alliance with the Sinaloa Cartel. Essentially, El Chayo received modest supplies of cocaine for local consumption, and in exchange the Sinaloa Cartel was free to use Michoacán, particularly the port of Lázaro Cárdenas, as a transit point for drugs. By this agreement El

Chayo abandoned the lucrative international drug trade, and reflecting the new focus, he adopted the name of Knights Templars (*Caballeros Templarios*) for his restructured organization in March 2011. As an additional precaution to disguise the transformation, El Chayo had faked his own death in a shootout with naval personnel the previous year on 9 December 2010, and a gullible Mexican government was broadcasting to the world the death of this fearsome cartel leader, even though authorities had not recovered his body.[16]

Successful as El Chayo had been in hiding his role in the creation of a new criminal organization, he still knew that he had to find a new lucrative activity to employ his followers. Clearly he and the other leaders were out of their depth in running an international drug cartel and they more realistically concentrated on what they knew best: Michoacán. In effect, in a masterful turnaround El Chayo reinvented the organization and transformed it from a drug cartel to a repressive governing structure. Ultimately the Knights Templars displaced or neutralized Mexican government authorities and set up their own parallel ruling structure in southern Michoacán with Apatzingán (Map 5) as their unofficial capital. The drug cartels used massive force in street battles against their rivals in the business but rarely against local inhabitants. Money, gifts, and donations had bought the sympathies and even the enthusiastic support of large segments of the local population. Instead, for the Knights Templars force was their principal instrument of domination, and money only served to cover their operating expenses. Intimidation and brutal terror became the favorite tactics of this savage organization.

In a region of massive unemployment and extensive migration to the United States, the Knights Templars with promises of high salaries and occasional booty easily recruited large numbers of followers. The key individual in their organization was the lookout (*puntero*) who constantly scanned the horizon for any signs of approaching troops; local policemen were no longer a concern because they had all either been bribed or more likely intimidated into obedience. The digital age had reached even this remote corner of Mexico, and the lookouts used not just their eyes to peer in the distance but also relied on electronic equipment and computers to intercept telephone conversations and digital messages. With this extensive network of paid informants, not surprisingly all the army campaigns in Michoacán had ended in failure. The only weak link in the network of lookouts was that the local population knew who worked for the Knights Templars.[17]

Paying the lookouts and the many gunmen (*sicarios*) who terrorized the local inhabitants was quite expensive for the Knights Templars who did not dispose of vast sums of money like the drug cartels. To collect funds, another network extorted payments from most of the local population. The members of this third branch of the Knights Templars initially were the same individuals who were lookouts and gunmen, but as the organization grew, separate networks appeared to extract the maximum amount of money from almost all inhabitants. The export and limes and avocados has constituted the most valuable economic activity in Michoacán. The Knights Templars initially demanded payments from large landowners and businesses in the towns but gradually they extended their demands and threats down to small landowners and even street vendors. As a first warning those who refused to pay had their establishments destroyed or burned down; the next time the gunmen came to kill family members. El Chayo and his associates became ingenious in finding new ways to extract money from local residents, and just to hold a job or to go to school required making a payment. Almost all town residents had to pay quotas to the Knights Templars.[18]

Soon the Knights Templars were not satisfied with just collecting protection payments from the owners of lime and avocado groves and went on to take full control of the properties. Lawyers even forced compliant authorities to register the transfer of property deeds to the Knights Templars. And the organization targeted not just large landowners but also took lands away from many small proprietors. The next ones to feel the burden were the harvesters in the lime groves, who had to pick the limes and then sell them to the Knights Templars at below market prices. A similar situation occurred with the avocado plantations: many of them remained abandoned because the owners had fled and the workers did not want to collect avocados for extremely low wages.[19]

By attacking almost all the social groups in Michoacán, the Knights Templars were laying the foundation for a multi-class opposition to emerge. Their excessive demands were making local inhabitants increasingly angry, but the automatic weapons of the ruling organization still inspired fear and submission, and the murder of citizens opposed to the Knights Templars terrorized many families.

The religious preaching of the evangelical El Chayo, who published a religious catechism and made his followers participate in long prayer sessions, helped the inhabitants tolerate the increasingly heavy economic burden of the Knights Templars. But in a very Mexican tradition, disputes over women finally turned the inhabitants against the criminal organization. The local population initially had welcomed the execution of rapists in previous years, but in a shocking reversal the Knights Templars began to kidnap and rape women by the end of 2012. Sometimes they demanded to have sex with the pretty daughters of ranchers and in other instances the Knights Templars simply took the women by force. In one case they kidnapped the pregnant niece of a doctor and she was never heard from again.[20]

The backlash began in a village 35 kilometers from Apatzingán on 24 February 2013. That day at 1000 hours an automobile using a loudspeaker summoned the inhabitants to a mass rally in the central plaza and over thirty showed up. These individuals were mostly lime gatherers who refused to pay the extortion quotas and as a first step they disarmed the local policemen who were tools of the Knights Templars. Similar uprisings erupted in nearby villages, and the original intention of these still amorphous groups was to operate independently and not to leave their villages. But when the gunmen counterattacked with their AK-47s and threatened to crush any resistance from individuals armed only with Mausers and shotguns, the Knights Templar seemed to be reestablishing their grip by means of executions and terrorism. Survivors recalled that the gunmen used AK-47s to spray any suspected target with clip after clip.[21]

As the clashes increased, the Peña Nieto administration misread the signs and concluded that this was simply more warfare among rival drug cartels. The usual solution was to send the army back to Michoacán, and a large deployment took place in May 2013. The arrival of the army kept the Knights Templars from crushing the incipient resistance movement, and community leaders knew that they had to act fast before the inevitable departure of the army units left them defenseless and exposed to bloody reprisals on the part of the Knights Templars. Businessmen and landholders contributed funds to purchase assault weapons and vehicles in Texas and provided the backing to organize the initial groups. At a later stage the Sinaloa Cartel offered funds and weapons mainly out of the desire to prevent the Zetas, the rival drug cartel, from moving in to fill the vacuum left by the collapsing Knights Templars. At the same time, at the local level community leaders decided to unite their different forces to create a regional self-defense movement (*auto defensas*) no longer limited to a single village. The consensus emerged that the self-defense forces had to drive out the Knights Templars from all the municipalities of Michoacán and had to eliminate all their leaders, including El Chayo whom the authorities believed was dead.[22]

The self-defense forces had no trouble recruiting willing residents from a wide specter of the population. Fear was a powerful driving force, and many self-defense volunteers confessed that they felt safer when on duty than when they traveled as unarmed private citizens. As a movement with many origins, initially no clear leader emerged, but a group of coordinators headed the volunteer groups in the common task of driving out the Knights Templars. Just as in Peru many inhabitants had turned against the impositions of Shining Path, so in Michoacán the citizens rebelled against the demands of the Knights Templars. The self-defense forces went on the offensive, and on 27 October 2013 over 300 followers made a first attempt to capture Apatzingán, but the Knights Templars repulsed this attack on their main bulwark.[23]

As the clashes between the self-defense forces and the Knights Templars multiplied in late 2013 and the first weeks of 2014, state officials were powerless to control the violence, and on 12 January 2014 the governor turned the matter over to the federal government. The next day the federal government took charge of maintaining order in the southern municipalities of the state and ratified this determination by using ten C-130 airplanes to fly in many soldiers. The administration of Peña Nieto still had not made up its mind about what to do with the self-defense forces, and as army patrols fanned out through the region, the likelihood of clashes increased. The troops were under orders to disarm all citizens, and because the Knight Templars traditionally went into hiding when troops arrived, the only ones left to disarm were the self-defense forces who saw no reason to hide from the army.[24]

When the soldiers tried to disarm the self-defense forces, some dangerous standoffs occurred, in several instances shots rang out, and in one case the soldiers killed three participants. But when the local inhabitants poured out in mass to embrace the self-defense volunteers and encircled the soldiers, the troops hesitated to open fire. Perhaps the peasant origin of the soldiers made them reluctant to fire on country people; for their part the veteran officers had never experienced similar spontaneous outpourings during the many anti-drug campaigns. In rapid consultations going up the chain of command, President Peña Nieto made the only logical decision of forming an alliance with the self-defense units against the Knights Templars. Negotiations with the leaders of the popular movement ensued, and on 27 January 2014 the government signed a formal agreement, but even before the alliance was formally drafted, the army was already cooperating closely with the self-defense volunteers.[25]

Under legislation dating back to the Mexican Revolution, the Defense

Ministry disposes of the authority to appoint civilians as rural guards in the countryside, but these appointments required background checks and the registry of all weapons. This legal formula of "rural guards" allowed the army to incorporate civilian forces as legitimate parts of the defense establishment. For the first time the soldiers came to enjoy abundant and specific information on who the Knights Templars were, and the volunteers easily rounded up all suspects and subjected them to interrogations to determine their degree of guilt. If the suspects had been only lookouts, the self-defense forces released them, and in a more controversial decision, sometimes recycled them back into the volunteer detachments. Because this process of vetting could take weeks if not months, the self-defense forces operated temporary detention centers. If the inquiries confirmed that the suspects had murdered or committed major crimes, then they were handed over to the federal authorities.[26]

The Knights Templars and its predecessor *La Familia Michoacana* had so corrupted government in the region that the army not only disarmed and removed all policemen but also most civilian officials. The state was rotten to the core, and only a mass purge could restore order. The offensive against the Knights Templars resumed in the first days of February 2014. Before the entry of the army, the heavy weapons of the Knights Templars had inflicted many casualties and in some clashes killed over a hundred followers of the self-defense forces. But now army vehicles rode ahead of the column and pulverized any points of resistance. Also, helicopter gunships provided effective air cover.[27]

Success came soon to the joint operations of the army and the self-defense forces, and the volunteers made a triumphal entry into Apatzingán, the former capital of the Knights Templars, on 9 February 2014. The volunteers then fanned out throughout the region as they tried to expel the Knights Templars from the remaining municipalities in Michoacán. Papa Smurf or *Papa Pitufo* (the nickname Estanislao Beltrán received because of his white beard and blue shirts) announced that the new policy was to continue patrols until all the Knights Templars' leaders had been captured or killed. In February Papa Smurf, an owner of lime groves himself, became the head of the self-defense forces. He got his wish when on 9 March 2014 navy personnel killed El Chayo, and this time unlike the fake death in 2010, the fingerprints on the body confirmed his identification.[28]

To carry out the process of transformation into the rural police, the leaders of the self-defense forces signed a second agreement with the central government on 14 April 2014. By this agreement the members of the self-defense forces promised to register their weapons with the army. Each weapon underwent ballistic testing for identification purposes, and each owner became responsible for any future use. The deadline for registration was 10 May 2014 and by then the persons of the self-defense forces qualified to become members of the new rural police in Michoacán. About 3000 individuals registered their arms before the deadline, but it was estimated that perhaps as many as 10,000 self-defense members preferred to hide their weapons in case the Knights Templars or another cartel might pose a danger again. The distrust of the authorities was understandable given the many years of neglect. For its part, the army and the federal police began arresting in June 2014 those self-defense members who had failed to register their weapons and were publicly brandishing them.[29]

As the threats to the region diminished, voluntary contributions declined, and finding funds for the self-defense members transformed into rural guards became a pressing question. Although the stated policy was to return to their rightful owners the lands confiscated by the Knights Templars, the self-defense forces were keeping some large lime groves and avocado plantations to generate funds. The owners of small plots recovered their properties, and most volunteers wanted to return to their normal lives, whether work or study. But as long as the threat of the Knights Templars lasted, farm hands felt the need to carry weapons while working in fields.[30]

Could the self-defense forces spread and provide the solution to Mexico's chronic lawlessness? The Knights Templars had extended to bordering states where the self-defense forces could help drive out the remnants of that criminal organization. But in states under the influence of drug cartels this solution was not possible. The fearsome arsenals and the private armies of the drug cartels are not the real obstacle, it is the money and prosperity that they bring to local communities. In Michoacán the Knights Templars attacked the economic interests of the entire society and provoked a multi-class backlash. The self-defense forces of Michoacán do not provide a model for the rest of Mexico, and only with other strategies can the Peña Nieto administration rein in the drug cartels.

12

Countries Without Armies

> *As I have said before, the foundation of all states is a good military organization, and where it is lacking there cannot be good laws or any other good thing.* —Niccolò Machiavelli[1]

In 1949 Costa Rica made history when it became the first country in the Western Hemisphere to abolish its army. Costa Rica did not stop there but became an advocate for eliminating armies not just in Latin America but throughout the world. One of the most forceful advocates for this pacifist policy was the Nobel Peace Prize winner Óscar Arias. Twice president of his small country, he strongly preached to anyone willing to listen about the evils of militarism. He also urged Western powers to stop giving economic assistance to poor countries, such as those in Africa that spent lavishly on supporting armies.[2] President Arias achieved his first great success in Panama when Costa Rica's southern neighbor abolished its army in December 1989. Shortly afterwards, this idealistic missionary of peace achieved his second great success when Haiti also dismantled its military in 1995. The momentum to abolish armies was spreading across the region, and unilateral disarmament no longer appeared as a utopian ideal.

Costa Rica: Creeping Militarization

Costa Rica abolished its army in 1949 primarily to bring internal peace. Until then the repeated military interventions in politics had kept the country in a state of turmoil and permanent instability. After 1949 the open political system and strong government intervention in the economy through state-owned firms and extensive regulation permitted the emergence of the largest middle class in Central America. A well-established electoral cycle and lack of voter fraud made elected officials very receptive to the needs of voters, and not surprisingly, citizens found guerrillas unappealing. Without having to face any insurgency, Costa Rica survived the Cold War years without the need for a large military establishment. And when Panama abolished its army in 1992, Costa Rica lost any remaining concern about any threat on its southern border.

A dark cloud still loomed on Costa Rica's northern border, however. Until 1979 the most despised military institution in all of Latin America had been the *Guardia Nacional* of the tyrant Somoza, but after his fall in July 1979, not even for a second did Nicaragua consider abolishing its army. Guns, uniforms, and soldiers were too deeply ingrained in the national identity of Nicaragua, and actually the military has been larger in the decades after Somoza than during his dictatorship. Managua repeatedly announced that its substantial army existed only for defense, and an almost farcical demonstration of that policy came in 2011. A misunderstanding over a border dispute, already submitted to international arbitration, led to Nicaragua mobilizing many army units north of the San Juan River (Map 10) in case Costa Rica should decide to occupy by force the disputed strip of land. While on the north side of the San Juan River thousands of Nicaraguan soldiers constructed bulwarks and set up heavy weapons and artillery to repulse any invasion, on the south bank the dozens of Costa Rican policemen could only watch with bewilderment. Finally the crisis, a classic tempest in a tea pot, blew over with not a single shot ever fired. The mobilization, however, had boosted the popularity of Sandinista President Daniel Ortega who went to the border to take personal command of the forces defending Nicaraguan territory.

Other than the latent danger from the Nicaraguan army, Costa Rica did not face any military threats for decades. Having a well-functioning republican system in Central America, Costa Rica seemed a natural U.S. ally, but its pacifist policies reflected an independent streak often at odds with U.S. imperial goals in the region. President Arias himself forced the U.S. government to stop the Contra War in southern Nicaragua in 1988, and when he returned to the presidency in 2006, U.S. officials expected him "likely to be at odds with the United States" because relations with Costa Rica although always friendly were "complex,"[3] an euphemism for not always following U.S. instructions. Quite naturally, during the first decade of the twenty-first century, Costa Rica never formed part of the trio (Colombia, El Salvador, and Guatemala) of unconditional U.S. allies in the Caribbean region.

Without any real external threat and enjoying internal tranquility, Costa Rica neglected its more than ten separate police forces for over thirty years. Slowly criminals began to take advantage of the weak law enforcement structure. In a first phase smugglers used the country as a transit point for bringing surplus weapons from the battlefields in Nicaragua to Colombia in exchange for cocaine. In the next step drug cartels relied on Costa Rica as an alternative to Panama for shipping Colombian cocaine to the United States. Because of the near absence of police in most areas of Costa Rica, the drug traffickers did not even have to spend a lot of money on bribing officials, as was the case in Mexico or other Central American countries. By 2009 even the Sinaloa cartel of Mexico had made Costa Rica a hub for its activities.[4]

The police of Costa Rica retained their reputation as being "the least corrupt in Central America, but they are significantly underfunded and under trained."[5] In reality the police were so inadequate as to be largely irrelevant to the criminal organizations spreading rapidly across the country. By 2009 the drug cartels were also using Costa Rica as a warehouse for storing large reserves of cocaine and not just as a transit point. In this way if authorities intercepted drug shipments, the drug cartels could still make their deliveries to U.S. consumers out of the ample cocaine stocks stored in Costa Rica. But the precaution was generally unnecessary, and reserves continued to pile up in Costa Rica. With so much cocaine sitting idle in warehouses and not earning any profits, drug traffickers decided to encourage domestic consumption by selling crack cocaine at local corner stores (*narcotiendas*) throughout the country.[6]

Soon a crime wave was sweeping Costa Rica, as drug users committed all sorts of felonies to obtain the money needed to support their addiction. In 2009 one person in every four households had been a victim of assaults, and the government seemed helpless to stop the rampage. The police did respond to violent crimes, but it was in a reactive mode of coming to deal with the cadavers and not having a clue as to how to prevent or even to investigate the murders. The police needed to develop a more proactive approach to protect citizens, but nothing in its tradition prepared the institution for this onslaught of drug-induced violence.[7]

Costa Rica was among the last countries to experience that wave of drug criminality already rampant throughout the region. When a country reaches this situation, the inevitable next step is to send soldiers to help the police regain control, but this possibility did not exist for Costa Rica. Without a military institution the country could not face the crime wave, but the pacifist tradition prevented establishing an army. President Óscar Arias remained in denial about the problem during the first two years of his term, and only in the middle of 2008 he finally accepted the reality of having to find a solution.⁸ The decision to increase the police budget brought the previously unpalatable consequence of having to turn to the U.S. government for assistance on how to fight the drug cartels. The Arias administration, in a remarkable reversal, stated in 2009 "Costa Rica needed all the help it could get, especially considering the increasingly high rate of both petty and violent crime in the country."⁹ The U.S. government, particularly after the Mérida Initiative, had set aside funds for helping Caribbean governments fight against drug cartels and other criminal enterprises. Only the U.S. government possessed the resources, the organization, and the expertise to help Costa Rica defend itself. But because of the militarization of U.S. foreign policy under George W. Bush, much of this material assistance and specialized training was available only through the U.S. military.

Costa Rica certainly faced a challenge in maintaining its pacifist stance. With the police at first it was easy, because the U.S. strongly supported a program to professionalize the force. The goal was to shift to community policing, that way uniformed personnel in close contact with residents and knowing what was going on in neighborhoods could intervene to prevent the outbreak of violent actions. The tactic was to stop small infractions before they escalated into major crimes. A new computer program of crime statistics pinpointed neighborhoods and blocks having the highest rate of violence, thus allowing police to concentrate personnel in these most critical areas. But the many police forces in Costa Rica were so outdated in their procedures that the program to professionalize the police required many years to implement. And the experience of other Central American countries showed that nothing was harder to accomplish than to convert police forces into efficient organizations. And most disturbing, as the U.S. trained elite units in the Costa Rican police forces to react to gang and drug violence, these special units increasingly acquired a military style. Almost imperceptibly at first, parts of the Costa Rican police were taking on the appearance of army combat units.¹⁰

Not just the land but also the air space and maritime waters of Costa Rica were very appealing to drug cartels. For air surveillance, Costa Rica possessed only civilian Cessnas more than a quarter of a century old, and most of those planes were no longer air worthy. The Costa Rican government asked for help to repair and refit these airplanes with new surveillance technology, and in a remarkable volte-face, even President Arias personally made a plea for a Black Hawk helicopter. Perhaps without realizing it, Costa Rica was taking the first steps toward the creation of an air force, because no matter how valuable a Black Hawk can be in drug interception programs, this helicopter is essentially a combat weapon.[11]

Drug cartels had found the territorial waters of Costa Rica even more attractive than the open skies of the country for bringing cocaine from Colombia to the United States. The large amounts of cocaine leaving the Pacific Coast of Colombia frequently passed through Costa Rican territorial waters. After the abolishment of its military in 1949, the few vessels of its small navy went into a Coast Guard. Costa Rica considered the Coast Guard as one of its police forces, but with most of its boats not seaworthy or otherwise unable to function, this unit was largely useless.

Both Costa Rican and U.S. officials devoted a lot of effort and thought to improve the Coast Guard. As was usual for U.S. officials throughout the region, they preferred to decide what vessels were best for the country. As part of the Mérida Initiative, Washington generously decided to give Costa Rica the most sophisticated patrol crafts in existence. In reality the Costa Rican Coast Guard only needed small patrol boats able to board suspicious vessels without causing damage. Instead "these high speed, high performance Enduring Friendship boats require extensive maintenance hours, both electronically and mechanically. They also require a significant amount of technical coxswain training for high speed maneuvering."[12] And as a reflection of the paternalistic attitude present in U.S. attitudes to the armed forces of the Caribbean, "This would be like giving a 15-year [old] a Ferrari."[13]

Increased funding from the government and U.S. military assistance began to transform the maritime unit. U.S. aid restored the many patrol boats previously idled in port, while U.S. Coast Guard officers for the first time in six years trained personnel on how to use the new technologies. The U.S. government also provided additional craft and contributed to the construction of new buildings to replace the old installations of the Coast Guard academy. As the Costa Rican Coast Guard increased in size and personnel, more and more it started to resemble a small navy. Just like the efforts to improve the police on land and sea were laying the foundations for an army and an air force, so the steps to rehabilitate the Coast Guard moved Costa Rica closer to having a navy.[14]

Costa Rica was finding it increasingly hard to remain a non-militarist country. Its government was putting to the test the proposition of whether having an army was a matter of choice and not of necessity. And while militarization quietly crept into Costa Rica in the twenty-first century, at least it could take comfort in knowing that it was not the only country trying to live without armed forces. Panama, its neighbor to the south, was engaged in a parallel experiment to try to live without armed forces.

Panama: The Struggle to Demilitarize

"Costa Rica certainly views its non-militarized southern neighbor as a kind of offspring to its own non-military example,"[15] and Panama was the second country in Latin America to try to follow the Costa Rican example. Panama was able to abolish its army, the Panama Defense Forces (PDF) only after the U.S. invasion of December 1989 removed the local dictator, Manuel Noriega, from power. Dismantling of the Panamanian military did not take place immediately, because the collapse of the PDF dumped almost all police functions on American troops in the weeks following the invasion. U.S. commanders felt that the thousands of PDF prisoners could be rehabilitated and recycled into a semi-military police force. In the rush to restore order to Panama City, President Guillermo Endara abolished the old PDF (numbering 15,000 men) on 24 December 1989 and as a stopgap measure created a new Panamanian Public Force (*Fuerza Pública Panameña*). Almost all of the initial 11,500 recruits had belonged to the former PDF.[16]

Washington and in particular the Chairman of the Joint Chiefs of Staff, Colin Powell, believed that the Panamanian officers were so hopelessly corrupt as to make counterproductive any attempt to preserve the PDF under a new name. The first director of the Panamanian Public Force had to resign after a few weeks when he could not explain deposits of $1.5 million made out to his wife's, his son's and his own bank accounts. When the new government announced that it was not going to purge former PDF members from the new Panamanian Public Force, outraged civilians loudly protested. Signs, posters, and bumper stickers spread throughout the city urging a "yes to police and no to army."[17]

At this very propitious moment, President Óscar Arias published in the *New York Times* an editorial proposing that Panama abolish its army. Translated into Spanish and reprinted throughout Latin America, his editorial cogently and concisely expressed the sincere desire of almost all Panamanians. The final endorsement came a few days later when in an editorial entitled "Panama Needs No Army," the *New York Times* solidly supported the proposal of Arias. These newspaper articles finished persuading almost all official Washington that a Panama without an army was in the best U.S. interest.[18]

As President Arias stressed directly, Costa Rica could only point the way, and it was up to Panamanians to decide what to do with their military. Colonel Eduardo Herrera Hassan, the second director, wanted to preserve an important role for the Panamanian Public Force. But when in July 1990 he tried to create a labor union to obtain collective bargaining rights for policemen, the next month the government forced him to retire and he left for Miami. After his dismissal, civilians and not uniformed personnel served as directors of the police force. The refusal of nearly 3000 former PDF members to accept invitations to join the new force indicated that recruitment was not going to be easy, yet the government still insisted on carrying out an intensive purge of the officer corps. All former PDF officers above the rank of major were fired from the Panamanian Public Force, as well as 51 percent of majors, 40 percent of captains, and 28 percent of lieutenants. All remaining personnel took a three-week training course on human rights sponsored by the U.S. Department of Justice, and the supposedly rehabilitated police sported new khaki uniforms. And in a complete break with past practice under the Noriega dictatorship, the new police did not administer its budget and instead, for the first time, its salaries and expenses were under the control of the civilian Comptroller General.[19]

U.S. officials were so confident about the rehabilitation of the former PDF personnel that they ended the last of the joint U.S.–Panamanian police patrols on 28 November 1990. In reality, as the purge of officers continued, deep discontent and anxiety was bubbling inside the police force. In October, when Colonel Herrera Hassan returned from Miami to refute charges of being involved in coup plotting, he was arrested and imprisoned at the barracks in Naos Island at the south entrance to the Panama Canal (Map 12). Rumors of coup plotting did not die down and, in a display of power by the civilian government, the Comptroller General delayed paying the salaries of the police in November.[20]

On 3 December 1990 around 1600 hours three former PDF soldiers who had not been incorporated into the new Panamanian Public Force hijacked a tourist helicopter in Panama City and flew to Naos Island to liberate Colonel Herrera Hassan. The guards meekly allowed the Colonel to board the helicopter and fly to a police station in Panama City. He soon was making frantic calls to all former PDF officers seeking their support for a coup. When he learned that the headquarters of the Panamanian Public Force was lightly guarded, at 2000 hours he set out with heavily armed followers wearing gas masks to occupy the building. The U.S. army advisers present, including two colonels, radioed for help from nearby Southern

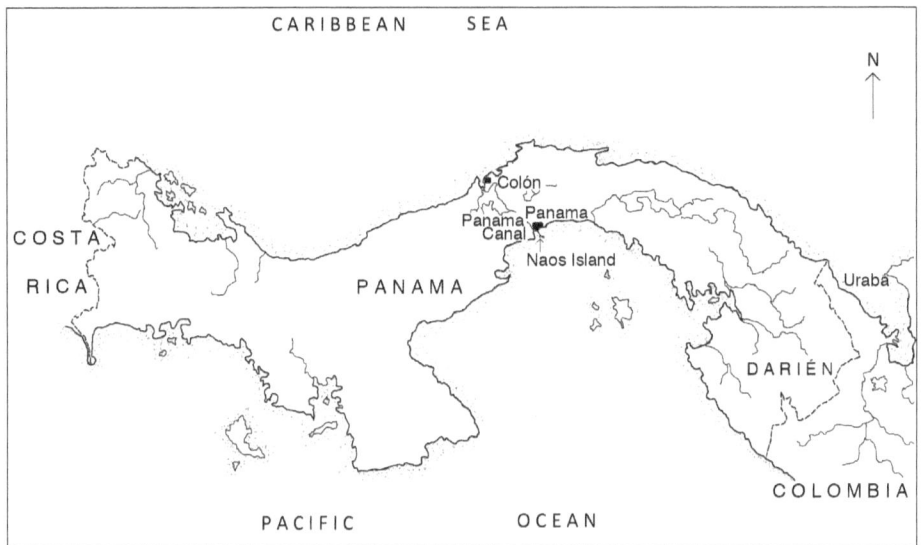

Map 12. Panama.

Command, and immediately U.S. troops were on their way. Any chance of rallying support gradually fizzled when Colonel Herrera Hassan realized that the police were waiting for the winner to emerge before openly backing him. As his coup attempt floundered, he tried to salvage the movement by taking his more than fifty armed followers on a protest march to the National Assembly to present a list of grievances at 2145 hours. By then five hundred U.S. soldiers surrounded the march, and they had authorization from the Endara government to disarm the rebels. After U.S. soldiers fired warning shots in the air and ordered all participants to drop their weapons, Colonel Herrera Hessan and most of his followers were handcuffed and taken to prison.[21]

This failed coup attempt finished souring the Panamanian public on having any military-style organization. The blatant failure of the Panamanian Public Force to stop Colonel Herrera Hessan and the embarrassing dependence on U.S. troops to preserve the civilian government made obvious the need for structural changes. The Endara government fired an additional 100 officers involved in the December coup attempt but had always hesitated to dismiss all former PDF personnel out of the valid fear that throwing these men into the ranks of the unemployed was a sure formula to create more unrest and disturbances. Actually, some rebels escaped the failed December coup and in 1992 began to carry out armed raids against civilian targets. At first U.S. officials and the Panamanian government feared that a guerrilla movement might be brewing, but concerns about an insurgency dissipated as the former rebels degenerated into criminals who robbed banks and assaulted private businesses for money. These criminal gangs had obtained automatic weapons from the looted arsenals and easily outgunned the police armed only with revolvers. Whenever police patrols clashed with these criminal gangs, heavily armed U.S. soldiers came to provide back up.[22]

In 1990 the U.S. government donated patrol cars, radios, and other equipment to the police force and also helped finance a new police academy as means to rehabilitate the Panamanian Public Force. But these measures were inadequate to stop the rampant crime wave sweeping Panama, and by 1992 the criminal gangs armed with assault weapons and riding fast cars were carrying out bank robberies and other violent actions. The police could not even pursue the criminals, because of 160 patrol cars the U.S. donated, lack of spare parts left only 58 in operation. The Endara administration seemed powerless to stop the criminals, and the bungling of the police reorganization left Panamanian citizens helpless in the face of the marauding gangs running wild in the cities.[23]

The crime wave gave the opposition party, the Democratic Revolutionary Party (*Partido Revolucionario Democrático*) or PRD, a great campaign issue to rally voters for the 1994 elections. The recovery of PRD was extremely remarkable, because this had been the political party of Ómar Torrijos and Manuel Noriega. The U.S. invasion of December 1989 smashed the PDF as a fighting force but left intact the vast grassroots organization of the PRD. The membership of this party consisted mainly of the dark-skinned working poor who made up the majority of the population in Panama. The white oligarchy of the Endara government had done nothing to break the grip of the PRD on the largest bloc of the Panamanian voters. And the imposition of privatization resulted in thousands of government employees losing their jobs and venting their anger at Endara. The party operatives carried out an intense campaign among the urban poor, and by emphasizing its original roots in the Torrijos era, largely succeeded in distancing PRD from both Noriega and the military. The party simultaneously campaigned for the election of its candidate Ernesto Pérez Ballesteros to the presidency and for the ratification of the constitutional amendments prohibiting the existence of any military establishment in Panama.[24] After winning the presidential election, Pérez Ballesteros kept the former PDF officers at a distance and "had forced the entire Noriega crowd below decks."[25]

The constitutional prohibition of 1994 seemed more than justified, because Panama, with army-free Costa Rica as its northern neighbor and separated from South America by some of the worst terrain in the world, did not seem to need an army. The only possible exception was the defense of the Panama Canal, but this was not a matter of immediate urgency, because until 31 December 1999 U.S. garrisons guarded the strategic waterway. Once attempts to extend the presence of U.S. troops into the twenty-first century failed, the agency operating the Canal created Special Forces to provide protection against any terrorist teams attempting to sabotage the locks or to disrupt the proper functioning of the waterway.[26]

The 1994 Constitution also authorized the National Assembly to pass laws establishing the police forces needed to maintain law and order inside the country, and the legislature duly created three new institutions. The largest and most important was the Panamanian National Police (*Policía Nacional Panameña*), which already in 2004 numbered 15,000 policemen, or about the same number as the abolished PDF at its maximum strength in 1989. The great majority of the police personnel no longer came from the discredited PDF because the government had purged almost all former officers from the ranks, and only enlisted men not directly connected to the abuses of the Noriega dictatorship remained in service. Although the purges of former PDF personnel ended with the election of Pérez Ballesteros to the presidency, they were still excluded not just from the new Panamanian National Police but also from holding any government positions. With lots of free time on their hands, the former PDF officers, the great majority of them of junior rank (major and lower), congregated at the offices of the PRD and planned a comeback after the year 2000. The former PDF officers devoted their superior organizational skills to put in long hours of political work among the masses. Having political mobilization as a peaceful outlet for the energies of former PDF personnel was a major factor in avoiding the problems Haiti experienced with its former army officers.[27]

The Panamanian government also inherited patrol boats and surveillance airplanes from the PDF. Because of the long coastlines of a country bordering on both the Atlantic and Pacific Oceans, selling off these air and naval assets and disbanding their hard-to-replace skilled personnel did not seem wise. But these military-like forces did not fit well with the mission of the Panamanian National Police, so the government decided to create separate institutions. The National Air Service numbering 700 individuals received the transport aircraft and helicopters of the former PDF, but by 2004 many of them were inoperable because of old age and poor maintenance. Of 14 helicopters, only five were airworthy and even these were too risky for U.S. Embassy personnel who were told not to board them. The National Maritime Service numbering 700 sailors received the decaying patrol boats of the former PDF. Its inefficiency provoked frequent public criticism and it soon earned a reputation as the worst law-enforcement agency in Panama.[28]

Promising a new and prosperous Panama, Martín Torrijos, the son of former Panamanian ruler Ómar Torrijos, won the presidential elections of 2004. His enthusiastic followers cheered wildly and honked car horns in a popular outpouring of emotion that had not occurred in the country for decades. When he took office on 1 July the most important item on his agenda was the enlargement of the Panama Canal to make possible the transit of the biggest vessels in the world. In a referendum, President Torrijos obtained approval for a construction project to enlarge the Canal, but his populist initiatives gradually withered. The initial idealism and illusions soon gave way to bitter disappointment.[29] And nothing damaged the reputation of the Torrijos administration more than its insistence on creating military-style institutions in Panama.

Both the Torrijos administration and U.S. officials underestimated the intense hatred that the PDF still generated among virtually all sectors of Panamanian society. Believing that the anti-military feelings had subsidized since 1989, the Torrijos administration proposed on 6 October 2005 to combine the National Maritime Service and the National Air Service into a new Coast Guard. Immediately the population feared that with this merger Torrijos was trying to remilitarize Panama. The use of the word "Guard" was unfortunate, because it brought back memories not only of the National Guard, the original name of the hated PDF but also the name of the intensely hated National Guard in Somoza's Nicaragua. The outpouring of hostile reaction to the proposal to combine these two services caught officials completely by surprise, particularly because Panamanians have a reputation for historical amnesia. But the hatred of all things military was so intense, that anything resembling the defunct PDF immediately provoked passionate hostility.

Contrasting the anti-military feelings in Panama with the pro-military sympathies in Nicaragua helps to explain this hostility. The Nicaraguans had loathed the National Guard so intensely that they were willing to revolt at any cost, and even unarmed citizens preferred to die rather than endure any longer that hated force of the despised tyrant Somoza. But once the National Guard was destroyed in Nicaragua in 1989, the thought of abolishing the military was simply inconceivable in that country. As a matter of fact the hatred against the National Guard was particularly intense because that institution had betrayed its patriotic roots by turning into a pliant tool to carry out bloody and savage repression for the tyrant. But the army was such an integral part of Nicaraguan national consciousness, that to destroy it meant undermining Nicaraguan identity. Instead the military had never held a key role in the formation of Panamanian nationalism, and the Panama Canal seemed to provide an adequate substitute symbol to foster a sense of Panamanian identity.

The proposal to combine the air and maritime services generated a very heated controversy, even though valid reasons strongly argued for this consolidation as a cost-cutting measure to improve the efficiency of both services. What intensified the passions was that this proposal surfaced just as the Torrijos government was giving many jobs in the bureaucracy to former PDF officers. As reward for their many years of dedicated field work to rehabilitate the PRD, the political party of Torrijos and Noriega, so many former PDF officers went to work for the government, including even a commander of the infamous Dignity Battalions or "Ding Bats," that the "joke making the rounds in Panama is that the only ex–PDF officer who hasn't found a job in the Torrijos government is Manuel Noriega."[30] Yet this widespread entry of former PDF officers into the government gave them a great opportunity to rebuild their lives and prevented the recurrence of revolts as happened in Haiti.

In addition, open displays of militarism reappeared under Torrijos and angered the Panamanian public even more. Already in 2004 the anthem of the abolished PDF was played in public ceremonies for the first time since the fall of Noriega in 1989. At parades policemen of tactical units appeared with their faces painted in camouflage colors and marched in full combat gear while carrying automatic weapons and rocket propelled grenades. Officers began using titles no longer existing such as the rank of colonel of the old PDF. The most egregious example was the chief of the National Maritime Service, who strutted around in the white uniform of a Rear Admiral, even though the rank of admiral no longer existed. And in a final confirmation of the government's intention, Torrijos appointed an active-duty uniformed officer as Director of the Panamanian National Police when the legislation stipulated that only a civilian could hold that office; the Torrijos administration ingenuously tried to deflect criticisms by stating that the new Director was holding the position only as an "interim" appointment and thus did not violate the letter of the law.[31]

The storm of criticism forced the Torrijos administration to abandon its proposal to combine the air and maritime services, but this was only a tactical withdrawal. The Torrijos administration regrouped and decided to try again with an even more ambitious agenda for militarization in 2008. The firestorm broke out when the Panamanian press claimed that the purpose of a recent trip to Washington, D.C., by Torrijos was to secure U.S. approval for the creation of armed forces in Panama. "The implied and often stated undercurrent of the anti-militarization camp is that there is a stealth project to remilitarize Panama being carried out by the U.S. and their allies" in the Torrijos administration.[32] And the fears of U.S. support were not totally groundless, because the Southern Command did support the reestablishment of the military. U.S. diplomats, who were much more sensitive to local concerns had the final word, and although they opposed the reestablishment of an army, they saw no danger in the modest proposals and considered them beneficial to improve police performance in Panama.[33]

In June 2008 the National Assembly granted President Torrijos extraordinary powers to legislate on the security services. Granting similar blanket authorization has been popular in Latin America with politicians who thus avoided having to cast potentially embarrassing votes. In late August the executive issued the measures reforming the security structure. In the most important innovation, one decree combined the National Air Service and the National Maritime Service into the new National Aero-Naval Service. In response to sensibilities about the word "Guard," the text avoided the emotionally loaded term of Coast Guard. Another decree created the National Frontier Service (SENAFRONT) as a separate agency to provide police protection in border regions. This unit already existed as a branch of the Panamanian National Police, but by making SENAFRONT an independent agency, police no longer had to rotate between service in the Darién jungle and street patrols in Panama City. In a separate decree, the executive obtained the authority to appoint uniformed personnel to hold the position of Director of the National Panamanian Police.[34]

These changes did not bring any notable reduction in the crime wave afflicting the country yet were of a sufficient scale to arouse the fears of Panamanians about militarization. Scandals about fraud and illicit payments helped to confirm the popular belief that the Torrijos administration wanted to reestablish the corrupt practices of the old PDF. The revelation that the "admiral" in charge of the maritime service had been working with drug cartels caused quite a sensation. The Torrijos administration hoped that the new National Aero-Naval Service, under the command of a supposedly honest aviation officer, would help to restore the tarnished prestige of those units. U.S. officials were not so sure, and even before the changes of August 2008, the U.S. government financed the improvement of the Riverine Patrol and Action Unit or UMOF, a force within the Panamanian Police. This riverine police proved quite effective in patrolling the many rivers of the Darién jungle, the easternmost region of Panama (Map 12).[35]

The success of the riverine police in far-distant Darién did not matter much to the Panamanian citizens experiencing in 2008 a sharp increase in violent urban crime. Recruitment for the poorly compensated police proved difficult, and the job no longer brought the power and opportunities for bribes as in the days of dictator Noriega. As the drug cartels transitioned their cocaine shipments from one long haul between Colombia and the United States to many short trips, Panama became a rest stop for the cocaine trade by 2008. As a consequence, crime violence and gang activity spread widely in 2009.[36] The Torrijos administration, already reeling from the remilitarization controversy and other scandals, dragged its candidates down to defeat in the next elections. The party of Torrijos and Noriega lost control of the National Assembly while the opposition candidate Ricardo Martinelli became the next president of Panama.

Upon taking office, President Martinelli ratified the measures the Torrijos administration had taken to restructure the police forces as indispensable first steps. If anything, the police needed more strengthening to cope with the crime wave in Panama's cities. What also was becoming obvious was the inability of SENAFRONT, the border police, to deal with the incursions of guerrillas from Colombia into the Darién jungle. The Revolutionary Armed Forces of Colombia (*Fuerzas Armadas Revolucionarias de Colombia*) or FARC for decades had used this frontier region as a transit point for smuggling weapons into the country. The FARC commander of the 57th Zone, known as "Silver" diversified into shipping cocaine to finance the arms purchases of the guerrillas. Occasionally FARC detachments clashed with SENAFRONT in Darién, and one serious engagement took place on 11 December 2006. Lack of helicopters prevented SENAFRONT from pursuing the retreating FARC detachment, and to try to solve this deficiency, the Panamanian government sought to lease helicopters from Colombia.[37]

Without an army, Panama had to turn both to Colombia and to the United States for help in coping with FARC. It was clear that SENAFRONT was no match for the Colombian guerrillas, because in spite of its military appearance it was at best a police force suited for patrolling and not for combat. To try to track down 57th Zone commander "Silver," SENAFRONT starting in 2009 had been carrying out a strategy with Embassy support, to cut off FARC supply routes, restricting the amount of food that enters the area, and increasing patrols in choke points to make it more difficult for FARC to operate ... it is the consensus view of Post Country Team that SENAFRONT does not have enough men, training, equipment or logistical capability to militarily confront FARC, and that the harassment strategy is the most forward leaning strategy that SENAFRONT can maintain.[38]

The U.S. Embassy in Panama planned to use the presence of large numbers of American troops on humanitarian missions in Darién from April to September of 2010 to launch an attack on "Silver." This U.S. deployment in Panama was the largest in nearly twenty years, and the U.S. Embassy under the guidance of DEA planned to use attack helicopters to strike a deadly blow on FARC. The U.S. Embassy did not want to realize that such a large display of American firepower was sure to anger many Panamanians.[39]

Unknown to the U.S. Embassy in Panama, a parallel process to take out "Silver" was taking place. When at a summit meeting Colombian President Álvaro Uribe Vélez offered to help Panama deal with FARC incursions, President Martinelli asked for cooperation with the Colombian police. The Director of the Colombian National Police Óscar Naranjo visited Panama on 9 December 2009, and after discussions with him President Martinelli authorized Colombian forces to attack FARC units which the authorities expected to gather in Darién for the holidays at the end of that year.[40] Because the U.S. government considered Panama as its exclusive preserve, "Embassy Panama is deeply concerned about this turn of events. A Colombian cross-border attack would not serve U.S. interest in Panama ... it would be unpopular with the Panamanian public and would arouse deep-seated Panamanian mistrust of Colombia and its intentions."[41]

This sudden awareness of local sensibilities was quite remarkable and actually was rather narrow because U.S. diplomats failed to grasp that Panamanian fears about U.S. intentions were greater than any possible mistrust of Colombia. For its part, the Colombian government as an unconditional ally of the United States was not willing to intervene without the full approval of Washington. The Director of the Colombian National Police

> repeatedly stressed that Colombian cooperation with Panama would be completely transparent and coordinated with the U.S. Government. The Government of Colombia would not take any action with Panama that would interfere with U.S. government operations in Panama or make the U.S. government uncomfortable. If the U.S. government did not completely accept the Colombian National Police working with Panama, Naranjo would not do so—the relationship with the U.S. government was too important.[42]

The U.S. Embassy in Panama concluded that having two uncoordinated plans directed from several capitals was too risky and decided as a precaution to withdraw from this scheme. And because Washington failed to take a leadership role, the inability to coordinate actions among the three countries left FARC free to operate at will in Darién.[43] Operations to destroy "Silver" inside Colombia failed, and Panama, without an army, could not even attempt a similar campaign. SENAFRONT was too weak to confront FARC but was sufficiently militaristic to scare Panamanian civilians. When the Martinelli administration brought SENAFRONT units to repress street protests in Colón (Map 12) and then deployed the border police in the parade for Independence Day on 4 November 2012, many Panamanians criticized this display of power. Panama did not possess an army, but its citizens remained bitterly opposed to anything that threatened to reconstitute the police forces into a military institution resembling the hated and abolished PDF.

Haiti: Flawed Demobilization

The third country having a population of more than a million inhabitants and lacking an army in the Caribbean region was Haiti. Since the establishment of the Haitian Republic in 1804 as the second independent country in the Western Hemisphere after the United States, the army always had been an essential component of Haitian life and national identity. But a history of repeated involvement in politics finally caught up with the Haitian army. On 6 January 1995 in a momentous decision, President Jean Baptiste Aristide by executive decree abolished the Haitian army of 6,200 officers and soldiers. In the past such an action not only was inconceivable but would have immediately produced a coup d'état to overthrow any civilian president foolish enough to challenge the Haitian army.[44]

Nothing happened to President Aristide because nearly 20,000 U.S. soldiers defended his regime. Previously President Bill Clinton had sent U.S. troops to occupy Haiti on 19 September 1994 precisely to restore Aristide to office after a military junta had overthrown the popular Haitian leader on 30 September 1991. The U.S. occupation, code named "Operation Restore Democracy" was really directed at the armed forces of Latin America and aimed to send the message that the international community no longer accepted military coups as a way to remove elected civilians from power. The entry of American troops in September 1994 marked the start of a nearly uninterrupted U.S. involvement of varying intensity in Haiti over the next twenty years.[45]

As has often been the case in many foreign interventions, the U.S. government could enter easily but found it hard to depart. As U.S. involvement in Haiti dragged on, Republicans in the United States and candidate George W. Bush enjoyed bashing the Clinton administration for its project of "nation building." And the United States could not have picked a worst place than Haiti to attempt nation building. Unlike the rest of Latin America, Haiti was the only country that was still overwhelmingly agrarian, with more than two-thirds of the population living in the countryside. The rural population did not produce profitable export crops but eked out a living in subsistence farming. Not surprisingly, Haiti was the most destitute country in the Western Hemisphere, exceeding by a wide margin Nicaragua, the second poorest country. The figures for Haiti were abysmally low even when compared with other island states in the Caribbean, and in the most significant comparison, Haiti's per capita income was a small fraction of that of the bordering Dominican Republic.[46]

The deep poverty and rampant misery throughout the Haitian countryside seemed to constitute irrefutable arguments to dismantle the Haitian army in January 1995. President Aristide originally had intended to maintain a small army of only 1500 soldiers and officers. But he always harbored a deep distrust of the officers who had overthrown him. When former President of Costa Rica Óscar Arias came to Haiti in November 1994 to propose the abolishment of the army just as Panama had recently done in December 1989, Aristide needed little persuading.[47] However, eliminating the army opened the country to two major problems.

The first problem came when the government placed the entire burden of maintaining order solely in the hands of a totally incapable police. Because of the blatant inefficiency and rampant corruption of the police, many of its functions by default had passed to the Haitian army during previous decades. Maintaining order in the countryside where the majority of the population lived had been primarily an army function, while soldiers stationed in the eastern border controlled the vital trade routes with the Dominican Republic. But starting in 1995 a new apolitical police force was supposed to replace the disbanded army. The government promulgated a law on 14 December 1994 creating a new Haitian National Police, and international assistance covered the start-up costs of the new law-enforcement institution. To join the ranks all applicants first had to graduate from a newly reconstituted police academy, but in an early indication of serious obstacles, all the former policemen failed the admissions exam. Finding suitable and qualified personnel for the new Haitian National Police became a frustrating task, and the goal of creating a force of 20,000 policemen proved excessively ambitious. The force barely totaled 5000 by the end of 1995, and the highest it ever reached was 10,000 policemen in the year 2000, a number still vastly insufficient to maintain order in the country. Scattered across the country in dilapidated stations and rarely armed with more than an old revolver, these policemen were usually the sole official presence in many neighborhoods and villages. And when the last American troops withdrew in January 2000 once the U.S. government tired of nation building, Haiti was rapidly reverting to a failed state.[48]

The second major problem was the appearance of a large number of unemployed and disgruntled former army personnel. Although the deposed junta leaders went abroad to enjoy a life of luxury on their foreign bank accounts, the majority of officers and the soldiers depended on the army for their livelihood in an economy offering no other alternatives for comparable employment. In spite of the acute shortage of policemen, the government had allowed only 1500 former soldiers to join the new Haitian National Police, leaving over 4000 officers and soldiers unemployed and feeling completely betrayed. And to make matters worse, even the presidential decree abolishing the army was illegal, because the Haitian Constitution, which the Americans claimed to have come to restore, directly stated that the army was an essential institution of the Haitian nation. U.S. and U.N. planners forgot the principle that it is harder to demobilize soldiers than to mobilize them. With such massive poverty in Haiti it seemed unjust to privilege former soldiers for preferential treatment in training, education, or receiving loans if not outright assistance. The simplest solution, giving a large severance payment to younger soldiers and awarding pensions to those near retirement age, seemed inconceivable to U.S. and U.N. officials often passionately determined to seek revenge for previous human rights violations.[49]

With nowhere else to turn, the unemployed soldiers knew that they only had to wait for the opportunity to strike once U.S. troops made their inevitable departure. The return to office of Jean Baptiste Aristide for another term as president in 2000 after he won disputed elections was too much for the former army officers who regarded Aristide as their worst enemy for having abolished the army in 1995. Quietly the former officers began obtaining armament and recruiting soldiers to strike a blow against Aristide. The central plateau of Hinche (Map 13) became their base of oper-

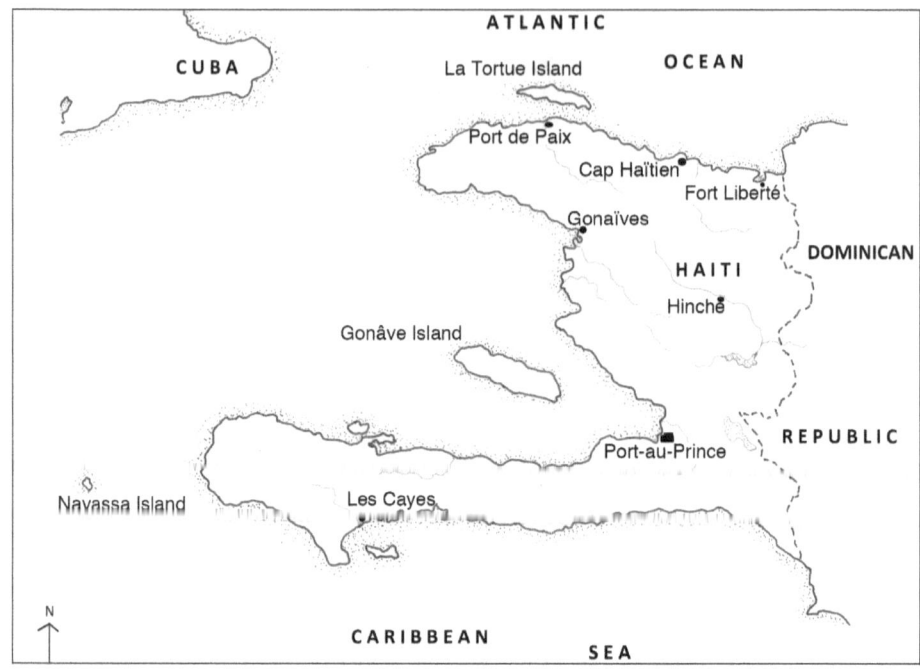

Map 13. Haiti.

ations, and the officers became the real authority in this remote region of Haiti. Demonstrations began in late 2003, and in February 2004 the protests became an open revolt when former army personnel pushed from Hinche to the coast and joined with disillusioned supporters of Aristide to take over the coastal city of Gonaïves.[50]

The revolt spread rapidly and soon the rebels captured Cap Haitïen until even President Aristide admitted that "the north of Haiti was out of control."[51] The police detachments, usually armed with no more than revolvers, were no match for the rebels who had acquired assault rifles. After sporadic shooting, the police detachments simply melted away when rebel groups appeared. On Monday March 1 the rebels entered Port-au-Prince and after taking over most of the capital, were heading to the presidential palace to demand the resignation of President Aristide and the restoration of the Haitian army. The Bush administration could not allow such a blatant repudiation of U.S. policy in the region and reluctantly sent American troops back to try to salvage the situation. French, Canadian, and Chilean troops promptly joined the effort, and soon the U.S. government obtained United Nations authorization to structure a peacekeeping force.[52]

The U.S. government rescued Aristide but realized that his rule was beyond saving. Aristide went into exile, and his departure met one of the two rebel demands. But the U.S. government vehemently opposed the second rebel demand that the Haitian army be reestablished and instead intensified the persecution of former soldiers. U.S. officials blamed the sudden collapse of the Haitian police on the presence within its ranks of 1500 former army personnel and in reprisal conducted an intensive purge that reduced the number of policemen to barely 3,000 by the end of 2004.[53]

For the former army personnel, the U.S. intervention meant a return to the previous waiting game. On one side, the rebels were not so foolish as to challenge in combat the vast firepower of the U.S. military. For their part American officers did not want to provoke the rebels into starting anything resembling the insurgency already bogging down U.S. troops in Iraq. A stalemate emerged as Haiti gradually drifted into a condition of lacking effective governing structures, while former army officers controlled regions such as Hinche and also had outposts outside the main cities of Haiti. These ex-officers pressed for the restoration of the Haitian army, a proposal that faced immense opposition from civilian elites. To bring pressure on the government, in 2006 a Haitian senator threatened to block approval of the national budget unless it included an entry for the reestablishment of the army.[54]

On 29 July 2008 former officers occupied abandoned military installations as the start of three days of protests demanding either the reestablishment of the army or the payment of their back salaries and pensions. Counterdemonstrations threatened to spiral out of control and to result in violent clashes, and just in time on 31 July Haitian officials persuaded the ex-officers to abandon the installations in exchange for sympathetic consideration of their needs. Even after a year the slow-moving government failed to respond to the demands, when the devastating earthquake of January 2010 shattered Haiti and in particular Port-au-Prince.[55]

The earthquake ended any chance of rebuilding the Haitian National Police as an effective force, but even before that natural catastrophe the institution had fallen hopelessly behind in trying to reach the goal of 12,000 policemen by 2012. During the January 2010 earthquake, at least 71 policemen died and 471 were missing. For the nearly impossible task of patrolling a ravaged Port-au-Prince only 4,235 policemen remained and less than that number for the rest of Haiti. The earthquake damaged the buildings of the police academy, and the government lacked funds to start another graduating class. The United Nations estimated that Haiti needed a police force of at least 20,000, but in September 2011 the number of policemen barely reached 10,000.[56]

By 2011 it was clear that all attempts to create an efficient police force in Haiti had ended in complete failure. As a realistic alternative President Michel Martelly floated a proposal to restore the army in September 2011. Of course, Haiti could not afford the $95 million dollar price tag of creating a new army, and for its implementation the proposal depended on contributions from foreign donors whether official or private. One part of the proposal was very promising because it called for the distribution of $15 million dollars among the former military personnel as payment for back salaries and pensions.[57]

At last many U.S. and U.N. officials realized that taking care of the former army personnel was the key to ending the periodic outbursts of violence in Haiti. Some of the $15 million dollars assigned for compensating former army personnel trickled in from foreign donors, but as usual, the Haitian government was slow in registering those eligible for the payments. Protests late in 2012 demanded a speedy processing of the compensation packages, and it seemed that after nearly twenty years, Haiti finally was implementing a difficult demobilization.[58]

As the former soldiers faded away, the need for having an army remained critical. The first contingent consisting of 41 army engineers began to operate under a newly created Defense Ministry in February 2014. They were construction workers in uniform who had received training in Ecuador. A second group of 200 students left to train in Brazil in June 2014 also as army engineers.[59] By beginning the reconstruction of its army with the least controversial elements, the Haitian government hoped to prevent the human rights abuses of the past. After two decades of enormous expenditures, the formula of having only a national police force had been tried only to have the country become a failed state as the Somalia of the Western Hemisphere. To function Haiti needed a set of national institutions, and it was hard to avoid the conclusion that an army formed an essential part of an effective governmental presence throughout its territory. Clearly Haiti had not come even close to matching the relative success of demobilization in Costa Rica and Panama.

13

An Inconvenient Ally: The Armed Forces of the Dominican Republic

As generous as Dominican Republic wants to be, it cannot assume Haiti's poverty, because it would fall to the same situation.—President Leonel Fernández[1]

The Military of the Dominican Republic

The U.S. government never accepted the military of the Dominican Republic. The preferred U.S. option for the Caribbean traditionally has been for a constabulary force combining police functions with rural patrols. In the Dominican Republic the chosen instrument was the National Guard the U.S. Marines created in the early 1920s. But when the dictator Rafael Trujillo transformed the National Guard into an army and created a separate national police force, the armed forces took permanent root in the Dominican Republic. Megalomania of the dictator certainly was a major factor in the creation of a separate army, but Trujillo also realized several things about his people that U.S. officials have never fully grasped. First uniforms and weapons were extremely popular in the Dominican Republic just as in Cuba, its closest cultural sibling. Secondly, Dominicans regarded the military as the indispensable bulwark to defend the country from a Haitian invasion. Trujillo was the first to tap into these two sentiments that survived long after assassination ended his rule in 1960. Until today the military have remained one of the most respected institutions and only the country's baseball players can rival its popularity.

What was most galling to U.S. officials was that without their approval the Dominican Republic preserved and frequently expanded its military under both Trujillo and his successors. For U.S. diplomats, the Dominican military was at best meaningless and "its primary function has been, and continues to be, to provide employment."² But whenever the U.S. government sensed an opportunity to push for the elimination or at least the reduction of the military, formidable barriers immediately appeared. "Reduction in the number of uniformed personnel is seen as politically difficult, in part because about 30 to 35 percent of the uniformed forces is actually employed in civilian government offices or in the private sector, often as private residence guards, chauffeurs, and other personal service assistants."³

Perhaps the last real opportunity to abolish the Dominican military in one blow came during the U.S. occupation of 1965–1966. But because right-wing elements inside the military were indispensable to support the U.S. occupation, reluctantly Washington had to ally itself with these officers and had no choice but to extend assistance to the generals and even to cover the payroll until the civilian government resumed normal functions. After the U.S. withdrawal, American officials sent the not-too-subtle message of channeling almost all assistance to the police as the chosen instrument to wage the campaign against Communism during the Cold War. The Dominican government continued to support its own military, but shortage of funds blocked most acquisitions of weapons and equipment.⁴

By 2004 the armed forces of the Dominican Republic at 42,000 members were, with the exception of Socialist Cuba, the largest in the Caribbean islands and in Central America. In comparison with a country the U.S. found most vexing, Venezuela had a surface area almost twenty times larger but an army only twice the size. But the large military of the Dominican Republic could be favorable to U.S. interests. The military made sure that peace and order prevailed in the country, thus making it suitable for trade, investment, and tourism. When in 2006 the violence from Jamaica and Haiti seemed to be spreading to the Dominican Republic, the army deployed 3000 soldiers to join the police in patrolling Santo Domingo, the capital (Map 14). The timely intervention by the army prevented the homicides from reaching the horrendous rates in nearby Caribbean countries and spared the U.S. government of having to deal with lawlessness and the collapse of authorities.⁵

Also, having a large military proved convenient for the George W. Bush administration when it was struggling to find allies to participate in the "Coalition of the Willing." In one of its few successes in Latin America, the Bush administration persuaded a pliant government to deploy a battalion to help in the occupation of Iraq in 2003. The success proved short-lived, however, and the Dominican public angrily opposed this involvement in a Middle Eastern country. In a futile attempt to influence the presidential

13. An Inconvenient Ally: The Armed Forces of the Dominican Republic

Map 14. Dominican Republic.

election of 16 May, the outgoing administration recalled Dominican troops from Iraq in April 2004. The nationalistic backlash was one of the factors contributing to the electoral victory of Leonel Fernández who took office as president on 16 August 2004.[6]

The deployment in Iraq had been a small distraction for the Dominican military whose fundamental mission always remained the defense of the country against Haiti. This reason became even more important with increases in the numbers of Haitians crossing the border illegally. Since the end of the nineteenth century Haitians have come to cut sugar cane during the harvest season, and since the 1960s they have also found work in construction, tourism, other agricultural sectors, and even in manufacturing.[7] Out of a population of nearly 10 million inhabitants in the Dominican Republic in the first decade of the twenty-first century, probably as many as a million were Haitians, and "the public perceives that there are too many Haitians in the country."[8] In reality, popular views were often in direct contrast with the attitudes of private businesses. Companies often preferred to hire illegal Haitians not just to exploit them by paying lower wages, but also because businessmen considered them more hard working and responsible than native Dominicans. Instead the traditional public perception, reinforced by the periodic recurrence of isolated actions such as one Haitian with a machete hacking to death a Dominican, was "to dismiss the Haitians as a savage people capable of unpredictable acts of cruelty."[9]

Anti-Haitian feeling was inevitable in a country celebrating as its national holiday the date of its independence from Haiti in 1844. But are countries perpetually trapped in their histories? The new president of the Dominican Republic, Leonel Fernández, with great optimism decided to make a first move to end the tense relations with the bordering country. He made a brief and low-key visit to Port-au-Prince (Map 13) on 12 December 2005, but shortly after he arrived at the presidential palace, an angry mob started to form outside. The protesters first lit bonfires, later burned tires, and appeared determined to kill Fernández. When the Haitian policemen melted away, the visitors decided to abandon the presidential palace at once. Four hovering Dominican helicopters fired machine guns to clear a path through the mob to allow the motorcade of Fernández to reach the airport. The visiting delegation grabbed the first available flight out of that dangerous capital.[10]

Wars have started over lesser incidents, but the peaceful president had no intention of using the riots as a *casus belli*. A chastened Fernández explained that "he had been the first Dominican president to visit the country in 65 years, … and perhaps the last for the next 100 years."[11] The news did cause an outrage in the Dominican Republic: "an event intended for amity instead played in Dominican media as an affront and an attack by Haiti on all things Dominican … the Haitians hate us, and those attacks were just more evidence of their anti–Dominican attitude."[12] The squalid poverty and the hopelessness Fernández observed in his short visit made a profound impression on him, so he continued to strive for improving relations with Haiti. The near killing of the Dominican president sent the Haitian elites and the government into panic, because the last thing they wanted was to face hostility from the wealthy and powerful neighbor. The Haitian government publicly apologized orally and in writing over the unfortunate incident, but the incoming president René Preval realized that this was not adequate satisfaction. As soon as he won the election, he broke precedent and made the Dominican Republic his first foreign visit as president-elect. President Fernández enthusiastically welcomed his Haitian counterpart, but aides clearly explained that Preval could not expect a return visit. In diplomatic terms this signified an insult, and the Haitian President had no choice but to swallow his pride and that of his country to compensate for the riots of Port-au-Prince. The visit to Santo Domingo was successful, and the Haitian President-elect received every possible honor; the fact that he was fluent in Spanish made the visit all the more successful. The Dominican people behaved in a completely courteous way to their Haitian guest as if trying to emphasize the contrast with the mobs on a rampage in Port-au-Prince.[13]

Presidential visits could not solve the problems of Haitians entering illegally the Dominican Republic. The illegal immigration grew rapidly after the disbandment of the Haitian army in January 1995. Up to that moment the Dominican army delegated to the Haitian army the task of reducing the flow of illegal immigrants. Haitian officers were extremely willing to please the Dominican army because carrying out this task gave them the opportunity to extort the Haitians trying to enter the bordering country. But once the Haitian army ceased to exist in 1995, the Dominican army was ill-prepared to fill the void because its training and equipment focused on stopping a conventional Haitian invasion and guerrilla infiltration. It did not have the abilities to police the many civilians who were supposedly crossing the border for trade and temporary employment but then overstayed their welcome and became illegal residents.[14]

President Leonel Fernández asked the U.S. government for advice on how best to stop the flow of illegal Haitians into the Dominican Republic, and in response to his request, a multi-agency task force prepared a report. Not surprisingly, U.S. officials promptly took advantage of the opportunity to make another attempt to reduce the Dominican army as a first step toward its abolishment. U.S. officials in the multi-agency task force proposed the creation of a separate civilian agency loosely patterned on the U.S. Border Patrol. The proposed agency also offered the attractive incentive of larger customs revenue from better border enforcement of existing laws. President Fernandez liked the idea of a specialized body to control the frontier, but he knew that in all matters concerning Haiti, the Dominican people trusted only the military. In 2006 he decided to create CESFRONT (*Cuerpo Especializado Fronterizo*) as an integral part of the Defense Ministry but not as a civilian agency. Uniformed personnel in the chain of command comprised the members of this new body, with 60 percent coming from the army and the remaining 40 percent divided equally between the navy and the air force.[15]

CESFRONT sent detachments throughout the border and concentrated its efforts on the four most heavily transited crossing points at Dajabón, Comendador, Jimaní, and Pedernales (Map 14). In September 2007 President Fernández suddenly ordered the immediate deployment of CESFRONT ahead of schedule when he received reports about the smuggling of U.S. rice into the Dominican Republic. The initial deployment seems to have gone well, except in Dajabón. At that site the soldiers began beating both Haitians and Dominicans trying to cross the border, and a video showed soldiers extorting payments from Haitians who wanted to enter the Dominican Republic. Complaints from civilian officials resulted in punishing and removing two officers and eight enlisted men who had committed those abuses. To prevent corruption and favoritism, CESFRONT adopted the practice of rotating all personnel in direct contact with civilians every 50 days, yet in spite of this precautions, soldiers could not resist demanding bribes from Haitians trying to cross the border.[16]

CESFRONT concentrated its efforts on Dajabón as the most important site between the two countries. Even though the easternmost suburbs of Port-au-Prince were less than 40 kilometers away from the Dominican border crossing at Jimaní, the proximity of towns of comparable size on both sides of the border made Dajabón the central point for commerce between the two countries. CESFRONT policed a curious trade on the border. As expected, the agriculturally prosperous Dominican Republic generally exported food staples and fresh produce to its neighbor. Haiti, besides its famous arts and crafts, exported clothing, liquors, and many industrial items presumably smuggled in from abroad for re-export into the Dominican Republic. Also smuggled in were some foodstuffs preferred by Dominican consumers, such as U.S. rice and Dutch cheeses. Market days on Monday and Fridays were particularly trying tests for CESFRONT, when 20,000 Haitians poured over the border into Dajabón to buy foodstuffs and sell merchandise. On those days the visitors did not need any border passes, because the troops set up a security ring around the town and then made sure that at 1600 hours all the Haitians returned back across the border.[17]

CESFRONT also helped to diffuse incidents on the border. When Haitian cattle rustlers kidnapped two Dominican construction workers, Santo Domingo sent additional troops to the border. Meanwhile CESFRONT obtained the release of the hostages in coordination with Haitian officials before the incident turned into an international crisis. Although U.S. officials preferred to have a civilian agency, CESFRONT was working, and to enhance its performance U.S. government employees came to teach the soldiers at a specially created training facility. American instructors imparted technical and investigative techniques, while the Dominicans also received courses in Creole, the dialect of French spoken in Haiti. And the additional training was certainly needed, because besides the usual smuggling of merchandise and the illegal trade in guns and drugs, human trafficking in women and children was becoming a more widespread criminal activity on the Haitian-Dominican border.[18]

CESFRONT was policing the Creole-speaking Haitians crossing the border and participated in the deportation of illegal residents, but the Dominican military always realized that security in the border required additional steps. CESFRONT was a complement to the bases and outposts existing in the border zones since the 1930s. Occasionally a surge in Haitian immigration threatened to overwhelm both CESFRONT and the regular garrisons, and in response the government hurriedly dispatched thousands of additional soldiers on temporary duty to man outposts on the frontier. A pathetic case came in June 2008, when food shortages in Haiti drove many starving peasants toward the border with the Dominican Republic, and to prevent a massive exodus the military reinforced its guard posts at all likely entry points.[19]

The military did not neglect taking more permanent measures when reports confirmed that illegal Haitian immigrants already occupied entire districts as squatters near Dajabón and Pedernales. The fertile soils of the Dominican Republic, in contrast to the deeply eroded slopes in Haiti, traditionally proved an irresistible attraction to starving peasants. As part of a "Dominicanization" program, Trujillo the dictator had replaced all French place names with Spanish words and expelled all Haitians in 1938. But because population density in the border provinces remained low, Haitians started to drift back after the death of the feared dictator in 1960. It became clear that patrols, checkpoints, and deportations were inadequate to keep out Haitian squatters from the border provinces. As a solution the army began a pilot project to turn reservists and active duty personnel into colonists. Soldiers who agreed to settle with their families in the border districts received considerable assistance from both private businesses and the military. The partnership between government and private business aimed to increase the number of Dominicans living in the fertile vacant lands near the border. To help the Dominican peasants already in the region, President Leonel Fernández issued a decree in 2009 requiring all graduates from the military academies to spend a year in the border provinces repairing private houses, fixing public installations, and teaching residents to read and write.[20]

This requirement of a year of service in the border for graduates of the military academies was one of many changes President Fernández was struggling to introduce in the armed forces. His attempt to imitate Hugo Chávez in Venezuela by giving the vote to soldiers and officers ran into obstacles.[21] President Fernández likewise was unsuccessful in trying to persuade the high command to reduce the excessive rotation of senior officers. A frustrated U.S. Embassy reported that five of six brigade commanders in the national army were replaced, although these commanders had only been in office six months. Poor continuity, as evidenced by these changes as well as the naming of a new CESFRONT commander, remains a systemic problem that hampers development of the armed forces as a professional institution and complicates Embassy's efforts to improve cooperation.[22]

Much more successful was the campaign of President Fernández to open up the armed forces to women. Until his presidency, very few females joined the military and they could serve only in the medical and administrative branches. Most countries of the world had followed a gradual approach of opening up the sectors of the military one at a time to women. This slow approach was not without contradictions, for how could a woman be unsuited one year for one sector and then the next year qualify as a candidate? President Fernández insisted on a simple and direct solution to female participation. In 2008 a new code of military discipline stipulated complete equality between the sexes in the military. It was then up to the individual women and their commanders to decide which individuals were best suited for a position.[23]

Reaching officer rank was not easy, and of the thirty women who entered the third mixed gender class of the military academy, only six completed their graduating. A first lieutenant made history in 2008 by becoming the first female to command a platoon of soldiers. Other women followed in commanding male soldiers, but the numbers of female officers remained very small. Only about two dozen graduates became officers and the highest rank females reached in the army was captain. Most of these women came from army families and lived in military neighborhoods, so their familiarity with the army life style probably gave them an advantage unavailable to women from civilian life. Although promotion of women to senior rank had a long way to go, at least the entry of women into the armed forces of the Dominican Republic has so far been free of the friction and controversies occurring in other parts of the world.[24]

The opening of the armed forces to women guaranteed that the officer corps always could draw on the talent pool of the entire population. In addition to having access to the best human resources, the military also needed to acquire new weapons to replace the armament usually dating back to the Trujillo years. Because the Dominican Republic imported all its weapons and ammunition, a long-held goal of the military was to establish a cartridge factory in the country. Trujillo himself tried to establish an arms industry but ultimately failed, and negotiations with Brazil were the

13. An Inconvenient Ally: The Armed Forces of the Dominican Republic 235

most recent attempt to attain this goal, but no concrete results emerged as of 2014.[25]

The negotiations for the munitions factory formed part of the deal for the Dominican Republic with its own funds to purchase Super Tucano airplanes from Brazil. The Super Tucano, a turboprop, was ideally suited for patrol functions and to intercept illegal civilian flights over the Dominican Republic. The airplane, although assembled in Brazil, contained many U.S. components and its sale required the approval of the U.S. government. The Dominican air force was practically without combat airplanes as the relics from the Trujillo era ceased to be operational. Normally the U.S. government did not want the Caribbean countries to have any combat airplanes even of such modest proportions as the Super Tucanos. The U.S. government had blocked the sale of Super Tucanos to Venezuela (chapter 6), but in the case of the Dominican Republic the astute President Fernández was able to overcome the opposition of Washington.[26]

The drug trade from Colombia had generally bypassed the Dominican Republic, but increased U.S. Coast Guard enforcement was encouraging drug traffickers to explore new routes by 2005. The Dominican Republic to the east was not the ideal location but did provide a serviceable alternative for drug traffickers. Fast airplanes loaded with cocaine flew safely over the Dominican Republic because drug traffickers knew that its air force required a long time to put into the air its slow-moving ancient aircraft. Because U.S. officials wanted to interdict the drug trade, they had no choice but to approve the use of American components in the Brazilian Super Tucanos. After having flown over South America, the new airplanes caused a very favorable impression in Santo Domingo. Large crowds turned out to participate at the formal inauguration ceremonies. The Super Tucanos were also well suited to patrol the border with Haiti.[27]

By the time the Dominican government devoted more efforts to stop drug trafficking, the problem had already contaminated both the armed forces and the police. Starting in 2009 and continuing through 2011, the government had to take strict disciplinary measures against uniformed personnel who had received bribes from drug traffickers. At the end of 2008 the government expelled 538 uniformed personnel including 92 officers of the rank of captain, majors, and colonels. By the end of 2009 the tally of expulsed officers and soldiers rose to 1500. The purge was still continuing in 2011 when 171 members of the police and the armed forces were expelled. The corruption seemed to have declined in 2012, not so much because of successful enforcement but because the drug traffickers were continuing their long-term practice of periodically rotating routes as better opportunities appeared to ship drugs to the United States.[28]

President Leonel Fernández began, and his successor Danilo Medina also undertook, a vast restructuring of the armed forces. Both presidents wanted not just a reshuffling of bureaus and commands but a fundamental change in the functioning of the military. By 2012 the armed forces had grown to 52,000 members, too large a number for the U.S. government. Although the popularity of the military precluded any reduction, significant structural flaws required correction. A first problem was that military personnel were so well ensconced throughout society, that it was not clear what they were doing and if they were performing military duties. Many active-duty personnel did not wear uniform and were paid by private businesses. Many colonels served as heads of security in large corporations, and other officers assigned their subordinate soldiers to work in private security firms supposedly to help them earn a better salary. While some high-ranking officers drove to work without chauffeurs or body guards, other officers put on the army payroll their civilian sons who worked as body guards and chauffeurs.[29]

A second structural problem was endemic to the military of Latin America: too many chiefs for too few Indians. With some exaggeration hostile journalists accused "the armed forces of having become an inverted pyramid over the previous decades, with more generals than soldiers, more lieutenants than sergeants, more sergeants than corporals, and more corporals than privates."[30] But the hyperbole was not completely unfounded when the Defense Minister revealed that the armed forces had 152 generals and around one thousand colonels for 52,000 uniformed personnel. The ratio of 1 general per 342 soldiers and 1 colonel for every 52 soldiers meant either rank inflation or more likely excessive promotion of unsuited officers.[31]

When the Defense Minister investigated the accusations, he found most disturbing the lack of leadership experience by officers. Most senior officers had never commanded even a platoon and had simply risen to high rank based on desk and bureaucratic criteria. Journalists were even harsher and denounced many of these generals as being overweight, pot-bellied, and unable to take to the field to lead troops. What these generals did was "sit in their homes without any command or responsibility of any kind, and their stars shine only when it is time to celebrate an event, such as national holidays or the holy mass commemorating the birthday of their branch."[32]

This top-heavy institution needed drastic pruning, and a six-year plan called for gradually reducing the number of generals to only 40 in the entire armed forces; twenty would belong to the army, and the remaining 20 were divided equally between the navy (with the equivalent rank of admiral) and the air force. The first step in the six-year plan called for limiting the number of generals to 60, and several new mechanisms helped to implement these reductions. The most significant was requiring generals to retire if they had not been selected from the list of candidates nominated to the president for appointment to top commands.[33]

In essence, the Dominican Republic was carrying out the long-overdue transformation of the armed forces. The military retained too many of the characteristics of the era of Trujillo, and the creature had outlived its creator for decades.[34] But the armed forces were a vital national institution in the life of the Dominican Republic, and by their existence they preserved stability and order inside the country and provided defense against an invasion from Haiti. The large armed forces were not only an integral part of the national identity of the Dominican people but also served as major factor restricting the ability of the U.S. government to impose its will on the internal affairs of the country.

14

Defiant Cuba

> *The military is generally better regarded in Cuba than the political institutions.*—U.S. Interests Section in Havana[1]

Cuba is the country that comes closest to satisfying the avowed goals of the U.S. government for all of Latin America. U.S. officials want to avoid lawlessness and to reduce criminal activities such as drug trafficking, armed gangs, death squads, illegal immigration, and the white slave trade. Paradoxically Cuba has come closest than any other Latin American country either to curtailing their incidence or sometimes even to eliminating them completely. The success of Cuba has been made possible by the absence of the acute income inequality, homelessness, and crushing poverty rampant in most other countries of the region. The Cuban people also enjoy educational and public health systems comparable to those in Western Europe and Canada. An extensive infrastructure of highways, railroads, ports, and airports reaches into virtually every corner of the island. And in perhaps the most vibrant political system in Latin America, citizens constantly participate in meetings to express their views and demand action from local, provincial, and national officials who have to cater to a well-informed electorate. As one Canadian scholar observed, "Cuba is replete with almost daily assemblies, meetings, and gatherings of various organizations to discuss and examine particular issues, in conjunction with the participation of government officials. This is part and parcel of the political process."[2]

In spite of these and many other accomplishments, the U.S. government has waged since the summer of 1959 a relentless campaign to overthrow the revolutionary regime existing in the island. Why this extreme hostility exists, at first glance appears irrational and inexplicable, when Cuba by its accomplishments amply qualifies to be an ideal ally of the U.S. government. Many experts blame Washington's hostility on political pressure from Miami Cubans who aspire to turn the clock back to 1958. While

Map 15. Cuba.

this unfounded nostalgia does contribute to the hostility, the Miami Cubans usually have been useful pawns in U.S. efforts to preserve and extend control over the informal empire in the Caribbean.

The real explanation for the hostility lies not in Miami but in the lack of U.S. influence over truly independent Cuba. In the rest of the Caribbean countries U.S. officials adroitly manipulate pressing issues, such as drug trade and human trafficking, not so much to solve them but to insert U.S. influence and control. The U.S. government can live with countries that fail to solve the problems, as long as U.S. officials are deeply involved in controlling and influencing Caribbean countries. Because Washington has grudgingly accepted the relative independence of Mexico, having Cuba follow the same path may not seem such an outrageous or dangerous step. But only because of Mexico's size, population, and long common border has the U.S. government impatiently put up with its annoying manifestations of independence. To allow small Cuba to follow the exasperating example of Mexico was to open the door for the rest of the small countries of the Caribbean to shape their own destinies outside of the orbit of U.S. influence.

Since 1959 the Cuban leadership with insightful prescience knew that the U.S. government was incapable of staying out of the local affairs of any Caribbean country. To ask Washington to respect Cuban or Caribbean independence was the equivalent of asking water to stop flowing downstream. Just as only dams can protect areas from ravaging floods, so only strong barriers can restrain world powers. The universal language of international relations is force, and Cuban leaders had no choice but to maintain a large military establishment to defend the Revolution against its sole foreign enemy. Nobody doubted that in any war the crushing superiority of the United States meant certain destruction for the revolution and seemingly made futile any resistance. Deterrence, or making too high the human, economic, and international costs of any unilateral invasion, remained the fundamental mission of the Cuban armed forces. But the large military establishment could be useful to preserve the revolutionary legacy in many other ways, as this chapter explains.

Preserving the Revolutionary Legacy

During the late 1980s, succession issues started to call the attention of foreign observers and, to a lesser degree, of Cuban leaders. A well-justified concern arose among citizens about how to preserve the legacy of the Revolution once its founders retired, became incapacitated, or died. When Cuba adopted a new Constitution in 1976, the institutionalization of the Revolution was already well underway. Its clauses contained mechanisms to assure the orderly transmission of power from the historic founders of the Revolution, in particular Fidel Castro and his brother Raúl, to younger leaders. The structural defect with this succession plan was the assumption that the Soviet Union and its allies would remain the most important economic and military partners of Cuba. That deeply held belief went unquestioned under the then reigning Marxist dogma.[3]

When the Soviet Bloc countries began to collapse like a row of dominoes, Cuba faced something much worse than a succession crisis and confronted a struggle for the very survival of the revolutionary regime. Because 85 percent of Cuba's commerce had been with the Soviet bloc, and 95 percent of the island's petroleum came from the Soviet Union, the end of this trade had catastrophic effects. Among other adverse results, Cuba ceased to earn valuable hard currency from the re-export of Soviet petroleum. Inside the island, the population experienced daily blackouts because fuel shortages nearly paralyzed electricity generation. Without fuel to operate agricultural machinery, food supplies dwindled, and the population faced the specter of hunger. In a shocking statistic, Gross Domestic Product shrank by at least a third between 1989 and 1994. Not just U.S. officials but also talking heads on television confidently predicted the collapse of the regime. Journalists were eager to proclaim the fall of another domino, and they avidly scanned the news searching for the first hints that the revolutionary regime was about to collapse. In such a supercharged atmosphere of expectation, for anyone even to suggest the possibility that the regime might not fall only provoked massive amounts of ridicule.[4]

While journalists, U.S. officials, and Miami Cubans fantasized about the imminence of a popular uprising, nobody in the island doubted the acute seriousness of the crisis. The mortal threat to the Revolution came from the economic collapse and no longer from a U.S. invasion. Tackling the issue directly, on 28 September 1990 Fidel Castro announced the start of a "Special Period" and promised massive structural changes so that Cuba could survive the disintegration of the Soviet Bloc.[5]

The first and foremost task was to achieve self-sufficiency in food. Under the trading arrangements with the old Soviet Bloc, each country cultivated those food crops enjoying a comparative advantage in its soils; now Cuba had to produce everything previously imported that could grow in its subtropical climate. Consumers had developed a taste for imported foods, and a vast information campaign was necessary to teach the population to enjoy nutritionally equivalent and locally grown fruits and vegetables; the best example was the partial replacement of wheat flour with *casabe* flour, the latter obtained from an abundant tropical tuber. The little fuel coming from domestic petroleum fields went to power agricultural equipment, yet even this proved inadequate, and the reliable oxen once again reappeared in Cuba's fields as harbingers of a Green Revolution and as a source of endless ridicule on the part of Miami Cubans.[6]

In a major permanent structural change and in a confession of failure, the government turned away from the state farms existing since 1960. At the start of the Special Period, most state-owned lands became very productive farmer cooperatives. The state farms that had controlled 70 percent of agricultural lands saw their share drop to about one-third. Many of the remaining state farms passed under the control of the army, which operated them with military discipline to obtain substantial increases in agricultural output. Because most of the conscripts were of peasant origin, their transition into farm laborers in the army farms proved easy. During the Special Period the conscripts often cut short their basic training or skipped it altogether to go to work directly on farms. Defense Minister Raúl Castro deployed nearly 70 percent of the soldiers to work as farm laborers, and they covered 90 percent of the dietary needs of the armed forces and also produced substantial amounts of food for civilian consumption. Making the military virtually self-supporting took a huge monetary burden out of the budget, so the government had the opportunity to channel its dwindling funds to other critical civilian expenses such as fuel.[7]

The fuel shortage was the major bottleneck blocking the economy. To solve the paralysis in public transportation, Cuba imported bicycles from China, and soon at least one-third of the population was riding bicycles. The potential of most of Cuba's hydroelectric sources had already been exhausted, while the adoption of alternate sources of power, such as wind and tide, remained expensive projects for a distant future. The exploitation of Cuba's petroleum deposits required decades of development and could not yield immediate results. The acute lack of fuel crippled Cuba's armed forces, units had to do initially with 30 percent of their fuel allotments, and by 1994 they were down to 10 percent. Any training requiring the consumption of fuel practically ended. Front-line pilots could no longer fly their required minimum number of hours, while reserve pilots saw their flight time decline from 200 hours to 20 hours annually. Cannibalization became the rule to obtain replacement parts to keep at least some aircraft, vehicles, and equipment operational. And as uniforms became scarce, even the soldiers started to look ragged.[8]

Tourism was already emerging as a lifesaver for the Revolution and as a way to mitigate the fuel shortage. Tourist dollars covered the costs of imported fuel, and military personnel adapted to a new civilian role. Combat pilots flew aerial sightseeing tours, naval personnel operated excursion boats, and soldiers served as drivers and guides for bus and automobile tours on land. The foreign language skills of many officers proved invaluable in making tourists feel even more welcome and comfortable during their stay in hospitable Cuba, and not surprisingly the military became the government partner in many of the joint ventures with foreign firms on tourism.[9]

The government slashed the military by more than half. The armed forces shrank to around 55,000 uniformed personnel, with the army having about 38,000 soldiers. In another cost-saving measure, the government reduced the duration of military service from 3 to 2 years, a period in tune with the prevailing European ideas of how long a conscript needed to stay in uniform to become a proficient soldier. Conscription became also less important to maintain the numbers of the regular forces, because reenlistments increased during the Special Period. The personal management style of Raúl brought him close to the troops as he routinely toured garrisons, listened to the soldiers, and sat down to eat with them at the mess halls.[10]

The Cuban armed forces, as was typically throughout Latin America, had a very large officer corps. However, in contrast to the rest of Latin America, the excessively large number was the result of a deliberate policy of having more officers available for any rapid mobilization of reserves. While the reserves had their own officers, any large scale mobilization to counter the threat of a U.S. invasion required additional professional officers to fill all the command positions. Cuba could not afford to maintain all these officers during the Special Period, but Raúl, who took great care of his officers, was determined to find them suitable postings. The Defense Ministry assigned civilian tasks in the economy to active-duty officers. Many enterprises previously considered hopeless became the responsibility of officers who remained in uniform and under military discipline. The assumed understanding was that if these officers wanted to maintain their rank and privilege, they needed to get the economy working again, and not surprisingly they poured body and soul into making sure that these state enterprises worked efficiently once again.[11]

The transition was not as traumatic for officers as it might seem at first glance. Raúl had insisted in the 1980s that commanders take management courses, and these acquired capabilities proved invaluable during the Special Period. The immense efforts of the officers and the support of civilians created very strong bonds and a sense of "we are all in this together" and reinforced the feelings of national unity. The freefall of the economy ended in 1994 when the first almost insignificant increase in Gross Domestic Product took place, and in 1995 the recovery was fully underway. Each year the Cuban people through their own hard efforts and sacrifices improved their condition until permanent relief came in 1999 when the alliance with the Venezuela of Hugo Chávez materialized. Enjoying petroleum from Venezuela at below market rates, Cuba was able to establish the foundations for a solid and prosperous economy.[12]

As some readers have already guessed, the shifting of uniformed personnel into economic functions dramatically reduced the combat capabilities of the armed forces. Never before had Cuba been so vulnerable to attack, and during those years the Cuban armed forces ceased to have a deterrent effect on the U.S. government. In an additional green light for an armed attack, the U.S. promise not to invade Cuba expired when the Soviet Union collapsed. Hubris became the best ally of revolutionary Cuba during the George Bush administration, which haughtily believed that Cuba would fall just like the dominoes of Eastern Europe and thus make unnecessary a U.S. invasion to accomplish regime change. This belief that the revolutionary regime was doomed to collapse lasted in the United States into the early twenty-first century. Actually, Cuba fell to its weakest condition and was most vulnerable to a U.S. invasion during the Bill Clinton administration, but good fortune smiled on the revolutionary regime at the time of its greatest need.

Bill Clinton was the rare U.S. president who was most reluctant to intervene or invade a Latin American country. He correctly concluded that these interventions ultimately were counterproductive to U.S. interests, as he verified when he took the bold decision to declassify the CIA documents on the disgraceful U.S. operation to overthrown the popularly elected Jacobo Arbenz of Guatemala in 1954. President Clinton was willing to invade Cuba only with unanimous international support or at least a massive internal uprising. He did not have to defend his reluctance very hard, because the preferred position from the propaganda perspective was to have the Cuban regime collapse on its own as had happened in Eastern Europe. But as the years started to slip by, impatience in the United States with the absence of a meltdown fueled the clamor for tougher economic sanctions. The new conventional wisdom became that the regime had been hanging on by a thread and that new sanctions, specifically the Torricelli Act of 1992 and the Helms-Burton Act of 1996, supposedly sufficed to push the revolutionary regime over the edge.[13]

In reality the Helms-Burton Act targeted the European trading partners of Cuba, and soon a deep rift emerged among Western countries over commercial relations with Cuba. Because the U.S. government has waived the most offensive provisions of the Helms-Burton Act, the only result of this punitive legislation has been to solidify the determination of the Cuban people to maintain their particular system. As part of the permanent propaganda campaign to discredit the revolutionary regime, the U.S. Congress asked for a formal report from the Pentagon on the magnitude of the military threat that Cuba posed to the United States. After careful study, the Defense Intelligence Agency concluded that Cuba no longer posed any type of military threat to the United States, but when this realistic conclusion was leaked prematurely to the press in March 1998, Miami Cubans were outraged and demanded a recitation of the usual propaganda attacks against the revolutionary regime.[14]

Under intense political pressure, Secretary of Defense William Cohen went into damage control and sent the report back the Defense Intelligence Agency. No civilian official should ever tell military officers to modify their strategic assessment, but Secretary Cohen insisted on fitting the conclusions to pre-conceived ideas. In an exemplary demonstration of military professionalism, the Pentagon returned its analysis virtually unchanged and just added a few throwaway sentences to sugar coat the conclusion—bitter for some—that Cuba did not pose any threat to the United States. Seeing that the officers were not going to bail him out of this political problem, Secretary of Defense wrote a covering letter essentially contradicting the conclusion of the professional Pentagon report. This embarrassing episode publicly exposed cracks inside the U.S. government over official policy to Cuba. To the dismay of politicians, the U.S. military did not want to invade and could not find any reason why Cuba should be invaded. On the contrary, officials from the Pentagon, the DEA, and other law-enforcement agencies have consistently pushed to increase cooperation in the pursuit of mutually shared goals such as counterterrorism and drug interdiction.[15]

The possibility of improving relations between Cuba and the United States suffered a major setback when George W. Bush became president in January 2001. One of his campaign promises to Miami Cubans had been to do something about Castro. His administration was busy preparing to overthrown the revolutionary regime when the 11 September 2001 attacks on New York City and Washington, D.C., diverted U.S. attention to the Middle East. The subsequent U.S. military occupations of Afghanistan and Iraq, and the determination of the Bush presidency to have the U.S. government act unilaterally to invade countries provoked a strong response from the Cuban people. In the first years of the Bush administration, Cuban citizens, in particular the young, began to report enthusiastically for additional duty in the reserves. Patriotism swept the island, and this unexpected wave of nationalist feelings allowed the Cuban government to begin the process of rebuilding its armed forces so that once again they could provide a deterrent effect should President Bush consider launching another invasion.[16]

Deterrence had been based on the principle of "the nation at arms" (*Guerra de Todo el Pueblo*), and obviously this defense policy became unworkable during the trying years of the Special Period. But the outbreak of popular enthusiasm for defending the island and the ongoing economic recovery gave a renewed urgency to the principle of "the nation at arms" as the best way to counter the aggressive threats from the Bush administration. The petroleum from Venezuela released many units from their economic functions and made possible the resumption of training and practice exercises. Engineers and mechanics showed great ingenuity to produce spare parts, and soon equipment and vehicles that had been considered lost were once again operational. Somehow Raúl scraped together funds to purchase only those spare parts simply impossible to manufacture at home.[17]

Fidel and Raúl decided that the time had come to resume large-scale military maneuvers as the best way to rally the people against the U.S. threat. For the first time since 1986, the government carried out a mobilization of all the reserves, and this operation called *Bastión 2004*, showed that the Cuban armed forces were definitely recuperating their former strength. When Cuban journalists asked whether it was not better to spend the money on the economic recovery, officials replied that the operation was budget neutral and in no way detracted from economic activities. Citizens closely followed *Bastión 2004* on TV and in the press, and just in Havana Province 684 umpires judged the performance of the troops. To make sure that the sometimes stone-deaf Bush administration received the message about the revival of defense capabilities, the Cuban government invited NATO officers as observers who were expected to share their impressions with the Pentagon. These NATO officers noticed that the mechanics maintained perfectly the Soviet-era armament and equipment and that all the soldiers displayed formidable discipline and impressive organization. And an aging Castro, in what turned out to be the last military operation of his career, watched the maneuver unfold before his eyes in large TV screens at his hidden command center. *Bastión 2004* turned out to be a great success, and Fidel was extremely pleased with the work of his brother Raúl in the Defense Ministry. There was a lot of truth in the statement that the armed forces under Raúl had saved the revolutionary regime during the Special Period.[18]

This final demonstration of his brother's extraordinary abilities confirmed Fidel's long-held position that if anything happened to him, his brother Raúl was the leader best suited to take the Cuban Revolution into the future. When illness forced Castro to relinquish power to his brother on 31 July 2006, the contrast between the reactions in Miami and in Cuba could not be greater. In Miami, hordes of residents poured into the streets to celebrate the fall of Castro, and their cars caused traffic jams and congestion throughout the city. Florida authorities braced for a tidal wave of refugees expected to descend upon the state, while TV commentators dusted off their old talking points and once again predicted the imminent collapse of the regime. Instead in Cuba, the transfer of power passed without a single incident, and when Raúl rode to work the next day, the only difference was that small groups of cheerful citizens spontaneously gathered to applaud his motorcade on his usual route. And meanwhile, the Roman Catholic bishops invited the faithful to join in praying for the recovery of Fidel and for the success of the new ruler.[19]

The incessant rumors about the imminent death of Fidel were proved wrong once again, but as his return to power became less likely, the regime felt a need to confirm its solidity and continuity. Raúl, who routinely consulted his physically ailing but mentally sharp older brother, decided to launch a major military parade on Armed Forces Day on 2 December 2006. Since *Bastión 2004* the very popular practice of conducting military parades had revived in the island, but these generally were small and local. In 2 December 2006 the armed forces paraded their restored equipment and a very few new pieces, while the soldiers marching in their new uniforms looked impressive. The French military attaché remarked that "the Cubans did a good job to keep the old Soviet era equipment running, not unlike Cuban mechanics keeping 1950s era U.S. model cars running."[20] The civilian population has always loved military parades, and nearly half a million citizens came to watch the troops march during this joyful event, and many viewers followed the nearly continuous coverage on television. The mobile missile launchers for anti-aircraft defense were always a hit with the public. The high point of the parade came when Cuban MIGs for the first time in decades carried out low flyovers to the enthusiastic cheers of the excited crowds.

Much as U.S. officials tried to belittle and ridicule the remarkable recovery of Cuba and its armed forces, the Bush administration grudgingly saw its chances for a quick and easy invasion of Cuba slipping away. Undeterred by the challenge and gloating after the apparently successful invasions of Afghanistan and Iraq, a triumphant Bush administration decided to crush Socialist Cuba once and for all. Taking the symbolism of the date of Cuba's Independence Day, on 10 October 2003 President Bush created a Commission for Assistance to a Free Cuba under the chairmanship of Secretary of State Colin L. Powell. The main charge of the Commission was to "help the Cuban people bring about an expeditious end of the Castro Dictatorship," and to accomplish that goal the Commission "would draw upon experts within our government to plan for Cuba's transition from Stalinist rule ... and to identify ways to hasten the arrival of that day."[21]

Aware that all previous attempts to overthrow the revolutionary regime had failed, the Commission believed that the solution lay in seeking "a more proactive, integrated, and disciplined approach to undermine the survival strategies of the Castro regime and contribute to conditions that will help the Cuban people hasten the dictatorship's end."[22] In open contrast to the Pentagon report of 1998 that saw no threat in Cuba, the Commission boldly stated the assumption that the regime was a threat to apparently everyone in the Western Hemisphere, including the people of the United States. But when the Commission laid out a course of action, "the report reads like an occupation manual" and proposed "the type of measures that would almost certainly prompt violence and social unrest."[23] In the new constitution for the country, the Commission wanted to eliminate all references to the armed forces. The end of conscription, the abolishment of the army and of the official party, and the privatization of all state properties, the same policies that yielded disastrous results in Iraq, were the measures the Commission wanted to replicate in Cuba. The Bush administration adopted the Commission's report as official policy and even went so far as to appoint a coordinator who presumably would be the U.S. proconsul in Cuba once the revolutionary regime fell from power.[24]

The 2004 Commission report was a propaganda windfall for the Cuban government and provided a lifetime supply of accusations against Washington. Many sections of the report were so extreme that they outraged even Miami Cubans. To correct the most glaring blunders and also to update it, the Bush administration decided to issue a sequel. A new Commission for Assistance to a Free Cuba, this time under the joint chairmanship of Secretary of State Condoleezza Rice and Secretary of Commerce Carlos Gutiérrez, issued a separate report in July 2006. Carlos Gutiérrez as a Cuban businessman was supposed to bring input from the Miami community.[25] The 2006 Commission for a Free Cuba not only buried the previous recommendation to abolish the armed forces but also blatantly pandered to the Cuban military as a way to undermine the regime and hasten its fall. No less than the 2004 Commission, the 2006 Commission considered Cuba to be a threat to the region because the regime still sought "to actively undermine U.S. interests."[26] But a new urgency motivated the 2006 Commission, and its primary goal was "to ensure that the Cuban transition is genuine and that the Castro regime's succession strategy does not succeed."[27] Believing that Cuba had "begun a gradual but intrinsically unstable process of succession,"[28] the 2006 Commission automatically ruled out Raúl Castro as capable of maintaining stability. The report had to include this doomsday scenario to fool Miami Cubans into believing that the moment was fast approaching for the inevitable collapse of the regime.

While U.S. officials engaged in ideological fantasies far removed from reality, Cuba calmly responded to the renewed threats. *Bastión 2004* had been such a success, that Raúl Castro scheduled another mobilization for November 2008. And as trial runs for the larger military exercise, Cuba carried out two smaller military maneuvers: Operation Caguairán in 2006 and Operation Moncada in November 2007. A political bombshell landed when in a letter dated 18 February 2008, right before the annual session of the National Assembly, Fidel Castro declared that he was not seeking reelection as president of Cuba and was retiring from public office. Adopting Fidel's recommendations, the National Assembly duly elected Raúl as the next president of Cuba on 24 February 2008. Although Fidel has continued to weigh in to a greater or lesser degree on major policy matters, Raúl was formally in charge of running the government. When a hurricane struck the island in December 2008, *Bastión 2008* had to be cancelled so that the government could deploy its forces for relief efforts. A rescheduled mobilization duly took place on 26–28 November 2009 under the new title of *Bastión 2009*.[29]

Fidel had always harbored resentment at the Soviet Union ever since the betrayal at the Cuban Missile Crisis of 1962, and any remaining sympathy vanished with the shoddy treatment Cuba received from a collapsing Soviet Union in 1990. But Raúl always had a particular affection for Russia, and he had developed a much closer relationship with Soviet officials in the past both in Cuba and during his extended vacations in the Soviet Union.[30] In one of his first moves as president on his own right, he decided to reach out to Russia to try to restore the close ties formerly existing between the two countries. Russia had previously been indifferent to Cuba, but the massive and harsh Western backlash against Russia's war in Georgia was a turning point in Moscow's foreign policy. Vladimir Putin "called for Russia to rebuild its positions in Cuba and other countries during an August 4 Presidium meeting."[31] The first major step of this new policy came when Russian President Dmitry Medvedev, as part of a goodwill tour of Latin America, spent two days in late November 2008 at Havana in response to many cordial invitations from Raúl.[32]

Now it was the turn for Raúl to make an official trip to Russia lasting from 28 January to 4 February 2008, the first high-level visit by a Cuban leader since 1985. This personal contact was a historic first: "Two such high-ranking visits in such a short space of time were unprecedented as this had never even taken place while Soviet-Cuban relations were in existence."[33] President Dmitry Medvedev warmly greeted the Cuban ruler, and the most important meeting of Raúl was with Putin, who at that moment was Prime Minister. A shared nostalgia for the Soviet period created strong bonds between the two leaders. Cuba negotiated a series of favorable trade agreements with Russia and also agreed to establish joint ventures in areas such as petroleum, nickel, and manufacturing with both private and state Russian firms. Cuba also ordered civilian aircraft to replace its aging Soviet fleet. High on Raúl's list was the acquisition of military equipment, but without the large funds of his Venezuelan ally Hugo Chávez, the Cuban leader could not afford any expensive orders. Nevertheless, by the middle of 2008 Cuba began to receive a trickle of spare parts and even some upgrades for its Soviet-era equipment and armament.[34]

Putin showed a strong personal interest in restoring ties with Cuba, and he was ready to make a major commitment when he became Russian president again in 2012. When Raúl returned to Russia on 12 July on the start of an official state visit, Prime Minister Medvedev "expressed joy for Raúl's visit to Moscow."[35] At his country residence in Novo-Ogaryovo, Putin and Raúl "embraced as old friends," and the Russian President further added "I want to ask you to convey the best wishes to our great friend Fidel Castro."[36] Besides the basic issue that Cuba could not yet afford new Russian armament to replace the aging Soviet weaponry, spare parts were no longer available for many of the older weapons and obsolete equipment.

Cuba attempted to circumvent this limitation by trying to acquire surplus Soviet equipment at discounted prices from other former Soviet bloc countries such as Bulgaria and Ukraine. But the U.S. government was very vigilant and promptly stopped those countries from selling any Soviet equipment to Cuba. In a particularly revealing example, the U.S. blocked the sale of spare parts for the Pechora air defense system in June 2009, thereby forcing Cuba to turn to the Soviet Union for newer but more expensive components. The violent opposition to the sale of defensive weapons was a position more natural for the aggressive Bush presidency than for the supposedly peaceful Barack Obama administration.[37]

Scarcity of funds to repair its military equipment forced Cuba to seek bargains for parts and services no longer available in Russia. It turned out that North Korea still possessed the technical skills to repair and restore old Soviet-era armament. The temptation was too great for Cuba to resist, particularly because North Korea accepted payment in sugar and other foodstuffs, thus avoiding the need to spend scarce hard currency. U.S. intelligence services detected the shipment of radar equipment, missiles, and two obsolete jet fighters aboard one transport ship, and U.S. officials arranged for impounding the cargo when the North Korean freighter was passing the Panama Canal in 2013.[38]

Cuba carried out *Bastión 2012*, another large scale maneuver, in November of 2012. The November 2012 exercise saw Cuba display many reconstructed and redesigned weapons of the Soviet era, but the mobilization had as its main goal testing the response capabilities of many units. The government decided to schedule another mobilization in 2016, and the coincidence of these dates with the U.S. presidential election cycle was not accidental. And when the crises in Crimea and the Ukraine broke out in 2014, the unconditional and complete Cuban support for the Russian actions created a very favorable impression in Moscow. During a visit to Havana in July 2014, Putin announced Russia's cancelation of Cuba's old debt with the Soviet Union, thus opening the door to strong economic links between the two countries. In June 2014 Russia established a station for its global position system in Cuba and, in a momentous decision, reopened the electronic eavesdropping post near Havana. This extensive facility had been abandoned since its closing in 2001. Quietly both countries intensified their military ties during 2014.[39]

Raúl had taken a major strategic gamble by assuming that the Obama administration had no intention of invading Cuba. But by the time the U.S. presidential election rolled around in 2016, Raúl would know whether a right-wing candidate had a chance of becoming president of the United States. In that case Raúl planned to turn to Putin to ask for special pricing on the new defensive weapons Cuba would desperately need to maintain the deterrence effect against the United States. And scheduling *Bastión 2016* in the last month of the Obama administration gave the perfect opportunity for the Cuban armed forces to mobilize in anticipation of an invasion from the United States.

The U.S. Vision for Cuba and the Caribbean

The U.S. government has considered the existing armed forces of Cuba to be excessively large but recognizes the need for the island to have a much smaller military. In the historical perspective, U.S. officials wrestled once again in the initial decades of the twenty-first century with the same issue their predecessors confronted in the early twentieth century. While the first experiment to create a Cuban army ended in complete failure, ever optimistic U.S. officials were confident that this second time around they would get right their efforts to construct a Cuban military appropriate to meet U.S. goals in the twenty-first century.[40]

An examination of their proposals provides not only a clue into U.S. thinking on Cuba but also opens a window into U.S. efforts to maintain its informal empire over the Caribbean. The rallying cry since the Cuban Revolution has been "no second Cuba," when a more realistic slogan would be "no second Mexico," a reference to the hands-off policy the U.S. government has been forced to follow toward internal affairs inside Mexico. This was the goal behind not just the unrelenting pressure to overthrow the Cuban regime but also behind U.S. support for the April 2002 coup against Hugo Chávez. The benefits of overthrowing Cuba are really tempting, because once its revolutionary regime falls, the task of riding herd over the client states in the Caribbean becomes incomparably easier. The local population would firmly assimilate the teaching that any resistance to the dictates of the United States is futile and actually outright suicidal.

Eager as U.S. officials were to shape new armed forces for Cuba, no actual implementation could begin until the revolutionary regime first fell from power. So confident were U.S. officials that the collapse of the Cuban regime was not just inevitable but also imminent, that they drafted a series of fascinating blueprints for shaping a future government and its armed forces. Although some of these documents remained secret, the U.S. government considered the propaganda value of these blueprints so great that it widely publicized most of them. And the assumptions and goals behind both the public and secret documents were practically identical.[41]

Captivated by the image of the collapse of Communist regimes in Eastern Europe, most U.S. officials confidently assumed that a similar phenomenon was certain to occur in Cuba. But by the first decade of the twenty-first century any chance of repeating the Eastern Europe revolts in Cuba had passed, and in reality that possibility had never existed. On the contrary, the disappointing performance of most Eastern European countries after the fall of Communism reinforced the conclusion that rampant corruption, widespread poverty, and income inequality were the inevitable outcomes should the Cuban revolutionary regime ever disappear.

No civilian opposition leaders existed in Cuba with a following capable of overthrowing the regime, and tacitly some U.S. officials began to accept the possibility that only a military coup could destroy the revolutionary regime. The 2004 Commission for a Free Cuba in one of its many major flaws failed to realize this possibility and, by declaring its intention to disband the institution, even antagonized the military. The 2006 Commission for a Free Cuba corrected this obvious blunder and held out the olive branch of saving a place for the armed forces in a provisional government. But because the military was already the most influential institution in the country, it was not clear why the Cuban officers would want to overthrow a regime that gave them so many privileges.[42] The only very remote possibility was that after the Castro brothers were no longer around to provide a moderating influence (something that at the earliest would take place around 2016) then a power struggle might break out among rival generals motivated by the ambition to become the ruler of the island. Only under that unlikely scenario was it conceivable to image an unscrupulous and power-hungry general seeking U.S. support to install a provisional government.

Since the mid–1990s some U.S. officers have repeatedly pressed for closer relations with their Cuban counterparts, but in stark contrast with the situation in Mexico, it is not Cuban generals but U.S. politicians who have blocked those approaches. The dream of U.S. officers is to succeed in establishing close ties to ambitious Cuban generals who might be tempted to seize power. Illusive as that goal seems, improving cooperation on law-enforcement issues such as drug trafficking, smuggling, and immigration brings immediate advantages to both countries. The Pentagon, its hands tied by American politicians, has turned to surrogates, such as NATO allies, to forge relations with the Cuban military. In particular, Britain and France, with U.S. blessing, have provided training and equipment to the Cuban armed forces for law enforcement in the airspace and territorial waters of Cuba. The U.S. government even has been able to slip a U.S. Coast Guard officer into Havana to enhance cooperation with the Cuban military. Cuban generals wanted a larger engagement with the Pentagon, but American politicians resolutely block any increase in ties with this perceived enemy.[43]

The Pentagon firmly recognized the need for armed forces to exist in a post-revolutionary Cuba, partly to avoid the collapse in Haiti after the demobilization of its army (chapter 12), and also to prevent the repetition of the disastrous results occurring in Iraq after U.S. forces disbanded its military. This conclusion reflected the realization that small armies such as those in Central America have been too weak to prevent the spread of violent gangs and drug cartels. Many areas of those countries, and even in Mexico, were largely out of the control of their governments. Ultimately this lack of control impacted negatively U.S. interests in the region, and the Pentagon did not want Cuba, so close to the United States, to fall into a pattern of lawlessness and disorder as was becoming the norm throughout Central American and the Caribbean.

More than in any other country in the region, a provisional government was very likely to face guerrilla warfare because of the very strong nationalist traditions of the Cuban people. Whether the U.S. government invaded Cuba or responded to urgent pleas for help from a provisional government, American troops were sure to get dragged into the island to put down an insurgency. The last thing the Pentagon wanted to do was to fight another insurgency in the guerrilla hide outs of the fabled Sierra Maestra and Escambray mountains. The best way to avoid losing American lives in a costly and unpopular intervention was to have a Cuban army adequate to the task. However, the fundamental question of why a revolutionary regime never had needed to use troops to prevent internal disturbances and instead a provisional government would require this military muscle to repress the Cuban people never seemed to have bothered U.S. planners.

According to the Pentagon, the Cuban armed forces, including the separate border guards and State Security units, would shrink under a provisional government from roughly 75,000 uniformed personnel to around 50,000.[44] This number was slightly lower than for the armed forces of the Dominican Republic, a country with a smaller population and less than half the size of Cuba. As the previous chapter explained, U.S. diplomats considered the armed forces of the Dominican Republic to be excessively large. U.S. officials have waged a relentless and unsuccessful campaign to reduce the armed forces of the Dominican Republic, and not surprisingly the Pentagon stipulated a much smaller military for Cuba.

More significant than the reduction in size was the elimination of key components of the armed forces of Cuba. The first forces slated for elimination were those comprising the vast air defense system scattered across the island. Because their sole purpose had been to defend the island from a U.S. invasion, American planners ingenuously reasoned that a provisional government no longer needed them and could convert these air defense units into engineer battalions. The abandonment of the strategy of "nation at arms" (*La Guerra de Todo el Pueblo*) meant the abolishment of conscription. The disappearance of the threat of a U.S. invasion made superfluous the militias numbering over a million individuals, and a provisional government could substitute them with very small army reserves. As part of the "nation at arms" strategy, the militias had complemented the Committees for the Defense of the Revolution, which existed in every block and village in Cuba. Just like the militias, these Committees also had to disappear, but the absence of these two mass organizations meant leaving a huge vacuum in authority throughout the country.[45]

The need to fill this void was a major reason why the Pentagon agreed to let Cuba have an army as necessary for repressing internal disorders and revolts. Seen from another perspective, the U.S. government intended to release the turbulent elements in the island and then imposed on a new army the obligation of keeping them under control. An even more fundamental flaw in the plan was the financing of a new military. Under Raúl the armed forces had become self-sustaining because of their participation in economic production, principally in growing food but also in other activities such as tourism and manufacturing. Because free-market principles considered the military with its stern discipline and rigid hierarchy to be unfair competition for the supposedly more efficient private businesses, the U.S. government required that the military abandon or privatize all its business activities and rely exclusively on government revenue to finance its operations. To expect the new pampered class of private businessmen to pay the taxes necessary to support an efficient military establishment is an assumption that goes against all the experience of similar business groups in Central America.[46] Consequently, low pay and inadequate funding meant poor performance and widespread corruption for the military and indeed for the entire provisional government.

The Cuban air force was a special case for the Pentagon. As chapter 6 explained, U.S. diplomats did everything possible to prevent Caribbean countries and even Mexico from acquiring jet warplanes. Cuba with the largest fleet of jet fighters in the region could not be an exception to this firm U.S. policy for long. The 2004 Commission for a Free Cuba left implicit that no powerful air force would be tolerated so near to the U.S. border. Making this statement was a blunder and only served to make the air force determined to fight to the death against any U.S. invasion. The 2006 Commission for a Free Cuba corrected this flaw, but it was up to the 2007 *Post-Transition Cuba* report to specify the details for a future Cuban air force. Because Cuba in reality is an archipelago of over 1600 islands, *Post-Transition Cuba* recognized that the obligation of reaching those many isolated islands made indispensable preserving intact the fleet of transport airplanes and helicopters, including the attack helicopters. A provisional government could probably count on U.S. assistance to replace the Soviet transport aircraft but probably not the attack helicopters.[47]

The response to the jet fighters was much more complex and disguised the Pentagon's healthy respect for the skilled Cuban pilots. Dismantling the also fearsome air defense system seemed easy to accomplish by transferring the personnel to other ground units, but disbanding the pilots and their jet fighters was a much trickier proposition. The pride of the Cuban air force was its less than a dozen jet fighters able to take on U.S. warplanes. Over decades of careful observation the Cuban air force had acquired secret techniques to delay the overwhelming aerial challenge from the United States, and this handful of planes was the first line of defense. Nobody doubted that these jet fighters would soon be shot down, yet pilots fiercely competed to fly these planes in the suicidal missions against massive U.S. airpower. Using the argument that supporting different airframes was too costly, the Pentagon was willing to allow Cuba to keep the 50 MIG 21s, obsolete planes of the 1960s no longer posing any risk to the United States.[48] But this major concession could only be temporary, and as cannibalization reduced the number of operational MIG 21s, U.S. diplomats, just as they did for Caribbean countries and even Mexico, were sure to block any attempts to buy replacements of even obsolete jet fighters.

The U.S. government had always condemned "the inordinately large size of the armed forces,"[49] therefore making all the more startling the recommendation "that the Cuban Navy should be dramatically expanded."[50] The measure of "coastline length" was a valid reason to justify Cuba's need for a navy as large as those of countries with comparable seashores, and the argument can be made that Cuba with many more islands than other countries actually requires an ever bigger navy. The U.S. government needed Cuba to perform law-enforcement functions over its maritime zones, and it was always a source of embarrassment to U.S. officials that the revolutionary regime actually performed those duties to a better degree than the client states in the Caribbean. Cuba was able to police its maritime zones effectively, even with a small navy, because of the reliance on the militias. Duty in uninhabited islands and patrols in open boats were routine assignments for the militia units from seashore communities.

Helicopters, patrol craft, and swift vessels fit well into the requirements for meeting U.S. interests, but Cuba's tiny Marine Corps did not make the cut as qualifying for survival. Numbering from 500 to 1000 Marines, this amphibious force had the capability of invading countries in the area but was too small to threaten any but the tiniest of the island nations of the English-speaking Caribbean. A revolutionary regime striving to cultivate the friendliest of relations with all the countries of the Caribbean never considered taking any hostile action with this small force of Marines. But the ideological baggage of the distant Cold War required disbanding this Marine Corps, even when those soldiers were ideal for conducting raids in the many isolated and often deserted Cuban islands.[51] Furthermore, as previous chapters had shown, U.S. officials had easily developed very close ties with the Marine Corps of Colombia and with the Marines in less friendly Mexico. Even accepting the questionable assumptions behind planning for the military of a transitional government, the abolishment of the Marine

Corps was contrary to U.S. interests and reflected the obsession with depriving Cuba of anything remotely resembling an offensive capability.

If by some miracle the U.S. government had the opportunity to refashion the Cuban armed forces, the army has comprised such an essential part of the national identity, that any reduction would face bitter opposition. At best the U.S. government would have to accept a military larger than that in the Dominican Republic but smaller than under the revolutionary regime. But without the militias, Committees of the Defense of the Revolution, and other mass organizations, a single military institution risked again the pattern of military involvement in politics, as was the case before the Cuban Revolution. Much as residents of Miami might welcome a return to an idealized past, the combination of an inadequate military, widespread poverty, income inequality, and corrupt politicians in a provisional government was sure to sink Cuba to that level of disturbances, lawlessness, and insecurity so rampant throughout the Caribbean region. Under the blueprints of the U.S. government, a post-revolutionary Cuba at best would become a playground for drug cartels as in Central America and at worst a failed state like nearby Haiti.[52]

15

Trying to Dismantle the Bush Legacy

> In light of the failure of CAFTA-DR to produce real gains in economic growth and job creation ... the region was losing patience with the Obama administration and expected the U.S. to move faster to change development policies, increase aid to the region, and establish a path to rapidly end the blockade of Cuba.—U.S. Embassy, Managua[1]

U.S. policy toward the Caribbean, Central America, and Mexico profoundly changed when George W. Bush replaced Bill Clinton as president of the United States on 20 January 2001. A survey of key actions amply demonstrates the contrast between the two administrations.

Friendship and a reluctance to intervene shaped the relations of the Clinton administration to the region, with the exception of Cuba. And even for Cuba, the President was responding mainly to political pressures from right-wing groups and Miami Cubans when he imposed new economic sanctions on the revolutionary regime. Yet in that case negotiation rather than confrontation was the preferred option, and the Clinton administration repeatedly postponed and tempered the harsh economic sanctions. And when Hugo Chávez became president on 2 February 1999, Clinton officials with curiosity watched unfold the new social experiment in Venezuela.

Not constrained by domestic ideological pressures over Venezuela, the Clinton administration was free to follow its preferred policy of letting the governments in the region explore solutions to the immense social and economic problems facing those countries. But when a crisis required direct action, the best example of the policy of friendship came with the financial bailout of Mexico in January 1995. At immense political risk, President Clinton used executive authority to provide funds from the U.S. Treasury to avert the bankruptcy of Mexico and to prevent the weakening of the U.S. financial system. The swiftness of the bailout and its complete success impressed President Ernesto Zedillo so much that, in gratitude, he did everything possible to strengthen his country's ties with the United States.

For the first time in history, a Mexican president ordered the armed forces to develop close working relationships with the U.S. military.

Without any doubt the Mexican bailout was the greatest achievement of the Clinton administration in Latin America, but to replicate that success when taking direct action in other countries proved to be a much more elusive task. In the case of Colombia, the Clinton administration drew up Plan Colombia to defeat Marxist guerrillas and destroy drug cartels. The original proposal called for massive U.S. investment in economic and social programs, but later drafts shifted gradually the emphasis to military support for the Colombian government. Reluctant to get involved in a land war, the Clinton administration excluded military support and largely limited Plan Colombia to assisting the police reduce the flow of cocaine flowing into the United States. But without a strong commitment to economic and social programs, this strategy failed to reduce the drug trade, while the absence of a military component made Plan Colombia ineffective against the Marxist guerrillas.

The inadequate results from Plan Colombia contrasted with the clear failure of the U.S. occupation of Haiti. A last-minute deal changed the intervention from a combat invasion to a peaceful occupation; in a very revealing detail, American paratroopers boarded airplanes assuming that they would jump into combat only to disembark unopposed to assume police functions. Although the initial entry of American troops into Haiti on 19 September 1994 was bloodless, everything else the U.S. government tried to do in that country was disastrous. Clearly Haiti had been the worst setback of the Clinton administration in Latin America. But because the country was the poorest in the Western Hemisphere and all previous interventions in Haiti had likewise ended dismally, the U.S. inability to transform the country passed largely unperceived.

Haiti has solidified its status as the only failed state in the Western Hemisphere and also revealed the limitations of a military solution. Indeed, the greater the direct intervention of the U.S. government and the less the economic assistance, the larger was the likelihood of failure. The lessons from the Clinton administration were that hostility and intervention were not effective while friendship and economic assistance were the best ways to enhance close relations with the region.

The incoming George W. Bush administration, motivated by ideological imperatives, rejected the obvious conclusions from the Clinton experience. Instead the new administration adopted a policy of confrontation and intervention and reduced the economic component to the minimum.

Already by the end of the twentieth century it had become obvious that the large countries of South America were most reluctant to join the North American Free Trade Agreement (NAFTA). Even before Hugo Chávez successfully led the efforts to block the approval of a Western Hemisphere free trade agreement in 2005, the U.S. government concluded that the proposal was dead. Previously U.S. diplomats had been working simultaneously on a much more modest alternative covering only the countries of Central America and the Caribbean. Among the island states of the West Indies only the Dominican Republic qualified to participate. The member states signed the treaty in 2004, and after ratification CAFTA-DR went into effect the next year.[2] Free trade can be beneficial but not when done independent of other essential measures such as massive capital investment, generous economic assistance, and extensive training. But for the ideologically driven Bush administration, the free trade agreement eliminated the need to consider the economic and social measures desperately needed to help poor countries escape their often grinding misery. Supposedly the unseen hand of the free market would do its magic and the region would almost miraculously escape poverty.

The unconditional allies of Colombia, El Salvador, and Guatemala heartily shared these conservative and free market ideas. The Bush administration found an ideological soul mate in President Álvaro Uribe Vélez and generously rewarded him by converting Plan Colombia into a massive assistance program for both the military and the police. And in Mexico the election of Vicente Fox in 2000 and then of his successor in 2006 meant that very conservative presidents were in office. For a while it seemed that the Bush administration was going to succeed in imposing its free market ideologies throughout the region.

However, because not all the governments were convinced that market mechanisms and free trade could miraculous transform the economies of the region, the Bush administration assumed the task of imposing regimes sharing these views about the curative powers of capitalism. Intervention to destroy regimes hostile to capitalism was not just a noble calling but an obligation. The administration espoused the goal of making the world safe for capitalism, an idea harking back to the earlier failed slogan of Woodrow Wilson of "making the world safe for democracy."

First on the hit list of the Bush administration was Cuba. Already during the electoral campaign the Republican candidate had promised to hasten the departure of Fidel Castro. Promptly upon taking office, the new administration increased the economic sanctions and did everything possible to persuade skeptical allies to end all trade with the revolutionary regime. But the U.S. invasion of Afghanistan and then of Iraq temporarily deflected the Bush administration from its stated goal of overthrowing Castro. Flushed initially with success at the start of the Iraq war, the Bush administration used the threat of invading Cuba as a way to provoke an uprising in the island. Because the resilience and endurance of the revolutionary regime surprised most Western observers, logically the next step for Washington to take was to invade the island. But as the U.S. occupations of Iraq and Afghanistan dragged on and turned into bloody quagmires for American troops, not surprisingly the Pentagon became most reluctant to take on another overseas commitment.

An undaunted Bush administration claimed to have found the explanation for the resilience of the revolutionary regime not in the determination of the Cuban people but in the assistance Hugo Chávez was providing from Venezuela. Consequently, or at least so the reasoning went, overthrowing Chávez was the essential prerequisite for destroying the Cuban regime. A gullible media in the U.S. eagerly depicted the elected Venezuelan president as a mad dictator capable of taking unpredictable actions. The CIA prepared to overthrow Chávez in the same way that it had done in Guatemala against Jacobo Arbenz in 1954, but the availability of a wealthy opposition made unnecessary a large U.S. involvement in the coup attempt of April 2002. The plotters knew that the Bush administration stood ready to assist the new provisional government in every way possible, but the utter failure of the coup left the U.S. with an invasion as the only way to eliminate the popular ruler in Venezuela.

Unilateral interventionism and the insistence on imposing right-wing ideology generated considerable resistance to the Bush administration not just in Latin America but also across the world. To its utter horror and dismay, after a few years the Bush administration began to confront the real possibility that leftist regimes were coming to power in Central America not by bullets but by votes in free and open elections. In spite of spending large amounts of money in those elections, the Bush administration could not convince the voters that capitalism rather than government intervention in the economy was the solution to the poverty and huge social problems of those countries. The first shock came when the Sandinistas won the elections in Nicaragua and Daniel Ortega took office as President on 10 January 2007. Right-wing conservatives in the United States regarded the Marxist-inspired party as a relic from the Cold War, and to have the Sandinistas return to office seemed inconceivable. Promptly U.S. diplomats began a campaign to discredit in every way possible the new Ortega presidency. The specter loomed that Nicaragua was joining Venezuela and Cuba in forming a new "axis of evil" in the Caribbean, but the Bush administration, drowning in foreign policy failures and the start of the Great Recession in 2007, could only direct words against the Sandinistas in Nicaragua.

Capitalism or neo-liberalism as it was then oddly called, was causing havoc in Central America, Mexico, and even Jamaica. Mexican drug cartels formed transportation and distribution networks stretching from the United States to Colombia and covering most of the Central America and Jamaica. Drug traffickers usually operated with near impunity in most of the countries, so gradually the cocaine problem became one no longer just of transportation but also of domestic consumption. Associated with this illicit trade but actually having different origins was the rise of violent gangs in most of the Central American countries from Guatemala to Panama. These gangs supplemented their participation in the drug trade with a wide range of criminal activities, including kidnappings and extortion.

Crime waves swept most of the countries, and in a tepid response the Bush administration obtained from the U.S. Congress passage of the Mérida Initiative, a campaign to help countries fight the epidemic of drug trafficking, particularly in Mexico. The allocated sums were too small to make more than a dent in the flow of drugs entering the United States. If anything, the Mérida Initiative was showing that government intervention rather than free market capitalism was the only way to start to solve the massive and overwhelming problems of the countries in the region.

By 2008 the accumulation of foreign and domestic failures was too much for the Bush administration to ignore, and American voters turned out of a sense of frustration to a newcomer on the political scene, Barack Obama, who was elected president on a wave of great expectations. Initial signs were encouraging, and in a fulfillment of a campaign promise, he lifted the restrictions the Bush administration had placed on travel and sending money to Cuba by Cuban-Americans. The ban on tourism and regularly scheduled direct airline flights remained in place, but Cuban-Americans could travel directly to Cuba on chartered planes. In a new innovation Americans could also ride those chartered jetliners when visiting the island on special study tours designed to learn about local culture and to meet with the Cuban people. The economic sanctions were still in force, but this initial easing created hope of establishing normal relations between the two countries and also of restoring friendly ties with Venezuela.[3]

Additional steps toward normalizing relations with Cuba and Venezuela had to wait until the Obama administration first decided upon a response to Central America's shift to the left. Was the coming to power of former Marxist guerrillas a fundamental threat to the United States? A quick examination of the leading figures in these leftist political parties showed that their ideology had mellowed considerably and actually resembled the views of the Obama administration. Many of the officials appointed to run the new governments turned to have business and engineering backgrounds and were in no way radical bomb throwers. By February 2009 at the latest the Obama administration concluded that the leftist governments of Central America posed no danger to the United States. In a clear statement to President Manuel Zelaya of Honduras about the new policy, "the Ambassador stressed that we had worked well on numerous issues with his government and that our concern was not whether he ruled from the left or right of the political spectrum."[4]

The Obama administration implemented the new policy of acceptance in Guatemala, El Salvador, and Nicaragua. The easiest case to solve was Guatemala, because even the Bush administration had concluded that the inauguration of Álvaro Colom as the first leftist president of the country in the twenty-first century did not pose a threat. It was easy for the Obama administration to ratify the policy of maintaining close relations with President Colom. Nicaragua was a tougher case, because the Bush administration had made clear its complete opposition to the return of the Sandinistas to power. But savvy U.S. diplomats were well attuned to the new winds prevailing in Washington, and gradually the alarmist reports about the revival of a totalitarian regime in Managua diminished and then faded away. The many concrete instances of cooperation between the military of Nicaragua and the United States finished burying any fears about hostility from the second poorest country in the Western Hemisphere.

Reversing U.S. policy toward Nicaragua was a major step toward returning to the policy of friendship reminiscent of the Bill Clinton administration. A first opportunity to start from scratch came when voters in El Salvador elected as president the candidate of the former guerrillas on 15 March 2009. Here the Obama administration could be proactive and not merely react to what the Bush administration had decided to do. The Obama administration extended the welcome mat to Mauricio Funes even before his inauguration on 1 June 2009, and the new president responded with sincere expressions that El Salvador intended to remain an unconditional ally of the United States. The attendance at his inauguration of Secretary of State Hillary Clinton, who wore a bright red dress matching the official color of the Salvadoran guerrilla party, was the most visible symbol of the start of a new relationship with the region.

Hopes were high that the Obama administration was returning to a policy of friendship with the region, but in retrospect Hillary in her bright red dress at the inauguration in San Salvador came to signify not the beginning of a new policy but the high point of the Obama's administration's efforts to dismantle the Bush policy of confrontation and hostility. Relations with Revolutionary Cuba, as has been the case since 1959, provided the most visible barometer on U.S. policy toward Latin America.

When the Obama administration lifted restrictions on travels and remittances by Cuban Americans to Cuba, expectations were high that these baby steps were merely the beginning of a long trip to restore normal relations with the United States. Just a few days after Hillary Clinton was strutting around in San Salvador, the Secretary of State made it absolutely clear that U.S. hostility toward Cuba had not changed in the least, in spite of considerable rhetoric and speculation.[5] The public banter even reached far away Bulgaria halfway around the world, and its government "in light of the willingness of the new administration in changing U.S. policy towards Cuba" was seeking guidance on a Cuban offer to purchase a Soviet-era air defense system.[6] Promptly the U.S. government stated its complete opposition to the transfer of any weapons to Cuba even if defensive. In effect, the Obama administration wholeheartedly embraced the extreme hostility of the Bush administration and left open as a final solution the possibility of launching an invasion to destroy what the U.S. government branded as a State Sponsor of Terrorism.

And in a startling similarity with the Bush administration, no sooner had Obama taken office when the U.S. government launched new programs to overthrow the Cuban government. Two of these programs, incongruously operated by AID, the Agency for International Development that is supposed to concentrate on economic and humanitarian assistance, have been uncovered. In the first program AID hired students from Costa Rica, Peru, and Venezuela who went to the island to recruit young Cubans to create opposition to the government. The Latin American students were supposed to provide information on HIV and other humanitarian causes but then gradually tried to turn Cuban youngsters into opponents of the regime.[7]

This program nicely complemented the second more ambitious project called Zunzuneo, a rough equivalent of a twitter. Rightfully the Cuban government had been most reluctant to open up the country to digital media because of fears that U.S. agencies could infiltrate these channels. As a tentative step, Raúl Castro authorized the use only of cell phones shortly after he became president. AID in a cloak-and-dagger operation obtained through a spy half a million phone numbers from the state telephone agency and then set up a shell company in Cayman Islands to start a primitive texting service relying only on the cell phones. This crude service came to have 68,000 subscribers by early 2012. Originally broadcasting harmless news about sports and music, Zunzuneo started to take a more political edge, and AID aimed to be able to summon "smart mobs" to bring pressure on the Cuban government to accept U.S. demands. Paradoxically, budget cuts forced AID to cancel the program in June 2012, when already suspicious Cuban authorities were trying to shut it down.[8]

The U.S. hostility came as no surprise to elder statesman Fidel Castro who from his retirement was among the first to state that the Obama administration was not abandoning the traditional U.S. desire to dominate the Caribbean region. Castro warned his very close friend Hugo Chávez to expect no modification in U.S. hostility toward Venezuela, but the ever optimistic and idealistic Venezuelan President ignored his friend's advice and believed a return to the policy of Bill Clinton was possible. Chavez went so far as to break protocol in a summit conference and approached Obama to congratulate him and to give the U.S. president a book explaining the real nature of the problems of Latin America. Both countries had been without high-level representatives since 11 September 2008 when Venezuela recalled its ambassador to Washington and expelled U.S. ambassador Patrick Duddy. After consultations with the Venezuelan government, President Obama took the unprecedented decision of returning the expelled Duddy back to Caracas, while the Venezuelan ambassador resumed his post in Washington, D.C. But barely had the U.S. ambassador returned to Caracas on 1 July 2009 amid favorable media coverage, when Secretary of State Hillary Clinton poisoned the well by granting a hostile interview to an opposition news channel. Not surprisingly efforts to resume a constructive dialogue with Venezuelan officials slowly crumbled.[9] A contrite Venezuelan President had to admit to his mentor Castro that the Cuban elder statesman had been right all along in predicting that the Obama administration would do little to change the Bush policy to the region.

In spite of the sincere attempts by Chávez to restore friendly ties with the United States at the start of the Obama administration, the belief spread in Washington that his mercurial personality was the real barrier to improving relations. The publicity stunts and his colorful antics had endeared him to the people of Venezuela and of many parts of Latin America, but Washington and most Western capitals considered them as being in bad taste and harmful to a healthy relationship. Thus, when a terminal illness brought the Venezuelan president to a premature death on 5 March 2013, expectations ran high that relations were sure to improve. A calm and less eccentric successor supposedly was more likely to restore at least a working relationship with the U.S. government. The newly elected president of Venezuela was the mild-mannered Nicolás Maduro who lacked the theatrical flair of his predecessor. But he was just as firm as Chávez in opposing U.S. imperial control, and already in March 2013 the new President expelled two U.S. military attachés accused of improperly intervening in Venezuelan internal affairs. Predictably the U.S. government responded with similar expulsions of diplomats from the Venezuela Embassy in Washington, D.C., and relations continued to deteriorate with both countries lacking ambassadors in each other's capitals since the departure of Ambassador Duddy in July 2010.[10]

President Maduro actually wanted to improve relations with the United States, and the new Secretary of State John Kerry picked up on this willingness when he publicly shook hands with the Venezuelan foreign minister on 5 June 2013. But it was too late for this attempt that would have been a great response to the gesture of Chávez in early 2009. Talks began about how to exchange ambassadors, but then the Edward J. Snowden crisis erupted and derailed a fragile arrangement. When Secretary of State Kerry tried to pressure his Venezuelan counterpart to deny asylum to the American whistleblower, official talks ground to a halt but were still not completely dead. The last blow came when the new U.S. ambassador to the United Nations during her senate confirmation hearings made harsh comments about the Venezuelan internal situation. In response, President Maduro had his foreign minister publicly announce in July 2013 that efforts to improve relations with the United States had come to an end. A downward spiral began when Venezuela expelled three U.S. diplomats on 30 September 2013, and several days later the State Department responded by expelling three diplomats from the Venezuelan Embassy in Washington. Relations with Venezuela have continued to deteriorate in 2014 with more mutual expulsions of diplomats and the involvement of U.S. officials in opposition demonstrations.[11]

It was hard to avoid noticing that both in 2009 and again in 2013 powerful groups inside the U.S. government did everything possible to sabotage attempts to improve relations with Venezuela. Actually the same pressures had already succeeded in blocking attempts to improve relations with Cuba. Only in Central America had the Obama administration reversed the policy of hostility to leftist governments, but after that step forward to better relations, the U.S. government largely abandoned attempts to dismantle the

Bush legacy. This has meant that as the Obama administration coasts toward the end of its second term in office, the Bush policy of hostility and domination has remained largely intact and become an integral part of U.S. foreign relations.

Expecting the Obama administration to return to the policy of Bill Clinton did not seem setting a high bar to measure success in the region. Yet this modest goal remained beyond the capacities of the Obama administration, and the last attempts under the new Secretary of State John Kerry have likewise not succeeded in dismantling the Bush legacy. Bill Clinton had been the maverick president who was most resistant to intervene in the affairs of the countries to the south and preferred to extend the hand of friendship. But once he left office in January 2001, the forces relentlessly driving the U.S. government to impose control over the Caribbean region tried to make up for the time lost under the Clinton presidency.

The 11 September 2001 terrorist attacks on the United States and the subsequent massive military involvement in the Middle East may seem to provide adequate reasons for the survival of a harsh and controlling U.S. policy toward the region. If such a permanent shift occurred in the initial years of the twenty-first century, then the failure of the Obama administration to dismantle the Bush policies toward the Caribbean region is not just understandable but also predictable. Waging war in the Middle East and a new mentality among Americans fearful of another terrorist attack seemed to justify a tough and controlling policy toward the Caribbean and Central American countries. But because no terrorist attack has ever come from this region, this argument does not prove a need for a harsh policy and only provides a justification for continuing with the Bush policies. In reality, the 9/11 terrorist attacks conveniently supported hostility toward Cuba and Venezuela as a way of imposing stricter controls over all the countries in the area. Because the terrorist threat is only a great pretext, the search to locate the forces driving U.S. policy toward Central America, the Caribbean, and Mexico must turn elsewhere. These structural currents run underground, and uncovering their exact nature poses a daunting task. Three explanations are possible for the U.S. policy toward the region.

Three Explanations

The chapters in this book have amply demonstrated that the United States has exerted great influence and even control over most the Caribbean,

Central America, and Mexico since the end of the Cold War. At the same time, the informal empire of the United States has not resulted in a significant improvement in the conditions of the majority of the population. Indeed, the living standards of most countries in this region are very low, and Haiti ranks among the poorest countries in the entire world. The political structures, although supposedly "democratic" with elections regularly occurring, have served in the majority of countries to perpetuate the privileges of small elites. In addition, gang violence and criminal acts routinely afflict many inhabitants of the region. Instead, those countries refusing to follow U.S. guidance have generally done better in most areas over the long run. The question inevitably arises of why the conditions of those countries following U.S. policies have remained so bad while at the same time U.S. influence has increased.

Three possible explanations are possible. The first and most charitable explanation claims that the U.S. government is genuinely committed to improving the lot of the countries under its protection, but that shortage of funds, the pervasive nature of the problems, and the ineffectiveness of government structures have meant that any substantial improvements require decades to achieve. The NAFTA and the CAFTA-DR treaties have tried to stimulate economic growth through free trade, while the Bill Clinton bailout of Mexico in 1994 showed that the United States was willing to give money to save a friend in need. The Bush administration with the Mérida Initiative indicated a commitment to offer Mexico equipment and training to fight the drug cartels. The Bush administration also expanded funding for Plan Colombia, and although the military dimension of the program dwarfed the economic assistance, the desire of the United States to help an unconditional ally in its war with Marxist guerrillas was clear. And programs such as Police Reestablishment in Colombia, besides helping in the war against guerrillas, also provided personal security for the first time to many inhabitants of rural areas.

Critics have often pointed that the programs of the Bush administration have been at best inadequate, usually useless, and at worst harmful. But whatever the shortcomings of the Bush-era programs, in contrast, the Obama administration did not even bother to launch comparable initiatives for the region. The Great Recession and the recovery efforts for the U.S. economy only partially accounted for the failure of the Obama administration to implement substantial programs in the region. On one level, the reason has been the difficulty of securing funds from the U.S. Congress for financing costly assistance programs in the region. In a strange paradox, American legislators have routinely authorized spending trillions of dollars to pour into the deserts of the Middle East but consistently have been most reluctant to approve millions of dollars to try to solve the problems of countries right in the U.S. doorstep. As will be recalled, even the Clinton bailout of Mexico was a purely executive branch action, because the U.S. Congress simply avoided having to take that politically risky action.

Consequently, the U.S. government, in spite of the best intentions, lacks the funds to shake the countries of the region from their chronic and massive structural defects of pervasive poverty and unemployment, rampant crime, and income inequality. All that U.S. officials can do is chip away at the problems and try to mitigate the worst consequences. And not much else seems likely until a combination of happy circumstances miraculously shakes ample funds out of a most hostile U.S. Congress to finance the long overdue investments in economic development and law enforcement.

Scholars have noted that since the end of the Cold War in 1990, despite reduced assistance, U.S. influence in the region has actually increased. The only major setback for the U.S. government has been the Chávez phenomenon in Venezuela after 1998. The collapse of the Soviet Union left the United States free to expand its control over the region, but these were only opportunities and not actual influence. So the paradox of increasing influence at a time of retrenchment in assistance seems at first sight hard to explain. A different scenario is needed to account for this unexpected outcome.

The second explanation argues that U.S. officials have deliberately used the problems of the region as the means to insert U.S. influence into those countries. If somehow magically those problems disappeared, then U.S. diplomats no longer enjoyed the justification to shape the internal affairs of those countries and thus lost leverage to pressure host countries. What became important for U.S. officials was to work alongside local officials in every major issue facing the country and not to worry that the problems never seemed to get resolved. Gradually the word spread among aspiring politicians that the most important vote in elections was that of the U.S. Embassy. Candidates for elected and appointed offices absorbed the lesson that U.S. blessing was indispensable. And most important, elected officials who in previous decades had even dared to challenge U.S. policies, now pleaded for U.S. help in solving the problems.

An excellent example of this complete turnaround was President Óscar Arias, who in 1986 forced the Ronald Reagan administration to end its support for the Contras in Costa Rica. This defiance was a bad precedent, but U.S. officials could not find a way to take reprisals against a recipient of the Nobel Peace Prize. Because of the spread of the drug trade in Central America, Costa

Rica asked for help from the United States, and in 2009 President Arias personally was pleading for a Black Hawk helicopter for his country. The transformation from defiance to a submission could not have been more complete. And best of all, the U.S. government, by letting free market forces do their work, did not have to lift a finger to convert Costa Rica into a client state.[12]

The United States has not enjoyed this control over all the countries, and the major exceptions have occurred in Mexico and Cuba since their revolutions, and of course in Venezuela since 1999. But even in the case of Mexico, its problems have served as an entry point to extend U.S. influence and control. And should a transitional government ever come to power in Cuba, the U.S. is ready with blueprints, as the previous chapter explained, to make sure that the inevitable and massive problems of a post-revolutionary Cuba guarantee permanent and widespread U.S. involvement in the internal affairs of the island. This second explanation is the one that best fits with the evidence presented in this book and in particular with the thousands of cables from U.S. embassies in the region and available in Wikileaks.

Harsh as the second explanation might seem, a third explanation takes a totally different approach to reach the most drastic conclusion of all. It is no longer the U.S. government using the problems of a region to extend its influence, rather it is the United States by its very existence being the cause of the worst problems afflicting the region. This idea made a brief appearance in the author's first book on the wars of Latin America when its conclusions noted that of the four countries suffering insurrections "all four countries experienced U.S. influence in varying degrees."[13] Those conclusions then went on to explain how the type of U.S. influence shaped individual insurrections, but the really fundamental question of what role the existence of the United States played remained unaddressed. The question haunted the next two volumes of the trilogy on the wars of Latin America, but the burden of having to trudge through battles in nearly continuous combat prevented studying the implications of this insight. Finally this last chapter can explore the question.

Not even whole volumes could hope to explain all the reasons that have made the United States the greatest, richest, and most powerful country in the world. Not surprisingly, even the most superficial examination reveals many profound differences between the United States and the countries of the Caribbean and Central America. The huge size, large population, and vast natural resources put the United States in a distinct category with the nearest rival being a Mexico trailing far behind in these characteristics. Because of this rich material base and an efficient tax collection system, the United States has been able to afford three branches in the national government (with even two houses in the Congress), a federalist system with considerable powers left to the fifty states, and many increasingly very expensive elections at all levels. The interaction of these many levels of government with an even larger private sector has proved to be a fruitful source of ideas, innovations, and productivity. Tension and disputes have constituted a normal and intrinsic part of the process. The positive results of these clashes demonstrate that in the United States the ideas of Niccolò Machiavelli about the fruitfulness of struggles between the wealthy and the people have eclipsed Karl Marx's gloomy predictions about class warfare.

Lacking the geographical scale and natural wealth of the United States, the countries of the Caribbean and Central America have been able to replicate only parts or segments of the U.S. model. In the case of Mexico, the attempt to mimic the federal system of independent states can only be described as an unstated failure. By not being able to reproduce the entire U.S. model, the countries of the region have succeeded in adopting only components easy to introduce such as free trade and have failed to adopt other essential practices such as an efficient tax collection system. But the adoption of only some elements means that they usually have harmful consequences. Even such apparently noble causes as "spreading democracy" can distort and pervert governments in many countries. Not only have the drug trade and lawlessness spread throughout the region, but these very problems serve to extend the influence and control of the U.S. government. In a nutshell, the United States is the source of most of the problems in the region, and then those issues serve as the entry for the U.S. government to try to find solutions. But the problems remain unsolved and the cycle of dependency continues to grow during the initial decades of the twenty-first century. Not surprisingly, Marx's ideas about the ruinous effects of class warfare clearly displaced the optimistic predictions of Machiavelli.

Having the countries of the region trapped in poverty and lawlessness does afford the United States the advantages of having cheap labor nearby. Whether working in factories producing for export to the American market or coming as illegal immigrants, the laborers of these countries obviously contribute to the profits of American corporations. And while these corporations are quite naturally strong advocates of free trade measures such as CAFTA-DR, they have no interest in asking the U.S. Congress to finance comprehensive assistance packages because if the poverty and lawlessness of those countries diminished, they would then lose their source of cheap labor so conveniently near to the United States and often inside in the form of illegal immigrants. At least a development program on the scale of the memorable Marshall plan to reconstruct a ravaged Europe has been needed for the region since the middle of the twentieth century. Actually, a comparable sum of assistance might be inadequate because so much of the task in the region consists of building from scratch rather than reconstructing as was the case in post–World War II Europe.

The police forces make an excellent example of the magnitude of the task facing the countries. In post–World War II Europe the restoration of pre-existing police forces proceeded smoothly, and the war-ravaged countries of Western Europe did not have to make major efforts to enjoy adequate police and judicial services. But in the countries of the Caribbean and Central America, police forces have been dismal failures at best. The only exemption has been the police of Colombia, which partly because of its French traditions, usually has provided a satisfactory performance. And as chapters 4, 10, and 11 abundantly showed, the most pathetic failure has been Mexico. After a century of efforts, the Mexican government has been completely unable to create efficient police forces, and rampant corruption remains the most notorious trait of that country's police. And without efficient law-enforcement agencies, how can a country pretend to develop its economy when its citizens are under constant danger of criminal activity?

In the well-known science-fiction series of *Star Trek*, the Prime Directive prohibits advanced civilizations from having any contact at all with peoples in a rudimentary stage of technological development. The well-reasoned argument was that even with the best and noblest of intentions, any involvement with less technologically advanced peoples inevitably brought unintended consequences always harmful and usually of a disastrous nature. And while it may be too soon to propose a Prime Directive of complete isolation for U.S. relations with the Caribbean and Central America, as the decades roll by without any substantial improvement, this policy option will have to be seriously considered.

A complete separation between countries is not as easy to accomplish as between planets in different solar systems, yet already elements of a modified Prime Directive are in place. The most notable example is Cuba thanks to the U.S. embargo. From a special perspective, the U.S. embargo has been the greatest gift to the Cuban people, because Cuba has been spared the most horrific consequences flowing from free and open intercourse with the United States. The revolutionary government closely limits the interactions between its citizens and the United States. Long distance helps to mitigate the spread of defects coming from the Western European countries and Canada, but in Cuba already some problems such as prostitution reappear when European and Canadian tourists freely enter the island without any prior controls. But the gang violence, the drug epidemic, the crime wave, the kidnapping, and the general lawlessness rampant in most of the countries of the region have largely been absent from Cuba.

Could it be that the United States is inherently an evil influence on the region? The experience of Haiti would seem to argue so, because in the country where the United States has been most deeply involved, the results could not have been more disastrous. It can be argued that almost anything U.S. officials try to do in Haiti turns out badly. Reversing the Cold War rhetoric, then it would seem that the evil empire was not the disappeared Soviet Union but after all was the United States itself. But this explanation assumes that the United States is inherently evil, when that is not true and when Haiti is such an exceptional case. However, the Haitian example best illustrates the real paradox that things, events, actions, and institutions are generally positive while occurring inside the United States but turn out to be disastrous if not downright evil when transplanted south of the U.S. border.

In some instances no doubt exists that the U.S. government and American citizens want to do harmful and even evil actions. The best example of U.S. government evil-doing is the well-publicized prisoner treatment at Guantánamo base. By exporting these detainees and their torture outside U.S. soil, the Bush administration could legally get away with actions that normally justified prosecution in war crimes tribunals. Extending the argument beyond Guantánamo, the principle is that in the region the U.S. government, private American citizens, and U.S. corporations are free to carry out actions that are completely unacceptable and illegal on U.S. soil. An indirect and tacit admission of this harmful behavior comes from the U.S. government when it insists on immunity from local prosecution for American soldiers in the Status of Force Agreements (SOFA). American soldiers receive the subliminal message that they can get away with almost anything as long as it happens only in the host country (an enlargement to the 10th power of the slogan "that what happens in Las Vegas stays in Las Vegas"). But as viewers of reality TV shows such as *Cops* know, even in Las Vegas the police and judges diligently crack down on criminal activity, so the SOFA example simply returns to the underlying argument that abuses and evil actions occur only when a part of the U.S. system is transplanted to another country.

To globalize the analysis, it can be argued that Western Europe and countries like Australia are not corrupted and degraded to the same degree despite extensive contacts with the United States. But countries in Western Europe have rich traditions of strong institutions to resist the most harmful

U.S. influences, and others such as Australia enjoy the safety of being a long distance away. But weak countries without adequate institutions and right next to the United States are completely unprotected from massive American influence and consequently are exposed to corruption and degeneration. Privatization removes the barriers sheltering inhabitants from U.S. corporations and weakens local institutions. At the individual level, the arrival of turbulent and sometimes unscrupulous American adventurers has been a source of turmoil in many countries. Instead in Cuba, where the visits of Americans are tightly controlled and channeled to cultural, artistic, and environmental sites, the government has been able to restrict the negative influence of unsavory individuals who would otherwise contaminate the Cuban people with defects and vices.

Large Mexico seems to be able to cope with free trade, although objections to these economic relations have not been lacking. In particular, privatization inevitably meant the rise of privileged billionaires in Mexico, a scenario likely to repeat itself in any post-revolutionary Cuba. For the smaller countries free trade has only helped a minuscule privileged minority enrich itself. Only strong government participation in the economy makes it possible for most countries to withstand the harmful pressures coming from the United States. Any type of solution necessarily requires a high degree of socialism, but only if leaders keep in mind the lessons of Hugo Chávez who insisted on preventing government institutions from falling under the control of elites.

Chapter Notes

Chapter 1

1. Bill Clinton, *My Life* (New York: Alfred A. Knopf, 2004), p. 642.
2. This and the next paragraph draw on Bart Jones, *¡Hugo! The Hugo Chávez Story from Mud Hut to Perpetual Revolution* (Hanover, NH: Steerforth Press, 2007), pp. 281–282, 377; Rory Carroll, *Comandante: Hugo Chávez's Venezuela* (New York: Penguin, 2013), p. 59; Steve Ellner, *Rethinking Venezuelan Politics: Class, Conflict, and the Chávez Phenomenon* (Boulder, CO: Lynne Rienner, 2009), pp. 110–112.
3. Harold A. Trinkunas, *Crafting Civilian Control of the Military in Venezuela* (Chapel Hill: University of North Carolina Press, 2005), pp. 208–214; Jones, *Hugo*, pp. 226, 238–240, 254–256; Cristina Marcano and Alberto Barrera Tyszka, *Hugo Chávez* (New York: Random House, 2007), pp. 130–131.
4. Brian A. Nelson, *The Silence and the Scorpion: The Coup Against Chávez and the Making of Modern Venezuela* (New York: Nation Books, 2009), p. 5; Jones, *Hugo*, pp. 251–254, 286–290, 274–275; Ellner, *Chávez Phenomenon*, pp. 197–198. Many scholars have read too much into the Venezuelan refusal to accept the assistance of Marine Corps personnel to help repair heavy damage from record floods near Caracas in December 1999. The comments of Chávez were actually diplomatic, and the real reason he did not want American military personnel was because the U.S. State Department required signing an innocent sounding "Status of Forces Agreement" essentially granting immunity from prosecution to American soldiers. As a fervent nationalist, Chávez could not accept having U.S. soldiers commit any crimes and leave the country without receiving punishment. Venezuela was joining Mexico and Costa Rica in the policy of refusing to allow the presence of foreign troops inside its territory. In contrast, almost all the client states in the Caribbean have signed Status of Forces Agreements. Readers may recall that the refusal of the Iraqi Parliament to renew the Status of Forces Agreement accelerated the departure of U.S. troops from that Middle Eastern country in 2011.
5. Jones, *Hugo*, pp. 296–303; Carroll, *Comandante*, pp. 71–72; Ellner, *Chávez Phenomenon*, p. 199. Just like in the case of Cuba in the summer of 1959, it would be very revealing to know the precise date when the Bush administration first considered overthrowing the Chávez regime and when exactly the U.S. government first decided to support a coup in Venezuela. But the answers may never be known, because U.S. officials have been more diligent in destroying incriminating evidence than were their predecessors on the decision to overthrow the revolutionary regime in Cuba in the summer of 1959. See René De La Pedraja, *Wars of Latin America, 1948–1982: The Rise of the Guerrillas* (Jefferson, NC: McFarland, 2013), pp. 86–87 and in particular p. 320n6.
6. Nelson, *Coup Against Chávez*, pp. 4–5; Jones, *Hugo*, pp. 294, 305.
7. Ellner, *Chávez Phenomenon*, pp. 112–114; Nelson, *Coup Against Chávez*, pp. 6, 135–136; Gregory Wilpert, "The 47-Hour Coup That Changed Everything," 2007 April 13, p. 1; Jones, *Hugo*, pp. 294, 305–308, 312. Most accounts do not emphasize the 49 decrees as the turning point in the Chávez presidency and its relations with the U.S. government. The decisive break with private business and organized labor took place in November 2001 and sometime afterward the Bush administration decided to support the overthrow of Chávez.
8. Wilpert, "47-Hour Coup," pp. 1–2; Jones, *Hugo*, p. 293; Steve Ellner, *Organized Labor in Venezuela 1958–1991* (Wilmington, DE: Scholarly Resources, 1993), pp. 181–184, 221, 227–231. Observers have overemphasized racism as a major source for the hatred of elites against the dark-skinned Hugo Chávez. In reality, racism is just one instrument in the tool kit of privileged groups. In Latin America the upper class has accepted even dark-skinned presidents, such as Fulgencio Batista in Cuba, as long as they were protecting elite privileges. While Chávez was preserving elite benefits and considering privatizations sure to favor the upper

class, the racism card remained absent. But after the 49 decrees of November 2001 Venezuelan elites incessantly used racism among other accusations to try to undermine the Chávez regime.

9. Marcano and Barrera Tyszka, *Hugo Chávez*, pp. 147–148; Eva Golinger, *The Chávez Code: Cracking U.S. Intervention in Venezuela* (Northampton, MA: Olive Branch Press, 2006), pp. 53–54. Ellner, *Chávez Phenomenon*, p. 114 calls the "prolonged alliance [...] unprecedented in Venezuelan history," and this alliance between the largest labor confederation and the main business association may well have been unique in Latin American history.

10. Jones, *Hugo*, pp. 309–310. The causes of the vitriolic hostility of private media owners against Chávez were varied; for a detailed discussion see Marcano and Barrera Tyszka, *Hugo Chávez*, chapter 13. Besides personality clashes between Chavistas and journalists, media owners missed the privileged status they enjoyed under previous presidencies. The loss of paid advertising from state agencies certainly hurt media profits, but this reprisal came after the break with Chávez had occurred and was not a real explanation. Media owners sympathized with fellow Venezuelan capitalists, who, however, proved more flexible in seeking accommodation with the Chávez regime than the permanently recalcitrant and inflexible media outlets. Even among elite groups media owners stood out by their intense opposition to Chavista populist attempts to increase the participation of previously ignored citizens.

11. Nelson, *Coup Against Chávez*, pp. 66–67; Jones, *Hugo*, pp. 255, 315; Ernesto Villegas Poljak, *Abril, golpe adentro* (Caracas: Correo del Orinoco, 2012), pp. 108-113.

12. Nelson, *Coup Against Chávez*, p. 278; U.S. Department of State, *A Review of U.S. Policy Toward Venezuela: November 2001-April 2002*, pp. 14–15. Even the Bush administration, in its whitewash of its involvement in the coup, admitted that the hostile U.S. Embassy reporting "underestimated President Chávez's popular support." Ibid., p. 39.

13. For Guatemala and the Bay of Pigs see De La Pedraja, *Wars of Latin America, 1948-1982*, pp. 34–45, 86–108. For Chile see Thomas C. Wright, *Latin America in the Era of the Cuban Revolution* (Westport, CT: Praeger, 2001), pp. 139–147. For the Brazil coup of 1964, the key book is Phyllis R. Parker, *Brazil and the Quiet Intervention, 1964* (Austin: University of Texas Press, 1979).

14. Golinger, *The Chávez Code*, pp. 47, 49, 58; Wilpert, "47-Hour Coup," pp. 18–19; Nelson, *Coup Against Chávez*, pp. 277–280; Jones, *Hugo*, pp. 302–304; U.S. Department of State, *A Review of U.S. Policy Toward Venezuela*, pp. 26–27.

15. Jones, *Hugo*, p. 303; U.S. Department of State, *A Review of U.S. Policy Toward Venezuela*, pp. 14, 40; Villegas, *Abril*, p. 422. For the U.S. intelligence failure during the Bay of Pigs invasion, see De La Pedraja, *Wars of Latin America 1948-1982*, pp. 88–91.

16. Nelson, *Coup Against Chávez*, pp. 24, 164–165; Villegas, *Abril*, pp. 84, 422.

17. Wilpert, "The 47-Hour Coup," p. 2; Jones, *Hugo*, pp. 204–205, 311–312.

18. Wilpert, "The 47-Hour Coup," pp. 2–3; Jones, *Hugo*, p. 315.

19. Nelson, *Coup Against Chávez*, pp. 13–14, 21–22; Wilpert, "The 47-Hour Coup," p. 3; Jones, *Hugo*, p. 315; Marcano and Barrera Tyszka, *Hugo Chávez*, p. 195.

20. This paragraph and the next draw on Nelson, *Coup Against Chávez*, pp. 20–21; Review of Dilan A. Nelson book by Gregory Wilpert, Amazon.com; Jones, *Hugo*, pp. 315–316; Venezuela, Asamblea Nacional, *Informe de la Comisión Parlamentaria Especial para investigar los sucesos de abril de 2002*, p. 23.

21. Nelson, *Coup Against Chávez*, pp. 14–15; Jones, *Hugo*, pp. 316–317; Francisco Toro, "The Untold Story of Venezuela's 2002 April Crisis," 14 April 2004, pp. 2, 7, Caracaschronicles.com; Villegas *Abril*, p. 56. A last chance for the police to stop the marchers came when they split into two columns: one proceeded in the tunnel and another in the parallel street to the north. See Map 2.

22. Nelson, *Coup Against Chávez*, pp. 232–233; Toro, "Venezuela's 2002 April Crisis," pp. 5, 13. Confusion exists about the deployment and movements of the tank column; I have found the most plausible version in Asamblea Nacional, *Informe de los sucesos de abril de 2002*, pp. 36, 39.

23. Asamblea Nacional, *Informe de los sucesos de abril de 2002*, p. 39.

24. Nelson, *Coup Against Chávez*, pp. 26–29; Jones, *Hugo*, p. 326. The decision to have the entire column march into Baralt Avenue is crucial to understand subsequent events. Other accounts, and even the map of Nelson, *Coup Against Chavez*, p. 10 have the march approaching Miraflores Palace simultaneously through Baralt Avenue, South Eight Street, and New Republic Viaduct, but Nelson, Ibid., p. 28 gives the crucial evidence when a marcher answers in response to the question of why not going up South Eight Street "Nobody's going that way." This point might seem quibbling over minutiae, but the implications are significant, because (1) turning first at Baralt Avenue was another indication that by the afternoon the number of marchers was considerably less than

those who had gathered early in the morning in front of the PDVSA building; (2) the leaders no longer had enough demonstrators to try to advance simultaneously along many streets and had to maintain a single strong column; (3) the leaders insisted on the turn at Baralt Avenue and did not want the marchers to veer off into other streets in the more direct routes to Miraflores Palace.

25. Wilpert, "47-Hour Coup," p. 5; Nelson, *Coup Against Chávez*, pp. 291-293. As proof of the government's bloody repression of the demonstration, anti-Chávez TV stations and hostile foreign media repeatedly broadcast a famous video of civilian Chavistas firing pistols at an undetermined target. After being arrested, these Chavistas at trial were acquitted of all charges because they had fired over an hour after the last civilian deaths occurred. Secondly, the distance from the site to the opposition demonstrators exceeded the range of their pistols. In all likelihood they were replying to shots from the rebel Metropolitan Police, but the video conveniently left out their target, so that the prosecution lacked evidence to convict them of killing anybody. For a careful examination of this video, see Jones, *Hugo*, pp. 327-329, 371.

26. Wilpert, "47-Hour Coup," p. 6; Jones, *Hugo*, pp. 326-327, 330-331; Nelson, *Coup Against Chávez*, pp. 132-134.

27. Jones, *Hugo*, pp. 332-333; Wilpert, "47-Hour Coup," pp. 10-11.

28. Jones, *Hugo*, pp. 334-340; Wilpert, "47-Hour Coup," p. 11; Nelson, *Coup Against Chávez*, pp. 154-158; Asamblea Nacional, *Informe de los sucesos de abril de 2002*, pp. 46-49.

29. Wilpert, "47-Hour Coup," pp. 10-11; Jones, *Hugo*, pp. 337-340; Villegas, *Abril*, pp. 109-113, 424.

30. Wilpert, "47-Hour Coup," p. 12; Nelson, *Coup Against Chávez*, p. 156.

31. Wilpert, "47-Hour Coup," p. 11. On the office seekers see also Nelson, *Coup Against Chávez*, p. 156 and Toro, "Venezuela's 2002 April Crisis," p. 14.

32. Toro, "Venezuela's 2002 April Crisis," p. 16; Asamblea Nacional, *Informe de los sucesos de abril de 2002*, pp. 77-81.

33. U.S. Department of State, *A Review of U.S. Policy Toward Venezuela*, pp. 73, 75; Golinger, *The Chávez Code*, pp. 44-45, 73-74; Nelson, *Coup Against Chávez*, pp. 183-187; Wilpert, "47-Hour Coup," p. 19.

34. David Adams and Phil Gunson, "The Unmaking of a Coup," *St. Petersburg Times*, 22 April 2002, p. 3; Toro, "Venezuela's 2002 April Crisis," pp. 14-15; Nelson, *Coup Against Chávez*, pp. 194-195; Ellner, *Chávez Phenomenon*, p. 117.

35. Wilpert, "47-Hour Coup," pp. 12-13; Toro, "Venezuela's 2002 April Crisis," pp. 16-17; Asamblea Nacional, *Informe de los sucesos de abril de 2002*, pp. 65-66.

36. Jones, *Hugo*, p. 356; Toro, "Venezuela's 2002 April Crisis," pp. 18-19; Asamblea Nacional, *Informe de los sucesos de abril de 2002*, pp. 68-71.

37. David Adams and Phil Gunson, "The Unmaking of a Coup," *St. Petersburg Times*, 22 April 2002, p. 5.

38. Wilpert, "47-Hour Coup," p. 12; David Adams and Phil Gunson, "The Unmaking of a Coup," *St. Petersburg Times*, 22 April 2002, p. 5; Jones, *Hugo*, pp. 351-352; Villegas, *Abril*, pp. 179-180, 426.

39. Nelson, *Coup Against Chávez*, pp. 277-281; Jones, *Hugo*, pp. 43, 366-368; U.S. Department of State, *A Review of U.S. Policy Toward Venezuela*, pp. 35-36, 43, 75.

40. Nelson, *Coup Against Chávez*, p. 198; Wilpert, "47-Hour Coup," p. 13; Jones, *Hugo*, pp. 352-353, 358; Toro, "Venezuela's 2002 April Crisis," p. 15; Villegas, *Abril*, p. 123.

41. Nelson, *Coup Against Chávez*, pp. 210-212; Jones, *Hugo*, p. 357.

42. Nelson, *Coup Against Chávez*, p. 226; Jones, *Hugo*, pp. 357-358; Wilpert, "47-Hour Coup," pp. 13-14.

43. Jones, *Hugo*, pp. 358, 363; Wilpert, "47-Hour Coup," p. 14. An English translation of the statement by General Vásquez Velasco is available in Toro, "Venezuela's 2002 April Crisis," pp. 17-18.

44. Jones, *Hugo*, pp. 353-355, 358-360, 363-364; Wilpert, "47-Hour Coup," p. 14.

45. Jones, *Hugo*, pp. 363-364; Wilpert, "47-Hour Coup," pp. 14-15.

46. Jones, *Hugo*, pp. 364-365; Wilpert, "47-Hour Coup," p. 15. The anti-Chávez Nelson, *Coup Against Chávez*, p. 226 admits that "the only people who were ever punished in connection with April 11 were the Metropolitan Police," but in other pages, such as p. 271 he gives the impression of extensive persecution of coup participants. My text reflects the nuanced presentation of Jones, *Hugo*, pp. 368-371. Even the seven fired PDVSA executives were reinstated, although at least one of them later betrayed this second opportunity Chávez gave them, see Ellner, *Chávez Phenomenon*, pp. 118-119.

47. Jones, *Hugo*, p. 365; Toro, "Venezuela's 2002 April Crisis," p. 20.

Chapter 2

1. Niccolò Machiavelli, *The Prince*, trans. W.K. Marriott (London: Everyman's Library, 1908), p. 21.

2. Steve Ellner, *Organized Labor in Venezuela, 1958-1991* (Wilmington, DE: Scholarly Resources, 1993), pp. xxiii, 82; Roberto Briceño-León, "Violencia, ciudadanía y miedo en Caracas," *Foro internacional* 47(2007): 557-558.

3. Harold A. Trinkunas, *Crafting Civilian Control of the Military in Venezuela; A Comparative Perspective* (Chapel Hill: University of North Carolina Press, 2005), p. 215; Bart Jones, *¡Hugo! The Hugo Chávez Story from Mud Hut to Perpetual Revolution* (Hanover, NH: Steerforth Press, 2007), pp. 231-232; Cristina Marcano and Alberto Barrera Tyszka, *Hugo Chávez* (New York: Random House, 2007), pp. 138-139.

4. "Transparency of Budgets/Military Spending," Cable from U.S. Embassy, Caracas, 23 November 2004, Wikileaks.

5. Jones, *Hugo*, pp. 231-232; "Keller Poll: Venezuelans Support Chávez and the Military Blindly," Cable from U.S. Embassy, Caracas, 13 May 2005, Wikileaks.

6. Trinkunas, *Military in Venezuela*, pp. 211-214; Jones, *Hugo*, pp. 255-256; Marcano and Barrera Tyszka, *Hugo Chávez*, p. 162.

7. Steve Ellner, *Rethinking Venezuelan Politics: Class, Conflict, and the Chávez Phenomenon* (Boulder, CO: Lynne Rienner, 2009), p. 111; Trinkunas, *Military in Venezuela*, pp. 211-214.

8. Jones, *Hugo*, p. 229; Trinkunas, *Military in Venezuela*, pp. 213-214; Marcano and Barrera Tyszka, *Hugo Chávez*, pp. 258-259.

9. Jones, *Hugo*, pp. 368-371; Trinkunas, *Military in Venezuela*, pp. 221-222, 228.

10. Ellner, *Chávez Phenomenon*, pp. 198-199. For the end of the insurgency in Venezuela, see René De La Pedraja, *Wars of Latin America, 1948-1982: The Rise of the Guerrillas* (Jefferson, NC: McFarland, 2013), pp. 182-185.

11. "How Real is Chávez's Oil Threat?" Cable from U.S. Embassy, Caracas, 4 March 2004, Wikileaks; "Transparency of Budgets/Military Spending," Cable from U.S. Embassy, Caracas, 23 November 2004, Wikileaks; Trinkunas, *Military in Venezuela*, p. 216.

12. Rory Carroll, *Comandante: Hugo Chávez's Venezuela* (New York: Penguin, 2013), pp. 142-143; Jones, *Hugo*, pp. 406-407, 437, 468-469. For Fidel Castro's speech declaring the Cuban Revolution to be socialist in April 1961 on the eve of the Bay of Pigs invasion, see De La Pedraja, *Wars of Latin America, 1948-1982*, pp. 94-95.

13. "Chávez Decrees Formation of Military Reserves," Cable from U.S. Embassy, Caracas, 11 April 2005, Wikileaks; "Chávez Calls for Citizen Soldiers," Cable from U.S. Embassy, Caracas, 18 May 2004, Wikileaks; Ellner, *Chávez Phenomenon*, p. 167.

14. "Chávez Decrees Formation of Military Reserves," Cable from U.S. Embassy, Caracas, 11 April 2005, Wikileaks.

15. "Chávez Decrees Formation of Military Reserves," Cable from U.S. Embassy, Caracas, 11 April 2005, Wikileaks.

16. "Venezuelan Military Doctrine Still Evolving," Cable from U.S. Embassy, Caracas, 5 July 2005, Wikileaks. See also *New York Times*, 11 June 2006.

17. "Electoral Logistics Continue to Favor Chávez," Cable from U.S. Embassy, Caracas, 27 October 2006, Wikileaks.

18. "Chávez Decrees Formation of Military Reserves," Cable from U.S. Embassy, Caracas, 11 April 2005, Wikileaks; "Russian Arms Arrive in Venezuela," Cable from U.S. Embassy, Caracas, 27 March 2008, Wikileaks; "Hugo Chávez Returns to Russia Seeking More Arms," Cable from U.S. Embassy, Moscow, 27 June 2007, Wikileaks.

19. "What to Expect from Chávez's Presidential Campaign," Cable from U.S. Embassy, Caracas, 21 February 2006, Wikileaks; Eva Golinger, *The Chávez Code: Cracking U.S. Intervention in Venezuela* (Northampton, MA: Olive Branch Press, 2006), pp. 33-34.

20. "Venezuelan Military Doctrine Still Evolving," Cable from U.S. Embassy, Caracas, 5 July 2005, Wikileaks.

21. "Chávez Speech Shows Military Cubanization Afoot," Cable from U.S. Embassy, Caracas, 7 January 2005, Wikileaks; "Venezuelan Military Doctrine Still Evolving," Cable from U.S. Embassy, Caracas, 5 July 2005, Wikileaks.

22. "Cuba/Venezuela Axis of Mischief: The View from Caracas," Secret Cable from U.S. Embassy, Caracas, 30 January 2006, Wikileaks.

23. "Chávez Speech Shows Military Cubanization Afoot," Cable from U.S. Embassy, Caracas, 7 January 2005, Wikileaks; "Cuba/Venezuela Axis of Mischief: The View from Caracas," Secret Cable from U.S. Embassy, Caracas, 30 January 2006, Wikileaks.

24. "The Venezuelan Military in Government and Society," Cable from U.S. Embassy, Caracas, 5 June 2006, Wikileaks; Marcano and Barrera Tyszka, *Hugo Chávez*, pp. 258-259.

25. "Venezuelan Military Morale Declining?" Secret Cable from U.S. Embassy, Caracas, 12 January 2006, Wikileaks; "Government of Venezuela Postures against Guerrillas and Alleged Subversion," Cable from U.S. Embassy, Caracas, 2 June 2005, Wikileaks.

26. "Venezuelan Military Morale Declining?" Secret Cable from U.S. Embassy, Caracas, 12 January 2006, Wikileaks.

27. "Zulia Governor Accused of Coup Plotting," Cable from U.S. Embassy, Caracas, 30 January 2006, Wikileaks; "Viva Zulia—Chávez Again Hits Rosales," Cable from U.S. Embassy, Caracas, 7 March 2006, Wikileaks; "BRV Backs Off Zulian Separatist Allegations for Now,"

Cable from U.S. Embassy, Caracas, 13 March 2006, Wikileaks.
28. "Viva Zulia—Chávez Again Hits Rosales," Cable from U.S. Embassy, Caracas, 7 March 2006, Wikileaks.
29. "Venezuela-Dutch Antilles: Some Ideas," Cable from U.S. Embassy, Caracas, 21 June 2005, Wikileaks.
30. "Carrier Visit to Benefit U.S. Policy Toward Venezuela," Secret Cable from U.S. Embassy, Caracas, 23 March 2006, Wikileaks.
31. "Ambassador's Additional Comment." In "Carrier Visit to Benefit U.S. Policy Toward Venezuela," Secret Cable from U.S. Embassy, Caracas, 23 March 2006, Wikileaks. "Ambassador's Additional Comment" is an indication of the exceptional importance of a particular issue or recommendation. Comments by Ambassadors are very scarce in the cables on Latin America available in Wikileaks. This cable was extremely sensitive and under normal circumstances would have been classified not just SECRET but also NO FORN; however, the possibility of wanting to share the message with Dutch or other allied governments may have persuaded U.S. diplomats to label the 23 March cable only SECRET.
32. "Electoral Logistics Continue to Favor Chávez," Cable from U.S. Embassy, Caracas, 27 October 2006, Wikileaks. Conversations during my study trip to Caracas of June 2007. Opposition leaders routinely denounced the sporadic involvement of soldiers in the voting process as part of a plot to steal the elections when the root cause was the traditional inefficiency of civilian employees.
33. "Election Update: Rosales Inroads, Still Uphill Battle," Cable from U.S. Embassy, Caracas, 17 November 2006, Wikileaks; "The Military's Increasingly Red Role in the Election," Secret/No Forn Cable from U.S. Embassy, Caracas, 21 November 2006, Wikileaks; Carroll, *Comandante*, pp. 142–144.
34. "Election Update: Rosales Inroads, Still Uphill Battle," Cable from U.S. Embassy, Caracas, 17 November 2006, Wikileaks; "The Military's Increasingly Red Role in the Election," Secret/No Forn Cable from U.S. Embassy, Caracas, 21 November 2006, Wikileaks.
35. "Election Update: Rosales Inroads, Still Uphill Battle," Cable from U.S. Embassy, Caracas, 17 November 2006, Wikileaks; Carroll, *Comandante*, p. 22; Jones, *Hugo*, pp. 446, 454–455.

Chapter 3

1. Honoré de Balzac, *Maximes et pensées de Napoléon* (Paris: Editions de Fallois, 1999), p. 83. Author's translation.
2. René De La Pedraja, *Wars of Latin America, 1982–2013: The Path to Peace* (Jefferson, NC: McFarland, 2013), pp. 157–159; Susan Peacock and Adriana Beltrán, *Hidden Powers in Post-Conflict Guatemala: Illegal Armed Groups and the Forces Behind Them* (Washington, D.C.: Washington Office on Latin America, 2003), p. 28; Jennifer L. Burrell, *Maya After War: Conflict, Power, and Politics in Guatemala* (Austin: University of Texas Press, 2013), pp. 35–36, 154.
3. "Guatemala Downsizes its Military," Cable from U.S. Embassy, Guatemala City, 6 November 2003, Wikileaks; "Guatemala Military Reservists to Aid Police," Cable from U.S. Embassy, Guatemala City, 24 February 2006, Wikileaks; "Guatemala reconoce la debilidad de su ejército," 23 October 2009, Notiamérica.com. For the genocide campaigns of the Guatemalan army, see De La Pedraja, *Wars of Latin America, 1982–2013*, pp. 14–19, 59–63.
4. Angelina Snodgrass Godoy, *Popular Justice: Violence, Community, and Law in Latin America* (Stanford: Stanford University Press, 2006), pp. 1–2; Burrell, *Maya After War*, pp. 117, 137, 143, 154; *Washington Post*, 24 February 2006. The omission of the crime wave sweeping Guatemala after 1996 is a major structural flaw in Peacock and Beltrán, *Post-Conflict Guatemala*.
5. "Guatemala Downsizes its Military," Cable from U.S. Embassy, Guatemala City, 6 November 2003, Wikileaks; Snodgrass Godoy, *Popular Justice*, pp. 54–55; Peacock and Beltrán, *Post-Conflict Guatemala*, pp. 63–67; Francisco Villagrán Kramer, *Biografía política de Guatemala: años de guerra y años de paz* (Guatemala: Editorial de Ciencias Sociales, 2004), pp. 401–402.
6. "Guatemala Downsizes its Military," Cable from U.S. Embassy, Guatemala City, 6 November 2003, Wikileaks; "Guatemala reconoce la debilidad de su ejército," 23 October 2009, Notiamérica.com.
7. "Scene Setter for March 5 Pcc on Guatemala," Cable from U.S. Embassy, Guatemala City, 1 March 2004, Wikileaks.
8. "Ambassador and Minugua Chief Tour D'Horizon," Cable from U.S. Embassy, Guatemala City, 10 January 2003, Wikileaks; "Letter from Guatemala (9)," Cable from U.S. Embassy, Guatemala City, 5 August 2005, Wikileaks.
9. "Government of Guatemala Accord on Deep Military Reductions," Cable from U.S. Embassy, Guatemala City, 30 March 2004, Wikileaks; "Breakfast with Berger: Progress on Military Downsizing," Cable from U.S. Embassy, Guatemala City, 10 June 2004, Wikileaks.

10. "Guatemala Downsizes its Military," Cable from U.S. Embassy, Guatemala City, 6 November 2003, Wikileaks.
11. "Guatemala Military Reservists to Aid Police," Cable from U.S. Embassy, Guatemala City, 24 February 2006, Wikileaks.
12. "Guatemalan Military Modernization Needed," Cable from U.S. Embassy, Guatemala City, 27 March 2006, Wikileaks.
13. "Breakfast with Berger: Progress on Military Downsizing," Cable from U.S. Embassy, Guatemala City, 10 June 2004, Wikileaks.
14. "Guatemalan Participation in Peacekeeping Operations is Growing," Cable from U.S. Embassy, Guatemala City, 10 December 2004, Wikileaks; "Guatemala Military Reservists to Aid Police," Cable from U.S. Embassy, Guatemala City, 24 February 2006, Wikileaks; "Guatemalan Soldiers Allegedly Beat Street Kids," Cable from U.S. Embassy, Guatemala City, 13 June 2006, Wikileaks; *Latin American Herald Tribune*, 10 February 2006; Kevin Lewis O'Neill and Kedron Thoman, eds., *Securing the City: Neoliberalism, Space, and Insecurity in Postwar Guatemala* (Durham: Duke University Press, 2011), p. 110.
15. "Office of Inspector General Inspection of Embassy San Salvador," Secret Cable from U.S. Embassy, San Salvador, 15 December 2005, Wikileaks.
16. For a full discussion of CAFTA-DR, see Mark B. Rosenberg and Luis G. Solís, *The United States and Central America: Geopolitical Realities and Regional Fragility* (New York: Routledge, 2007), pp. 83–96.
17. "El Salvador: The Case for Major Non-NATO Ally Status," Confidential/No Forn Cable from U.S. Embassy, San Salvador, 5 November 2007, Wikileaks. Paragraph also draws on Elana Zilberg, *Space of Detention: The Making of a Transnational Gang Crisis Between Los Angeles and San Salvador* (Durham: Duke University Press, 2011), pp. 43, 48 and Ellen Moodie, *El Salvador in the Aftermath of Peace* (Philadelphia: University of Pennsylvania Press, 2010), p. 79.
18. "Your Visit to El Salvador," Cable from U.S. Embassy, San Salvador, 28 September 2007, Wikileaks; De La Pedraja, *Wars of Latin America, 1982–2013*, p. 156.
19. "El Salvador's Proposed Changes for SOFA," Cable from U.S. Embassy, San Salvador, 11 January 2007, Wikileaks; "El Salvador Agrees to SOFA," Cable from U.S. Embassy, San Salvador, 1 March 2007, Wikileaks.
20. "Scenesetter for Visit of Salvadoran President Saca," Secret Cable from U.S. Embassy, San Salvador, 20 February 2007, Wikileaks; "El Salvador to Stay in Iraq," Cable from U.S. Embassy, San Salvador, 30 April 2007, Wikileaks.

21. "Your Visit to El Salvador," Cable from U.S. Embassy, San Salvador, 28 September 2007, Wikileaks; Congressional Research Service, *El Salvador: Political, Economic, and Social Conditions and Relations with the United States*, 3 April 2006, p. 5.
22. "El Salvador: The Case for Major Non-NATO Ally Status," Confidential/No Forn Cable from U.S. Embassy, San Salvador, 5 November 2007, Wikileaks. The only Latin American country to have acquired Major Non-Nato Ally status was Argentina in 1998. The threat of repealing that status was a bargaining chip for U.S. officials, see Heritage Foundation, "Argentina No longer Deserves to Be a Major Non-Nato Ally of the U.S.," 26 November 2012.
23. "Violence Spirals in El Salvador, Government Grasps for Solutions," Cable from U.S. Embassy, San Salvador, 17 February 2006, Wikileaks.
24. "Annual OSAC Crime/Safety Report—El Salvador," Cable from U.S. Embassy, San Salvador, 9 January 2006, Wikileaks; De La Pedraja, *Wars of Latin America, 1982–2013*, pp. 156–157; Charles T. Call, "Democratisation, War and State-Building: Constructing the Rule of Law in El Salvador," *Journal of Latin American Studies* 35(2003): 833; Canada, Immigration and Refugee Board, *El Salvador: The National Civilian Police* (Ottawa: Immigration and Refugee Board, 1998), pp. 2–3, 14–15.
25. "Salvadoran Law Enforcement Overwhelmed by Violent Street Crime," Cable from U.S. Embassy, San Salvador, 13 September 2006, Wikileaks. Perhaps because of a typing error in the original, the cable gives the total of 572 but adding the numbers gives 575.
26. This paragraph and the next draw on "Violence Spirals in El Salvador / Government Grasps for Solutions," Cable from U.S. Embassy, San Salvador, 17 February 2006, Wikileaks and Laura Pedraza Fariña, Spring Miller, and James L. Cavallaro, *No Place to Hide: Gang, State, and Clandestine Violence in El Salvador* (Cambridge: Harvard Law School, 2010), pp. 52–53.
27. Moodie, *El Salvador in the Aftermath of Peace*, pp. 68, 154–155; Call, "War and State-Building," pp. 835–836, 843; Pedraza Fariña, Miller, and Cavallaro, *No Place to Hide*, pp. 37–38; Zilberg, *Space of Detention*, pp. 40–41; Canada, *The National Civilian Police*, p. 6.
28. Oscar Bonilla, "El caso de El Salvador," in José Raúl Peralta, ed., *Reforma de las fuerzas armadas en América Latina y el impacto de las amenazas irregulares* (Washington, D.C.: Woodrow Wilson International Center, 2008), pp. 18–19; Zilberg, *Space of Detention*, p. 46; Pedraza Fariña, Miller, and Cavallaro, *No Place to Hide*, pp. 38–40; Call, "War and State-Building," pp. 839–843.

29. Zilberg, *Space of Detention*, p. 46.
30. "Salvadoran Law Enforcement Overwhelmed by Violent Street Crime," Cable from U.S. Embassy, San Salvador, 13 September 2006, Wikileaks; Bonilla, "El caso de El Salvador," p. 18.
31. "Scenesetter for Visit of Salvadoran President Saca," Secret Cable from U.S. Embassy, 20 February 2007, Wikileaks.
32. "Prospects for Colombia's Guerrilla War," Cable from U.S. Embassy, Bogotá, 8 July 1986, Wikileaks.
33. Ibid.
34. "The Guerrilla Situation: Preparing for War?" Cable from U.S. Embassy, Bogotá, 9 June 1987, Wikileaks; "Prospects for Colombia's Guerrilla War," Cable from U.S. Embassy, Bogotá, 8 July 1986, Wikileaks; CIA, "Colombian Counterinsurgency: Steps in the Right Direction," 26 January 1994, National Security Archive.
35. William Avilés, *Global Capitalism, Democracy, and Civil-Military Relations in Colombia* (Albany: State University of New York Press, 2006), pp. 102–104; De La Pedraja, *Wars of Latin America, 1982–2013*, pp. 193–194, 201–202.
36. Mario A. Murillo, *Colombia and the United States: War, Unrest, and Destabilization* (New York: Seven Stories Press, 2004), pp. 19–20, 127–128; De La Pedraja, *Wars of Latin America, 1982–2013*, pp. 200–202; Avilés, *Civil-Military Relations in Colombia*, pp. 129–130. The original proposal calling primarily for economic and social development was transformed into Plan Colombia concentrating on police and military assistance.
37. "Plan Colombia Phase II," Cable from U.S. Embassy, Bogotá, 18 February 2004, Wikileaks; Murillo, *Colombia and the United States*, pp. 20–21, 128; Adam Isacson, *Consolidating "Consolidation"* (Washington, D.C.: Washington Office on Latin America, December 2012), pp. 2–3; Avilés, *Civil-Military Relations in Colombia*, pp. 130–131.
38. "Operation Seventh of August Winds Down," Cable from U.S. Embassy, Bogotá, 14 September 2001, Wikileaks; De La Pedraja, *Wars of Latin America, 1982–2013*, pp. 206–207; James F. Rochlin, *Social Forces and the Revolution in Military Affairs: The Cases of Colombia and Mexico* (New York: Palgrave Macmillan, 2007), pp. 44, 46, 51–52.
39. Murillo, *Colombia and the United States*, pp. 126–128; Isacson, *Consolidating "Consolidation,"* pp. 2–3; Rochlin, *Social Forces*, p. 44.
40. Quotation and rest of paragraph from: "Plan Colombia Implementation Round-up, January 2005," Secret Cable from U.S. Embassy, Bogotá, 18 February 2005, Wikileaks and for rest of paragraph.
41. Overlooking no detail for running a competent organization, Admiral Luis Fernando Yance Villamil made it clear that overweight officers lost all chances at earning future promotions; see "Plan Colombia Implementation Round-up, January 2005," Secret Cable from U.S. Embassy, Bogotá, 18 February 2005, Wikileaks.
42. "Plan Colombia Implementation Round-up, March 2005," Cable from U.S. Embassy, Bogotá, 8 April 2005, Wikileaks; "Clipped Wings: Colombia's Air Asset Limits," Secret Cable from U.S. Embassy, Bogotá, 17 June 2005, Wikileaks; "Plan Colombia Monthly Highlights—August," Secret Cable from U.S. Embassy, Bogotá, 29 September 2005, Wikileaks.
43. "Plan Colombia Implementation Round-up, March 2005," Cable from U.S. Embassy, Bogotá, 8 April 2005, Wikileaks; "Clipped Wings: Colombia's Air Asset Limits," Secret Cable from U.S. Embassy, Bogotá, 17 June 2005, Wikileaks.
44. "Counter-Terrorism Wrap-Up," Cable from U.S. Embassy, Bogotá, 6 January 2005, Wikileaks.
45. *El Tiempo*, 28 April 2005.
46. "General Myers Meets with General Ospina," Cable from U.S. Embassy, Bogotá, 25 April 2005, Wikileaks.
47. "Plan Colombia Monthly Highlights—August," Secret Cable from U.S. Embassy, Bogotá, 29 September 2005, Wikileaks.
48. "Plan Patriota Phase 2b: Status Report," Cable from U.S. Embassy, Bogotá, 7 December 2005, Wikileaks; "Plan Colombia Monthly Highlights—August," Secret Cable from U.S. Embassy, Bogotá, 29 September 2005, Wikileaks; *El Tiempo*, 12 November 2005; "Joint Task Force Omega," 24 March 2013, Wikipedia.
49. Murillo, *Colombia and the United States*, pp. 144–146; "Plan Colombia Monthly Highlights—August," Secret Cable from U.S. Embassy, Bogotá, 29 September 2005, Wikileaks; Rochlin, *Social Forces*, pp. 51–52; Avilés, *Civil-Military Relations in Colombia*, p. 134.
50. "Plan Patriota Phase 2b: Status Report," Cable from U.S. Embassy, Bogotá, 7 December 2005, Wikileaks.

Chapter 4

1. "Engaging the New Mexican Administration on Security and Counterterrorism," Cable from U.S. Embassy, Mexico City, 15 June 2006, Wikileaks.
2. *New York Times*, 24 October 1995.
3. U.S. Department of the Army, *Army Country Profile—Mexico, Part I (U)*, April 1993, p. 31, Electronic Briefing Book No. 120, National Security Archive, George Washington

University (henceforth NSA); Craig A. Deare, *U.S.-Mexico Defense Relations: An Incompatible Interface* (Washington, D.C.: Institute for National Strategic Studies, 2009), pp. 2–3.

4. J. Jesús Esquivel, *La DEA en México: Una historia oculta del nacrotráfico contada por los agentes* (Mexico City: Grijalvo, 2013), p. 97; *U.S. News & World Report*, 11 December 2000; My conversations with Mexicans in the 1960s.

5. Jeffrey Davidow, *The Bear and the Porcupine: The U.S. and Mexico* (Princeton: Markus Wiener Publishers, 2007), p. 173. Paragraph also draws on: U.S. Department of the Army, *Army Country Profile—Mexico, Part I (U)*, pp. 15–16, 19, 26, 42, NSA; Stephen J. Wager, "Basic Characteristics of the Modern Mexican Military," in David Ronfeldt, ed., *The Modern Mexican Military: A Reassessment* (San Diego: Center for U.S.-Mexican Studies, 1984), p. 90; Edwin Lieuwen, *Mexican Militarism* (Albuquerque: University of New Mexico Press, 1968), p. 120; George W. Grayson, *Mexico: Narco-Violence and a Failed State?* (New Brunswick, NJ: Transaction Publishers, 2011), p. 159; Deare, *U.S.-Mexico Defense Relations*, p. 5.

6. The percentage of the budget for the two ministries declined slightly in the twenty-first century but more so for the Navy Ministry (down to around 25 percent) and less for the Defense Ministry (down to around 68 percent) mainly because the government created a separate budget line for the expenses of the armed forces social security institute; see Iñigo Guevara Moyano, *Adapting, Transforming, and Modernizing Under Fire: The Mexican Military 2006-2011* (Carlisle, PA: U.S. Army War College, 2012), p. 6.

7. René De La Pedraja, *Wars of Latin America, 1899-1941* (Jefferson, NC: McFarland, 2006), pp. 277–278; Roderic Ai Camp, *Generals in the Palacio: The Military in Modern Mexico* (New York: Oxford University Press, 1992), pp. 202–207, 249–260.

8. The PRI's monopoly on elections conveniently solved politicians' dilemma of wanting a long political career but not being able to run for reelection. In exchange for complete loyalty, the PRI nominated politicians to a succession of different elected positions, thus complying with the constitutional prohibition on reelection to the same office.

9. Grayson, *Narco-Violence*, pp. 132–135. In an additional restriction on police, municipal forces may only patrol, arrest, and respond to citizen complaints. Detective work, investigations, and the collection of evidence all fall under the purview of the state police except when the crimes also violate federal law, and in those cases the federal police forces have primary jurisdiction. See Ricardo C. Ainslie, *The Fight to Save Juárez: Life in the Heart of Mexico's Drug War* (Austin: University of Texas Press, 2013), p. 34.

10. "Mexico: More Interagency Cooperation Needed on Intelligence Issues," Secret/NoForn Cable from U.S. Embassy, Mexico City, 10 November 2009, Wikileaks.

11. Alfredo Corchado, *Midnight in Mexico: A Reporter's Journey through a Country's Descent into Darkness* (New York: Penguin, 2013), pp. 42–45.

12. Grayson, *Narco-Violence*, pp. 132–139. The evidence from former DEA agents in Esquivel, *La DEA en Mexico*, passim totally refutes the claim in Davidoff, *Bear and Porcupine*, pp. 46–47, who probably was following the official line, that "much of the information was developed by Mexican authorities and passed to the DEA office in Washington." But already in p. 47 and again in p. 59 of his book the former U.S. ambassador implies that U.S. agencies operated their own networks of agents.

13. "The Mexican Army—Still Passive, Isolated, and Above the Fray," Cable from U.S. Embassy, Mexico City, 11 May 1995, Electronic Briefing Book No. 120, NSA; Cable from U.S. Embassy, Mexico City, 13 April 2007, Wikileaks; Mexico, Instituto Nacional de Estadística y Geografía, *Anuario Estadístico de los Estados Unidos Mexicanos 2012* (Mexico: INEGI, 2013), Table 9.37.

14. "The Mexican Army—Still Passive, Isolated, and Above the Fray," Cable from U.S. Embassy, Mexico City, 11 May 1995, NSA. See also for paragraph, Esquivel, *La DEA en Mexico*, pp. 124–129.

15. Jesús López González, *Presidencialismo y fuerzas armadas en México, 1876–2012* (Mexico City: Gernika, 2012), pp. 232–233, 236–237; Lynn Stephen, *Zapata Lives! Histories and Cultural Politics in Southern Mexico* (Berkeley: University of California Press, 2002), pp. 140–145.

16. López González, *Fuerzas armadas en México, 1876-2012*, pp. 233–234, 237–238.

17. "The Mexican Army—Still Passive, Isolated, and Above the Fray," Cable from U.S. Embassy, Mexico City, 11 May 1995, NSA. For Álvaro Obregón's purchases from the United States to put down the De La Huerta Revolt, see De La Pedraja, *Wars of Latin America, 1899-1941*, pp. 281–282. Just as the arrival of American ammunition and weapons in 1923–1924 brought few results from a military perspective, so in 1994 the failure of the U.S. government to supply equipment and weapons had no military effect. But in both cases, the psychological and political impact was immense.

18. López González, *Fuerzas armadas en México, 1876-2012*, p. 255; "The Mexican Army—Still Passive, Isolated, and Above the Fray," Cable from U.S. Embassy, Mexico City, 11 May 1995, NSA; *Los Angeles Times*, 25 October 1995.

19. Susan Kaufman Purcell and Luis Rubio, eds., *Mexico Under Zedillo* (Boulder, CO: Lynne Rienner, 1998), p. 110; Bill Clinton, *My Life* (New York: Alfred A. Knopf, 2004), pp. 641-642.

20. Kaufman Purcell and Rubio, eds., *Mexico Under Zedillo*, p. 111; Clinton, *My Life*, pp. 642-645; *New York Times*, 14 March 1995.

21. Grayson, *Narco-Violence*, p. 226; *New York Times*, 29 December 1997.

22. International Consortium of Investigative Journalists, "U.S.-Trained Forces Linked to Human Rights Abuses," 12 July 2001, Internet; *New York Times*, 11 October 1995, 2 May 1996; Donald E. Schulz, *Between a Rock and a Hard Place: The United States, Mexico, and the Agony of National Security* (Carlisle, PA: U.S. War College, 1997), p. 14.

23. Grayson, *Narco-Violence*, p. 226; *Los Angeles Times*, 25 October 1996; *New York Times*, 2 May 1996; Deare, *U.S.-Mexico Defense Relations*, p. 3.

24. "Migration, Poverty, and Other Dynamics in Chiapas," Cable from U.S. Embassy, Mexico City, 26 October 2007, Wikileaks; López González, *Fuerzas armadas en México, 1876-2012*, pp. 233, 238, 240-242; Carlos Montemayor, *La guerrilla recurrente* (Mexico: Debate, 2007), pp. 41-42, 49; Stephen, *Cultural Politics in Southern Mexico*, pp. 199-211.

25. Carlos Montemayor, *Chiapas: la rebelión indígena en México*, 2d ed (Mexico: Joaquín Mortiz, 1998), pp. 77, 232; López González, *Fuerzas armadas en México, 1876-2012*, pp. 257-258; James F. Rochlin, *Social Forces and the Revolution in Military Affairs: The Cases of Colombia and Mexico* (New York: Palgrave MacMillan, 2007), p. 119.

26. "EPR-GOM Dialogue Helping to Keep the Peace," Cable from U.S. Embassy, Mexico City, 6 August 2008, Wikileaks; López González, *Fuerzas armadas en México, 1876-2012*, pp. 257-258, 289. The fullest scholarly account in English on EPR and the revolt in Guerrero is Rochlin, *Social Forces*, pp. 122-127.

27. U.S. Department of the Army, *Army Country Profile—Mexico, Part I (U)*, p. 28, NSA; López González, *Fuerzas armadas en México, 1876-2012*, pp. 239-240.

28. Yvonne M. Dutton, *U.S.-Mexico Extradition and Cross-Border Prosecution* (San Diego: Trans-Border Institute, 2004), pp. 1-2; *New York Times*, 2 May 1996.

29. When Secretary of Defense William Perry made the public announcement that U.S. and Mexican forces were going to conduct joint maneuvers, a firestorm of protest swept Mexico. The outrage over the presence of U.S. troops on Mexican soil forced President Zedillo to cancel the naval maneuvers that had always been intended to take place in the high seas; see Schulz, *Between a Rock and a Hard Place*, p. 13. Needless to say, Mexico since the Mexican Revolution has remained adamant in its refusal to sign a Status of Forces agreement authorizing armed U.S. soldiers to enter the country under complete immunity from local prosecution. Costa Rica since 1949, Cuba since 1959 and Venezuela since 1999 have been the other countries to refuse to grant immunity from prosecution to armed U.S. troops.

30. Davidow, *Bear and Porcupine*, pp. 93-94; Jorge Luis Sierra Guzmán, *El enemigo interno: Contrainsurgencia y fuerzas armadas en Mexico* (Mexico: Universidad Iberoamericana, 2003), p. 292; *New York Times*, 23 September 1995, 2 May 1996.

31. Davidow, *Bear and Porcupine*, p. 94. Paragraph also draws on Sierra Guzmán, *El enemigo interno*, pp. 292-293 and Schulz, *Between a Rock and a Hard Place*, p. 14.

32. *New York Times*, 1995 September 23.

33. The specifics of the operation for the arrest of Gulf Cartel chief Juan García Ábreu are highly disputed; see Esquivel, *La DEA en Mexico*, pp. 176-185.

34. This quotation and the next from Schulz, *Between a Rock and a Hard Place*, p. 19.

35. López González, *Fuerzas armadas en México, 1876-2012*, pp. 301-303; Esquivel, *La DEA en Mexico*, pp. 120-122; Anabel Hernández, *Los señores del narco* (Mexico: Grijalbo, 2010), pp. 210-212; Petter Watt and Roberto Zepeda, *Drug War Mexico* (London: Zed Books, 2012), pp. 128-132.

36. Grayson, *Narco-Violence*, pp. 75-76.

37. *New York Times*, 2 November 2002; Grayson, *Narco-Violence*, 76-77; *U.S. News & World Report*, 11 December 2000.

38. Grayson, *Narco-Violence*, 75-77.

39. Hernández, *Los señores del narco*, p. 274; López González, *Fuerzas armadas en México*, 1876-2012, pp. 300-301.

40. Ainslie, *The Fight to Save Juárez*, p. 33; Hernández, *Los señores del narco*, p. 436; "Calderon's Security Cabinet," Cable from U.S. Embassy, Mexico City, 11 December 2006, Wikileaks.

41. José Reveles, *El Cártel incómodo: El fin de los Beltrán Leyva y la hegemonía del Chapo Guzmán* (Mexico: Grijalbo, 2010), pp. 28-29, 57-63; Hernández, *Los señores del narco*, pp. 207-209. A plausible version claims that to shift the blame on someone else for his partic-

ipation in the murder of the Roman Catholic cardinal, the Lord of the Skies (Amado Carrillo Fuentes) betrayed El Chapo Guzmán to the authorities in 1993 and then as a partial compensation arranged for the latter's transfer to Puente Grande in 1995.

42. Hernández, *Los señores del narco*, pp. 237-261.

43. This paragraph and the previous one rely on Hernández, *Los señores del narco*, pp. 261-322, Reveles, *El Cártel incómodo*, pp. 60-69, and Ioan Grillo, *El Narco: Inside Mexico's Criminal Insurgency* (New York: Bloomsbury Press, 2011), pp. 92-93.

44. Hernández, *Los señores del narco*, pp. 359-363; Grayson, *Narco Violence*, p. 60; Grillo, *El Narco*, pp. 93-94.

45. Hernández, *Los señores del narco*, pp. 359-363; Ainslie, *The Fight to Save Juárez*, p. 13; U.S. General Accountability Office, *Drug Control: U.S. Assistance Has Helped Mexican Counternarcotics Efforts* (Washington, D.C.: Government Printing Office, 2007), pp. 16-17.

46. "President Fox's National Security Legacy," Cable from U.S. Embassy, Mexico City, 25 October 2006, Wikileaks; López González, *Fuerzas armadas en México, 1876-2012*, p. 282; Hernández, *Los señores del narco*, pp. 363, 405-406.

47. Grayson, *Narco-Violence*, pp. 104-105.

48. "President Fox's National Security Legacy," Cable from U.S. Embassy, Mexico City, 25 October 2006, Wikileaks; Instituto Nacional de Estadística y Geografía, *Anuario Estadístico 2012*, Table 9.38; Grillo, *El Narco*, p. 94.

49. Hernández, *Los señores del narco*, pp. 403-406, 412-413; Grayson, *Narco-Violence*, p. 153; Grillo, *El Narco*, pp. 98, 213; Diego Enrique Osorio, *La guerra de los Zetas; viaje por la frontera de la necropolítica* (Mexico: Grijalvo, 2012), pp. 155, 199.

50. López González, *Fuerzas armadas en México*, 1876-2012, p. 283; Ainslie, *The Fight to Save Juárez*, pp. 12-14.

51. Hernández, *Los señores del narco*, pp. 414-420.

52. Grayson, *Narco-Violence*, p. 119; Hernández, *Los señores del narco*, p. 422. By keeping the Pentagon in the dark about the troop deployments to the north, the Mexican army could not take "advantage of U.S. expertise or intelligence," to strike blows against the drug cartels; see GAO, *Drug Control*, p. 30.

Chapter 5

1. Sun Tzu, *The Art of War*, trans. Thomas Cleary (Boston: Shambala Editions, 1988), p. 168.

2. *Semana*, No. 1242 of 19 February 2006 and No. 1253 of 26 February 2006.

3. "Promising New Colmil Leadership off to Rocky Start," Cable from U.S. Embassy, Bogotá, 5 October 2006, Wikileaks; "La hora de Juan Manuel," *Semana*, No. 1258 of 11 June 2006.

4. "Colmil Penetrates FARC Heartland in La Macarena," Secret Cable from U.S. Embassy, Bogotá, 29 August 2006, Wikileaks.

5. "Colmil: Joint Task Force-Omega onto a More Aggressive Footing," Secret Cable from U.S. Embassy, Bogotá, 27 December 2006, Wikileaks.

6. "Colombian Military: Joint Task Force-Omega onto a More Aggressive Footing," Secret Cable from U.S. Embassy, Bogotá, 27 December 2006, Wikileaks.

7. "Government of Colombia Wealth Tax Boosts Defense Budget," Cable from U.S. Embassy, Bogotá, 23 February 2007, Wikileaks; "Más plata para la guerra," *Semana*, No. 1278 of 28 October 2006; "Ambassador Discusses Wealth Tax, Human Rights with Defense Minister Santos," Cable from U.S. Embassy, Bogotá, 21 December 2006, Wikileaks. The armed forces grew from a strength of 158,000 in 2002 to 260,000 in 2006, while the police grew from 104,000 in 2002 to 134,000 in 2006. The number of soldiers per 1000 inhabitants stood at 2.08 in 1980 and 2.86 in 1990; see "Why January 2006 Is Different from January 2005," Cable from U.S. Embassy, Bogotá, 6 January 2006, Wikileaks and Diego Otero Prada, *Las cifras del conflicto colombiano*, 2d ed (Bogotá: Indepaz, 2007), pp. 50-51.

8. This paragraph and the next relies on "Milgroup Bogota (2)—Closing the Colombian Army's Leadership Gap," Cable from U.S. Embassy, Bogotá, 29 June 2007, Wikileaks.

9. *Semana*, 27 October 2010.

10. "River Wars in FARC's Heartland," Cable from U.S. Embassy, Bogotá, 16 February 2007, Wikileaks.

11. "Ambassador Discusses Wealth Tax, Human Rights with Defense Minister Santos," Cable from U.S. Embassy, Bogotá, 21 December 2006, Wikileaks; *El Tiempo*, 21 December 2004; *El Comercio*, 2 March 2008.

12. *El Tiempo*, 7 February 2008; *Semana*, No. 1345 of 9 February 2008.

13. "Country Team Assessment on Submarine Combat Systems for Colombia," Cable from U.S. Embassy, Bogotá, 26 February 2008, Wikileaks.

14. "Colombian Military: Joint Task Force-Omega onto a More Aggressive Footing," Secret Cable from U.S. Embassy, Bogotá, 27 December 2006, Wikileaks; "Air Mobility Key to Success in Colombia," Cable from U.S. Embassy, Bogotá, 30 May 2008, Wikileaks.

15. "UAVs—Eyes in the Sky for Colmil Operations," Secret Cable from U.S. Embassy, Bogotá, 20 December 2006, Wikileaks.
16. "Terrorist Takedowns Reflect USG-GOC Intelligence Cooperation," Secret Cable from U.S. Embassy, Bogotá, 5 June 2006, Wikileaks.
17. "UAVs—Eyes in the Sky for Colmil Operations," Secret Cable from U.S. Embassy, Bogotá, 20 December 2006, Wikileaks.
18. "Plan Colombia: Nationalization of FMF/DOD Funding," Cable from U.S. Embassy, Bogotá, 28 March 2008, Wikileaks; "Economic Slowdown Prompts Cuts to Colombian Defense Budget," Cable from U.S. Embassy, Bogotá, 29 January 2009, Wikileaks.
19. "Colombia Conflict Update—3 Quarter 07," Cable from U.S. Embassy, Bogotá, 23 October 2007, Wikileaks.
20. "Colombia Conflict Update for July-August 2008," Secret Cable from U.S. Embassy, Bogotá, 22 December 2008, Wikileaks.
21. "Colombian Military Offensive Gets Results," Cable from U.S. Embassy, Bogotá, 6 March 2009, Wikileaks.
22. "The Military Moves Further Red," Cable from U.S. Embassy, Caracas, 11 July 2007, Wikileaks; "Chávez's Way Ahead: Words to Deeds," Cable from U.S. Embassy, Caracas, 12 January 2007, Wikileaks.
23. "The Military Moves Further Red," Cable from U.S. Embassy, Caracas, 11 July 2007, Wikileaks; "Outgoing Defense Minister Questions Socialist Plans, Armed Forces Gets New Leadership," Cable from U.S. Embassy, Caracas, 27 July 2007, Wikileaks; "General Baduel Arrested," Cable from U.S. Embassy, Caracas, 7 October 2008, Wikileaks; "General Baduel Arrested Again," Cable from U.S. Embassy, Caracas, 3 April 2009, Wikileaks.
24. "Outgoing Defense Minister Questions Socialist Plans, Armed Forces Gets New Leadership," Cable from U.S. Embassy, Caracas, 27 July 2007, Wikileaks.
25. "Venezuelan Army Hosts U.S. Military Delegation," Cable from U.S. Embassy, Caracas, 4 April 2007, Wikileaks; "Chávez Insulates Venezuelan Armed Forces from Undesirables," Cable from U.S. Embassy, Caracas, 4 March 2008, Wikileaks.
26. "Venezuela and Cuba—Confederation?" Cable from U.S. Embassy, Caracas, 15 October 2007, Wikileaks; *El Universal*, 7 December 2006.
27. "Chávez's Way Ahead: Words to Deeds," Cable from U.S. Embassy, Caracas, 12 January 2007, Wikileaks.
28. *New York Times*, 3 March 2008.
29. "Minimal Bolivarian Republic of Venezuela Armed Forces Response to Deployment Order," Secret/No Forn Cable from U.S. Embassy, Caracas, 4 March 2008, Wikileaks.
30. "No New Troop Movements Reported," Secret Cable from U.S. Embassy, Caracas, 7 March 2008, Wikileaks.
31. "No New Troop Movements Reported," Secret Cable from U.S. Embassy, Caracas, 7 March 2008, Wikileaks.
32. "Chávez Strikes Conciliatory Tone, Restores Relations with Colombia," Secret Cable from U.S. Embassy, Caracas, 11 March 2008, Wikileaks.
33. "Chávez Appoints New Vice President, Defense Minister, and Environment Minister," Secret Cable from U.S. Embassy, Caracas, 27 January 2009, Wikileaks.
34. "Chávez Laying the Groundwork for Confrontation with Colombia," Secret Cable from U.S. Embassy, Caracas, 10 November 2009, Wikileaks; "GoC Calm as Tensions Rise with Venezuela, but Economic Uncertainty Remains," Cable from U.S. Embassy, Bogotá, 31 July 2009, Wikileaks; "Venezuela Issues Almost All Pending Diplomatic Visas," Secret Cable from U.S. Embassy, Caracas, 25 March 2009, Wikileaks.
35. "Chávez Laying the Groundwork for Confrontation with Colombia," Secret Cable from U.S. Embassy, Caracas, 10 November 2009, Wikileaks.
36. "Rampant Crime in Venezuela—Chávez's Achilles Heel?" Cable from U.S. Embassy, Caracas, 24 August 2009, Wikileaks; "Chávez tackles Rising Crime with the Bolivarian National Police," Cable from U.S. Embassy, Caracas, 26 January 2010, Wikileaks.

Chapter 6

1. U.S. Department of State, *Papers Relating to the Foreign Relations of the United States, 1895, Part I* (Washington, D.C.: Government Printing Office, 1896), p. 558.
2. René De La Pedraja, *Wars of Latin America, 1948-1982: The Rise of the Guerrillas* (Jefferson, NC: McFarland, 2013), pp. 35–36, 38, 61, 63.
3. René De La Pedraja, *Wars of Latin America, 1982-2013: The Path to Peace* (Jefferson, NC: McFarland, 2013), pp. 111–114; Francisco Barbosa Miranda, *Historia militar de Nicaragua: Antes del siglo XV al XXI* (Managua: Hispamer, 2010), pp. 441–442.
4. "Codel Nelson Meets with Vice President Morales," Cable from U.S. Embassy, Managua, 28 February 2007, Wikileaks; Mark. B. Rosenberg and Luis G. Solís, *The United States and Central America: Geopolitical Realities and Regional Fragility* (New York: Routledge, 2007), pp. 58–59.
5. "Scene Setter for Secretary Powell's Visit,"

Cable from U.S. Embassy, Tegucigalpa, 22 October 2003, Wikileaks.

6. "Chairman of Joint Chiefs of Staff Myers Meets with Honduran Minister of Defense," Cable from U.S. Embassy, Tegucigalpa, 15 August 2003, Wikileaks; "Scene Setter for Secretary Powell's Visit," Cable from U.S. Embassy, Tegucigalpa, 22 October 2003, Wikileaks.

7. "Scene Setter for Secretary of Defense Rumsfeld," Cable from U.S. Embassy, Managua, 22 September 2006, Wikileaks. The threat to reduce military assistance is not in the Wikileaks cables but appears in Barbosa Miranda, *Historia militar de Nicaragua*, p. 442. Perhaps the ultimatum was mentioned in one of the Top Secret cables not included in Wikileaks; or a U.S. diplomat might not have been present when Collin Powell, or one of his assistants, issued the threat informally. Another possible explanation exists for the apparent omission in the written record: diplomats, not wishing to admit the failure of negotiations, traditionally have been reluctant to leave a paper trail about ultimatums.

8. "Scene Setter for Secretary of Defense Rumsfeld," Cable from U.S. Embassy, Managua, 22 September 2006, Wikileaks; "Herty Seeks Cordial, Constructive, Cooperative Relations with the United States," Cable from U.S. Embassy, Managua, 2 February 2006, Wikileaks.

9. "Moving Closer," Cable from U.S. Embassy, Managua, 3 May 2006, Wikileaks.

10. "Opposition Assembly Caucus Leaders Seek Training," Cable from U.S. Embassy, Managua, 21 February 2007, Wikileaks.

11. "Wha/cen Director Feeley Visits Nicaragua," Cable from U.S. Embassy, Managua, 26 March 2008, Wikileaks.

12. "Is There a Doctor in the House?" Cable from U.S. Embassy, Managua, 23 November 2007, Wikileaks; "Wha/cen Director Feeley Visits Nicaragua," Cable from U.S. Embassy, Managua, 26 March 2008, Wikileaks.

13. "Ambassador Meets with Foreign Minister Santos," Cable from U.S. Embassy, Managua, 19 September 2008, Wikileaks.

14. "No Short-Term Deliveries from Chávez's Arms Deal," Cable from U.S. Embassy, Caracas, 16 September 2009 and "A View of Venezuelan Arms Purchases from Russia," Cable from U.S. Embassy, Caracas, 5 November 2004, Wikileaks; René De La Pedraja, "Statesmen and Air Power in Latin America, 1945–2010," in Robin Higham and Mark Parillo, eds., *The Influence of Airpower Upon History: Statesmanship, Diplomacy, and Foreign Policy Since 1903* (Lexington: University of Kentucky Press, 2013), pp. 168–169.

15. "Aircraft Contract Rumors Resurface," Cable from U.S. Embassy, Caracas, 27 September 2004, Wikileaks.

16. "Better Late Than Never: Russian Choppers Arrive in Venezuela," Cable from U.S. Embassy, Caracas, 4 April 2006, Wikileaks.

17. "A View of Venezuelan Arms Purchases from Russia," Cable from U.S. Embassy, Caracas, 5 November 2004, Wikileaks; "Venezuela Looking to Buy Brazilian Light Attack Aircraft," 12 February 2005, BBC Monitoring Americas.

18. "Aircraft Contract Rumors Resurface," Cable from U.S. Embassy, Caracas, 27 September 2004, Wikileaks.

19. "A View of Venezuelan Arms Purchases from Russia," Cable from U.S. Embassy, Caracas, 5 November 2004, Wikileaks.

20. "Spain/Venezuela: Reaction to Denial of License for Tech Transfer," Cable from U.S. Embassy, Madrid, 12 January 2006, Wikileaks.

21. "La venta de barcos a Chávez dividió al Gobierno español," *El País*, 9 December 2010.

22. "Lula in Caracas—Parsing a Strategic Alliance," Cable from U.S. Embassy, Brasilia, 16 February 2005, Wikileaks; "Brazil's Lula Gets Intense Re: U.S.–Brazil Relations," Secret/No Forn Cable from U.S. Embassy, Brasilia, 31 July 2009, Wikileaks; "Venezuela Looking to Buy Brazilian Light Attack Aircraft," 12 February 2005, BBC Monitoring Americas.

23. "Brazil's Lula Gets Intense Re: U.S.–Brazil Relations," Secret/No Forn Cable from U.S. Embassy, Brasilia, 31 July 2009, Wikileaks.

24. "No Surprises in Bolivarian Republic of Venezuela Response to Counterterrorism Designation," Cable from U.S. Embassy, Caracas, 17 May 2006, Wikileaks. The U.S. embargo caught the Venezuelan Military Acquisition Office (VMAO) in Miami with a modest inventory of U.S. parts and equipment in its warehouse and other purchased items on the way, but zealous U.S. officials refused to allow shipment to Venezuela. This office was the traditional channel the Venezuelan armed forces used to purchase military equipment, field gear, and armament from American factories. The U.S. Embassy in Caracas wrote: "We should conclude the VMAO drama as soon as possible. It is a net loser for the U.S. government. To Venezuela, the rest of the world, and even inside the United States, retaining material already purchased by the Bolivarian Republic of Venezuela and offering no compensation will appear to be expropriation without compensation. [...] The longer this issue stretches out, the more damage we do to ourselves. As best as we can tell from here, there is no truly sensitive equipment in the VMAO, and little that could be classified as lethal. I hope the policymakers will direct the lawyers to find an argu-

ment permitting the immediate export of everything in this warehouse that is not subject to absolute legal prohibition." "Action Request: BRV Queries on Military Acquisitions," Cable from U.S. Embassy, Caracas, 28 February 2007, Wikileaks.
25. "Spanish Minister of Defense on U.S. Arms Transfer Policy to Venezuela," Cable from U.S. Embassy, Madrid, 26 September 2006, Wikileaks.
26. *New York Times*, 28 July 2006.
27. "Russia Delivers Sukhoi Fighter Jets to Venezuela," 2 December 2006, BBC Monitoring Former Soviet Union.
28. "Venezuelan Air Force Announces Russian Aircraft Purchase," Cable from U.S. Embassy, Caracas, 10 December 2007, Wikileaks; "Russian Arms Arrive in Venezuela, Air Defense Planned," 27 March 2008; *El Universal*, 7 December 2006.
29. "Venezuelan Air Force Announces Russian Aircraft Purchase," Cable from U.S. Embassy, Caracas, 10 December 2007, Wikileaks.
30. *Los Angeles Times*, 27 September 2008.
31. *New York Times*, 13 September 2009; *Asia News Monitor*, 14 September 2009.
32. "Russia's Anticipated Transfer of Igla-S to Venezuela," Secret/No Forn Cable from U.S. Embassy, Moscow, 14 February 2009, Wikileaks.
33. Ibid.
34. "Government of Colombia Asks for U.S. Government Help to Persuade European Union and Russia to Suspend Arms Sales to Venezuela," Cable from U.S. Embassy, Bogotá, 30 July 2009, Wikileaks; "Russia's Anticipated Transfer of Igla-S to Venezuela," Secret/No Forn Cable from U.S. Embassy, Moscow, 14 February 2009, Wikileaks.
35. *Listin Diario*, 17 August 2011.
36 "Mexican Finance Secretary Discusses Reform Agenda, Lays Out Proposals for Increased U.S.–Mexico Cooperation," Cable from U.S. Embassy, Mexico City, 24 January 2007, Wikileaks. Paragraph also draws on Iñigo Guevara Moyano, *Adapting, Transforming, and Modernizing Under Fire: The Mexican Military 2006–2011* (Carlisle, PA: U.S. Army War College, 2011), p. 6.
37. Jorge Luis Sierra Guzmán, *El enemigo interno: Contrainsurgencia y fuerzas armadas en México* (Mexico: Universidad Iberoamericana, 2003), pp. 289–290; Roberto Badillo Martínez, *El libro verde de las fuerzas armadas* (Mexico: Bredición, 2012), p. 87; George W. Grayson, *Mexico: Narco-Violence and a Failed State?* (New Brunswick, NJ: Transaction Publishers, 2011), p. 226.
38. "Government of Mexico Backing Off Purchase of Russian Fighter Aircraft," Secret Cable from U.S. Embassy, Mexico City, 25 January 2007, Wikileaks.
39. Guevara Moyano, *The Mexican Military 2006–2011*, pp. 6, 25; "Government of Mexico Backing off Purchase of Russian Fighter Aircraft," Secret Cable from U.S. Embassy, Mexico City, 25 January 2007, Wikileaks.
40. "Israeli Relations with Mexico," Secret/No Forn Cable from U.S. Embassy, Mexico City, 18 July 2008, Wikileaks.
41. "Russian Foreign Minister Lavrov Visit to Mexico: Constructive and Non-ideological," Cable from U.S. Embassy, Mexico City, 19 February 2010, Wikileaks.

Chapter 7

1. Niccolò Machiavelli, *Il principe, la mandragola, lettere* (Roma: Armando Curcio Editores, 1966), p. 107. Translation by Beatriz De La Pedraja.
2. This quotation and the next come from "Air Mobility Key to Success in Colombia," Cable from U.S. Embassy, Bogotá, 30 May 2008, Wikileaks. U.S. diplomats labeled this cable "unclassified" thus facilitating its widespread circulation without having to go through the lengthy and cumbersome procedure of declassification.
3. "Human Rights in Colombia—Widespread Allegations of Abuses by the Army," Cable from U.S. Embassy, Bogotá, 27 July 1990, National Security Archive, George Washington University, Washington, D.C.
4. "Minister of Defense Invites Assistant Secretary Shattuck to Visit Colombia," Cable from U.S. Embassy, Bogotá, 21 October 1994, National Security Archive.
5. "Cashiered Colonel Talks Freely about the Army He Left Behind," Secret Cable from Defense Attaché, Bogotá, 24 December 1997, National Security Archive.
6. The U.S. military developed a specialized vocabulary to describe its many quantitative measurements. "Takedowns," which refers to how many enemy were killed in a surprise engagement, is a good example. This could be a valid indicator for measuring success under some circumstances, but inevitably it reflected the assumption that killing more guerrillas was the way to achieve victory. If a war of attrition was the only way to win, then using "takedowns" as one of the quantitative measurement seemed reasonable, but usually in warfare attrition has been only one of many factors in determining the final outcome. In some cases killing more of the enemy actually went against the goal of winning, and in the Colombian case a perverted distortion resulted in a mad scram-

ble to find bodies for "takedowns" and other metrics on enemy dead.

7. *Semana*, No. 1271 of 8 September 2006, No. 1272 of 16 September 2006, No. 1291 of 3 February 2007, No. 1343 of 26 January 2008. Accounts occasionally mention photographing the victims to be able to claim them as kills. Presumably collections of these photographs with their accompanying reports existed either in battalion headquarters or in Bogotá. Because both regional commands and Bogotá traditionally have discarded all their military records after a few years, officers already had plenty of time, motive, and opportunity to destroy all the evidence.

8. "Ministry of Defense Commission Finds Army Corruption and Lack of Oversight Contributed to Extrajudicial Killings," Cable from U.S. Embassy, Bogotá, 5 November 2008, Wikileaks; "Ambassador Brownfield Presses Colombian Army Commander on Human Rights," Cable from U.S. Embassy, Bogotá, 20 May 2009, Wikileaks; Diego Otero Prada, *Las cifras del conflicto colombiano*, 2d ed (Bogotá: INDEPAZ, 2007), p. 271.

9. "Ministry of Defense Commission Finds Army Corruption and Lack of Oversight Contributed to Extrajudicial Killings," Cable from U.S. Embassy, Bogotá, 5 November 2008, Wikileaks.

10. Some guideposts exist for the estimate of 10,000 as the number of "False Positives." The first is the military's claim to have killed on the average 1400 guerrillas annually between 1997 and 2008; the annual average is probably too high for the twentieth century but because it is almost certainly too low for the twenty-first, the 1400 figure seems a reasonable annual average for the ten-year period. The second is the figure of 400 as the annual average of military deaths in the same ten-year period. See Otero, *Las cifras del conflicto*, pp. 179–180, 271–272. Assuming that one soldier died for each guerrilla killed (instead of 3.5 guerrillas killed for each soldier who died), then the number of guerrilla casualties each year averaged at around 400, a figure probably still too high. Subtracting 400 from 1400 left 1000 as the average annual number of "False Positives" between 1997 and 2008, while multiplying by ten gave the 10,000 figure. If anything, the figure may be too low, because "False Positives" were widespread before 1997, and most disturbingly, continued for years after the scandal broke out in 2008. The number of around 1000 civilians killed annually resembles the figures for the more quiet years of *La Violencia* (1948–1966), but because the population of Colombia had more than doubled by the twenty-first century, the death of generally forgotten individuals or of poor peasants was not immediately noticed and could easily be missed in a country traditionally very conflict prone. See René De La Pedraja, *Wars of Latin America, 1948–1982: The Rise of the Guerrillas* (Jefferson, NC: McFarland, 2013), pp. 13, 31–33, 49.

11. "Colombian Military Commander Resigns, General González Peña Named as Replacement," Secret Cable from U.S. Embassy, Bogotá, 6 November 2008, Wikileaks; "Colombia's Army Chief Resigns in Wake of Scandal," 5 November 2008, CNN.com.

12. "Ambassador Brownfield Presses Colombian Army Commander on Human Rights," Cable from U.S. Embassy, Bogotá, 20 May 2009, Wikileaks; "Army Inspector General Still Blocked, But New Minister of Defense Sparks Hope," Cable from U.S. Embassy, Bogotá, 21 August 2009, Wikileaks.

13. "Minister of Defense Overrules Army Commander's Transfer of Key Human Rights Official," Cable from U.S. Embassy, Bogotá, 17 November 2009, Wikileaks; "Witnesses Killed, Investigators Questioned in False Positive Cases," Secret/No Forn Cable from U.S. Embassy, Bogotá, 19 February 2010, Wikileaks.

14. "Army Inspector General Still Blocked, But New Minister of Defense Sparks Hope," Cable from U.S. Embassy, Bogotá, 21 August 2009, Wikileaks; "Witnesses Killed, Investigators Questioned in False Positive Cases," Secret/No Forn Cable from U.S. Embassy, Bogotá, 19 February 2010, Wikileaks.

15. "2009 Stalemate Continues: Colombia Conflict Update for July-September 2009," Secret/No Forn Cable from U.S. Embassy, Bogotá, 16 November 2009, Wikileaks.

16. "¿Alguien espió a los negociadores de La Habana?" *Semana*, 3 February 2014; "Andrómeda la sala de inteligencia que enreda al Ejército," *El Espectador*, 4 February 2014. For a discussion of why army officers need the guerrilla war to continue, see De La Pedraja, *Wars of Latin America, 1948–1982*, p. 338n33.

17. "Caso 'chuzadas' el misterio de la casa gris," *Semana*, 3 February 2014.

18. "Fuerzas oscuras están detrás de las chuzadas a negociadores de paz," *El Espectador*, 4, 23 February 2014; *El Tiempo*, 5 February 2014; *Semana*, 14 May 2014.

19. For the army's efforts to derail peace efforts under President Belisario Betancur, see De La Pedraja, *Wars of Latin America, 1948–1982*, pp. 128–130.

20. *Semana*, 7 February 2014. The DAS scandal began when *Semana* magazine revealed in February 2009 that the domestic intelligence agency had engaged in extensive illegal espionage of political opponents. In historical perspective, the DAS episode foreshadowed the

later scandal of Military Intelligence over Operation Andromeda.
21. *El Espectador*, 4 February 2014; *El Tiempo*, 4 February 2014.
22. While it is tempting to blame the drug trade for the unusual wealth of retired army officers, in reality this was rarely the case because the police had primary jurisdiction over the drug trade. Instead for wealthy police officers the prevailing assumption has been that they received payments from the drug cartels.
23. *Semana*, 16 February, 3 March 2014; *El Tiempo*, 16 February 2014; *El Espectador*, 3, 5 March 2014. For the turbulent history of the armed groups in the Sinú and San Jorge rivers, see René De La Pedraja, *Wars of Latin America, 1982-2013: The Path to Peace* (Jefferson, NC: McFarland, 2013), pp. 134–141, 217–224.
24. *El Espectador*, 17, 19 February 2014. Authorities were also investigating 823 NCOs and 2,908 soldiers for the False Positives. Only 89 of the accused belonged to the navy, and the grand total for the army including sundry categories was 4,173 uniformed personnel. Just this high number of accused powerfully suggests that 4,173 army personnel could not have murdered only 5,000 civilians. Colombians have been reluctant to admit the large number of innocent civilians killed as False Positives. Even highly critical *Semana* magazine (18 February 2014) accepts the low figure of 4,000, while the Mothers of Soacha claim only 5000. The figure of 10,000 or more victims of army massacres is very damming evidence for those who argue that Colombians are by nature brutally savage and predisposed to commit the most heinous crimes. The self-denial about these embarrassing figures of False Positives is completely understandable.
25. "Exclusivo: Los negocios en el Ejército," *Semana*, 16 February 2014.
26. *El Tiempo*, 16 February 2014; "Exclusivo: Los negocios en el Ejército," *Semana*, 16 February 2014.
27. *El Espectador*, 16 February 2014; "Bienes de los generales vinculados en escándalo de corrupción," *El Espectador*, 2 March 2014; *Semana*, 18 February 2014.
28. "Por chuzadas a Santos, Ejército inicia investigación interna," *El Espectador*, 24 February 2014; "Casos de corrupción no son una práctica generalizada: comandante del Ejército," *El Espectador*, 4 March 2014.
29. "Mañana será otro día para la paz," *El Espectador*, 8 March 2014.
30. "Madres de Soacha al Señor Juan Manuel Santos Presidente de la República," Bogotá, 20 February 2014.
31. *Washington Post*, 27 July 1999.
32. "Crime in Jamaica," Cable from U.S. Embassy, Kingston, 6 October 2008, Wikileaks.
33. *Philadelphia Tribune*, 23 May 2010; *Washington Post*, 27 July 1999; Anthony Harriott, *Organized Crime and Politics in Jamaica: Breaking the Nexus* (Kingston: Canoe Press, 2008), p. 37; Hume N. Johnson, *Challenges to Civil Society: Popular Protest and Governance in Jamaica* (Amherst, NY: Cambria Press, 2011), pp. 251–252; Anthony P. Maingot and Wilfredo Lozano, *The United States and the Caribbean: Transforming Hegemony and Sovereignty* (New York: Routledge, 2005), p. 138.
34. "Crime in Jamaica," Cable from U.S. Embassy, Kingston, 6 October 2008, Wikileaks; "Jamaica: Scene Setter for Rear Admiral Joseph Kernan," Cable from U.S. Embassy, Kingston, 7 April 2009, Wikileaks; National Geographic, "Drugs, Inc.: Jamaican Gangs, Guns, and Ganja," broadcast August 2013; Harriott, *Crime in Jamaica*, pp. 21, 43–44; Johnson, *Governance in Jamaica*, pp. 257–258. Members of the Shower Posse bragged about having missiles in their arsenal, but almost certainly they meant the fairly easy to procure Rocket Propelled Grenades (RPGs).
35. "Crime in Jamaica," Cable from U.S. Embassy, Kingston, 6 October 2008, Wikileaks; "Jamaica: Scene Setter for Rear Admiral Joseph Kernan," Cable from U.S. Embassy, Kingston, 7 April 2009, Wikileaks; National Geographic, "Drugs, Inc.: Jamaican Gangs, Guns, and Ganja," broadcast August 2013; *Philadelphia Tribune*, 23 May 2010. Followers of Christopher "Dudus" Coke claimed that the don pioneered bringing cocaine from Colombia through Jamaica; see Discovery Channel, "Gangsters, America's Most Evil; The Kingston Kingpin: Christopher 'Dudus' Coke," broadcast November 2013. While he may have been the primary player in setting up that route, the use of Jamaica as a base for sending Colombian cocaine to the United States was well established under his father.
36. "Jamaica: Crime, Corruption, and Economy," Cable from U.S. Embassy, Kingston, 12 March 2008, Wikileaks; "In Its First Three Months: The New Anti-Corruption Branch of the Jamaica Constabulary Force Arrests 27 Officers and Two Civilians," Secret Cable from U.S. Embassy, Kingston, 24 April 2008, Wikileaks; Harriott, *Crime in Jamaica*, p. 24; Hume, *Governance in Jamaica*, pp. 252–253, 260; Maingot and Lozano, *The United States and the Caribbean*, pp. 134–135, 139–140. Actually the tourists were not always innocent bystanders, and young Americans were responsible for introducing ecstasy into the island. Jamaican gangs scrambled to supply this new drug, but as of 2013 the demand has not spread to Ja-

maican drug addicts and has remained limited to the wild party scene of foreign tourists frolicking on the island.

37. "Jamaica: Crime, Corruption, and Economy," Cable from U.S. Embassy, Kingston, 12 March 2008, Wikileaks; "Acting NAS Director Calls on Chief of Defense Staff and Commissioner of Police," Cable from U.S. Embassy, Kingston, 3 April 2008, Wikileaks; "Jamaica: Scene Setter for Rear Admiral Joseph Kernan," Cable from U.S. Embassy, Kingston, 7 April 2009, Wikileaks; National Geographic, "Drugs, Inc.: Jamaican Gangs, Guns, and Ganja," broadcast August 2013; Johnson, *Governance in Jamaica*, p. 254. To minimize the role of the army, the government attempted to maintain the image that crop eradication was principally a police activity, when in reality, as the Wikileaks cables show, the army was conducting most of the eradication in the countryside.

38. "Crime in Jamaica," Cable from U.S. Embassy, Kingston, 6 October 2008, Wikileaks.

39. "Acting NAS Director Calls on Chief of Defense Staff and Commissioner of Police," Cable from U.S. Embassy, Kingston, 3 April 2008, Wikileaks.

40. "Jamaica's Election-year Violence," Cable from U.S. Embassy, 23 January 2007, Wikileaks; Harriott, *Crime in Jamaica*, p. 25.

41. "In Its First Three Months: The New Anti-Corruption Branch of the Jamaica Constabulary Force Arrests 27 Officers and Two Civilians," Secret Cable from U.S. Embassy, Kingston, 24 April 2008, Wikileaks; "Jamaica: Input for Caribbean Security Assessment," Cable from U.S. Embassy, Kingston, 29 May 2008, Wikileaks.

42. "Crime in Jamaica," Cable from U.S. Embassy, Kingston, 6 October 2008, Wikileaks.

43. "Minister of National Security Plans to Issue Retraction Regarding Support for Police Accused of Misconduct," Cable from U.S. Embassy, Kingston, 1 June 2009, Wikileaks; *Wall Street Journal*, 26 May 2010; Johnson, *Governance in Jamaica*, pp. 165, 267.

44. "Minister of National Security Plans to Issue Retraction Regarding Support for Police Accused of Misconduct," Cable from U.S. Embassy, Kingston, 1 June 2009, Wikileaks.

45. *Wall Street Journal*, 23 May 2010; *Philadelphia Tribune*, 23 May 2010; Johnson, *Governance in Jamaica*, pp. 101–102, 163.

46. *New York Times*, 24 May 2010; *Wall Street Journal*, 26 May 2010.

47. Johnson, *Governance in Jamaica*, p. 164; *New York Times*, 3, 23 June 2010; *Wall Street Journal*, 26 May 2010; National Geographic, "Drugs, Inc.: Jamaican Gangs, Guns, and Ganja," broadcast August 2013; Discovery Channel, "Gangsters, America's Most Evil; The Kingston Kingpin: Christopher 'Dudus' Coke," broadcast November 2013.

48. *New York Times*, 1, 3 June 2010; National Geographic, "Drugs, Inc.: Jamaican Gangs, Guns, and Ganja," broadcast August 2013; Johnson, *Governance in Jamaica*, pp. 164–165; Discovery Channel, "Gangsters, America's Most Evil; The Kingston Kingpin: Christopher 'Dudus' Coke," broadcast November 2013.

49. *New York Times*, 3, 23 June 2010; *Philadelphia Tribune*, 2 October 2011; Discovery Channel, "Gangsters, America's Most Evil; The Kingston Kingpin: Christopher 'Dudus' Coke," broadcast November 2013.

50. *New York Times*, 1 June 2010; *Christian Science Monitor*, 29 December 2011; *Philadelphia Tribune*, 8 January 2012.

Chapter 8

1. Mark B. Rosenberg and Luis G. Solís, *The United States and Central America: Geopolitical Realities and Regional Fragility* (New York: Routledge, 2007), p. 59.

2. For the Contra War of the 1980s, see René De La Pedraja, *Wars of Latin America, 1982–2013: The Path to Peace* (Jefferson, NC: McFarland, 2013), pp. 3–10, 67–75, 106–115.

3. "Ambassador Discusses Bilateral Concerns with Nicaraguan Ambassador-Designate Cruz," Cable from U.S. Embassy, Managua, 18 January 2007, Wikileaks.

4. "The Sandinista Government 60 Days Out—and Our Responses," Cable from U.S. Embassy, Managua, 5 March 2007, Wikileaks.

5. Ibid.

6. "Acting Minister of Defense Plans Ministry Purge, Ideological Oversight of Armed Forces," Cable from U.S. Embassy, Managua, 18 June 2007, Wikileaks.

7. "Codel Nelson Meets with Vice President Morales," Cable from U.S. Embassy, Managua, 28 February 2007, Wikileaks; "Ortega Appoints Another Pseudo-Minister of Defense," Cable from U.S. Embassy, Managua, 10 May 2007, Wikileaks; "Central American Security Requirements: Nicaragua," Cable from U.S. Embassy, Managua, 14 December 2007, Wikileaks. The army in 1999 numbered 14,083 individuals, including 2,016 officers and 1,374 civilians. See Francisco Barbosa Miranda, *Historia militar de Nicaragua: antes del siglo XV al XXI* (Managua: Hispamer, 2010), p. 423.

8. "The Sandinista Government 60 Days Out—and Our Responses," Cable from U.S. Embassy, Managua, 5 March 2007, Wikileaks.

9. "Nicaragua SOFA Letter: Potentially Difficult Road Ahead," Cable from U.S. Em-

bassy, Managua, 25 March 2008, Wikileaks; "Nicaragua SOFA—Next Steps to a Longer-Term Agreement," Cable from U.S. Embassy, Managua, 24 September 2008, Wikileaks.

10. "Death to the Yanqui—Ortega Lambastes U.S. at Nicaraguan Army's 30th Anniversary Celebrations," Cable from U.S. Embassy, Managua, 17 September 2009, Wikileaks.

11. "Colom Announces His Cabinet," Cable from U.S. Embassy, Guatemala City, 11 January 2008, Wikileaks.

12. "Wha/cen Director Focuses on Security during Visit to Guatemala," Cable from U.S. Embassy, Guatemala City, 10 November 2008, Wikileaks.

13. Ibid.

14. "2009 in Review; What to Watch for in 2010," Cable from U.S. Embassy, Guatemala City, 17 February 2010, Wikileaks; U.S. House, Committee on Foreign Affairs, Subcommittee on the Western Hemisphere, *Guatemala at a Crossroads* (Washington, D.C.: Government Printing Office, 2009), pp. 10–14; *Wall Street Journal*, 13 May 2009; *Los Angeles Times*, 23 May 2009.

15. For the many bizarre twists in the Rodrigo Rosenberg case see the excellent article by David Grann, "A Murder Foretold: A Reporter at Large," *The New Yorker*, 4 April 2011.

16. "2009 in Review; What to Watch for in 2010," Cable from U.S. Embassy, Guatemala City, 17 February 2010, Wikileaks.

17. "Rogue Elements of Guatemalan Military Selling Weapons to Narcos," Secret Cable from U.S. Embassy, Guatemala City, 8 June 2009, Wikileaks; *Latin American Herald Tribune*, 4 June 2009; *Los Angeles Times*, 4 June 2009; Hal Brands, *Crime, Violence, and the Crisis in Guatemala: A Case Study in the Erosion of the State* (Carlisle, PA: U.S. Army War College, 2010), pp. 14, 17–19.

18. "Ambassador and President Colom Review Security Issues in Puerto Barrios," Cable from U.S. Embassy, Guatemala City, 1 October 2008. Wikileaks.

19. "While Stealing Cocaine, Police Confronted by Attorney General's Office and Army," Secret Cable from U.S. Embassy, Guatemala City, 7 August 2009, Wikileaks; Kevin Lewis O'Neill and Kedron Thoman, eds., *Securing the City: Neoliberalism, Space, and Insecurity in Postwar Guatemala* (Durham: Duke University Press, 2011), p. 63.

20. Subcommittee on the Western Hemisphere, *Guatemala at a Crossroads*, p. 45.

21. "Ambassador and President Colom Review Security Issues in Puerto Barrios," Cable from U.S. Embassy, Guatemala City, 1 October 2008. Wikileaks.

22. "U.S. Government Assistance Positively Impacting Poptún; Playa Grande a Playground for Narcotraffickers," Cable from U.S. Embassy, Guatemala City, 21 October 2009, Wikileaks. U.S. Special Forces used the Poptún base to train the Kaibiles.

23. "Southcom Visit Provides Opportunity to Showcase U.S. Government Assistance and Areas for Improvement," Cable from U.S. Embassy, Guatemala City, 23 December 2009, Wikileaks; "U.S. Government Assistance Positively Impacting Poptún; Playa Grande a Playground for Narcotraffickers," Cable from U.S. Embassy, Guatemala City, 21 October 2009, Wikileaks; Subcommittee on Western Hemisphere, *Guatemala at a Crossroads*, p. 45.

24. O'Neill and Thomas, *Securing the City*, pp. 60, 129, 138; "Guatemala Shooting Raises Concerns about Military's Expanded Role," *New York Times*, 21 October 2012.

25. Mike McDonald, "Guatemala's Military Man, Nicaragua's Revolutionary," *Americas Quarterly*, Winter 2012, pp. 28–32; "Guatemala Shooting Raises Concerns about Military's Expanded Role," *New York Times*, 21 October 2012.

26. "A Pragmatic Shift or Merely a Tactical Move to Win?" Cable from U.S. Embassy, San Salvador, 24 June 2008, Wikileaks; *Los Angeles Times*, 26 June 2008.

27. "Early Impressions of Funes Government," Secret/No Forn Cable from U.S. Embassy, 28 July 2009, San Salvador, Wikileaks; "Funes Government Split Personality," Secret/No Forn Cable from U.S. Embassy, San Salvador, 5 June 2009, Wikileaks.

28. "Salvadorans Withdrawing from Iraq," Cable from U.S. Embassy, San Salvador, 23 December 2008, Wikileaks; *Los Angeles Times*, 26 June 2008, 17 March 2009; Linda Garrett, *A New Chapter for El Salvador: The First Hundred Days of President Mauricio Funes* (Washington, D.C.: Center for Democracy in the Americas, 2009), pp. 5–16.

29. "Your Visit to El Salvador," Secret Cable from U.S. Embassy, San Salvador, 8 May 2009, Wikileaks; *Washington Post*, 2 June 2009; "Early Impressions of Funes Government," Secret/No Forn Cable from U.S. Embassy, 28 July 2009, San Salvador, Wikileaks; Linda Garrett, *Expectations for Change and the Challenges of Governance: The First Year of President Mauricio Funes* (Washington, D.C.: Center for Democracy in the Americas, 2010), pp. 22–24.

30. "Government of El Salvador Discusses Transition and Pathways with Assistant Secretary Shannon," Cable from U.S. Embassy, San Salvador, 7 April 2009, Wikileaks; Garrett, *A New Chapter for El Salvador*, p. 3; *New York Times*, 25 March 2010.

31. "Your Visit to El Salvador," Secret Cable

from U.S. Embassy, San Salvador, 8 May 2009, Wikileaks; Garrett, *A New Chapter for El Salvador*, p. 12.

32. "Funes usó propuesta de 1998 para ascender a Munguía Payés," 10 August 2009, El Faro.net; Garrett, *A New Chapter for El Salvador*, p. 12.

33. "New Minister of Defense Seeks to Continue Military Cooperation with the U.S.," Cable from U.S. Embassy, San Salvador, 16 June 2009, Wikileaks; "Cantidad de militares ha crecido un 57 por ciento en primer gobierno de izquierda," 7 December 2011, El Faro.net.

34. "Funes Government Split Personality," Secret/No Forn Cable from U.S. Embassy, San Salvador, 5 June 2009, Wikileaks.

35. "Your Visit to El Salvador," Secret Cable from U.S. Embassy, San Salvador, 8 May 2009, Wikileaks; "Funes Government Split Personality," Secret/No Forn Cable from U.S. Embassy, San Salvador, 5 June 2009, Wikileaks; "Early Impressions of Funes Government," Secret/No Forn Cable from U.S. Embassy, San Salvador, 28 July 2009, Wikileaks. Persistent U.S. diplomats did not lose sight of their target and kept pressing Funes administration officials on the need to remove Manuel Melgar from office.

36. "Arms Dealer Says Police Moving Away from U.S. Weapons," Cable from U.S. Embassy, San Salvador, 28 August 2009, Wikileaks.

37. Ibid.

38. "Fraser Visit Highlights Strong U.S.–Government of El Salvador Military Relationship," Cable from U.S. Embassy, San Salvador, 17 September 2009, Wikileaks; "Military Anti-Crime Plan: A Gamble for Funes," Secret/No Forn Cable from U.S. Embassy, San Salvador, 2 November 2009, Wikileaks; "Funes aprueba uso de la Fuerza Armada para combatir la violencia y la delincuencia," 3 November 2009, El Faro.net.

39. "Cantidad de militares ha crecido un 57 por ciento en primer gobierno de izquierda," 7 December 2011, El Faro.net; "Fraser Visit Highlights Strong U.S.–Government of El Salvador Military Relationship," Cable from U.S. Embassy, San Salvador, 17 September 2009, Wikileaks; "Military Anti-Crime Plan: A Gamble for Funes," Secret/No Forn Cable from U.S. Embassy, San Salvador, 2 November 2009, Wikileaks; "Scene Setter for Department of Defense Visit to El Salvador," Secret Cable from U.S. Embassy, San Salvador, 23 February 2010, Wikileaks.

40. "Cantidad de militares ha crecido un 57 por ciento en primer gobierno de izquierda," 7 December 2011, El Faro.net.

41. "Fraser Visit Highlights Strong U.S.–Government of El Salvador Military Relationship," Cable from U.S. Embassy, San Salvador, 17 September 2009, Wikileaks.

42. "Funes decide enviar tropas a Afganistán," 18 August 2011, El Faro.net.

43. *Voices in the Border*, 11 November 2011; "Security Minister Leaves Amid Public Discontent," *Economist Intelligence Unit*, 6 December 2011; "Funes nombra a un general como director de la Policía Nacional Civil," 24 January 2012, El Faro.net; "Militares activos, militares retirados," 25 January 2012, El Faro.net.

44. "Negociación con pandillas," 26 March 2012, El Faro.net.

45. "La lista de peticiones que las pandillas hicieron al gobierno," 17 July 2012, El Faro.net; *Washington Report on the Hemisphere*, 3 May 2012.

46. *Christian Science Monitor*, 16 January 2003; *Washington Report on the Hemisphere*, 3 May 2012; "La lista de peticiones que las pandillas hicieron al gobierno," 17 July 2012, El Faro.net; "Military Anti-Crime Plan a Gamble for Funes," Secret/No Forn Cable from U.S. Embassy, San Salvador, 2 November 2009, Wikileaks.

47. "Scene Setter for Department of Defense Visit to El Salvador, March 8–9," Secret Cable from U.S. Embassy, San Salvador, 23 February 2010; *Washington Post*, 16 July 2012; *Washington Report on the Hemisphere*, 12 April 2012.

48. "Sánchez Cerén reitera ingreso de país a Petrocaribe," 19 February 2014, laprensagrafica.com; "Sánchez Cerén pide a Arena trabajar en un pacto de nación," 10 March 2014, Elfaro.net; Daily TV news broadcasts of Telemundo, March 2014.

49. *Los Angeles Times*, 13 March 2014; *Miami Herald*, 7 July 2014; "Sánchez Cerén mantendrá en su gabinete al ministro de la tregua," 21 May 2014, El Faro.net.

Chapter 9

1. Honduras, Comisión de la Verdad y la Reconciliación, *Para que los hechos no se repitan: Informe* (Tegucigalpa: Comisión de la Verdad y la Reconciliación, 2011), p. 5; (henceforth Comisión de la Verdad, *Informe*).

2. "Scene Setter for Secretary Powell's Visit," Cable from U.S. Embassy, Tegucigalpa, 22 October 2003, Wikileaks; "Fiscal Year Enhanced International Peacekeeping Capabilities Nomination for Honduras," Cable from U.S. Embassy, Tegucigalpa, 29 October 2003, Wikileaks.

3. Comisión de la Verdad, *Informe*, p. 203; "Honduras Ready to Move Forward on Regional Disarmament; Will Not Negotiate Bilaterally," Secret cable from U.S. Embassy, Tegucigalpa, 4 September 2003; "Fiscal Year Enhanced International Peacekeeping Capabilities Nomination for Honduras," Cable from U.S. Embassy, Tegucigalpa, 29 October 2003, Wikileaks. For the

Hundred Hours' War, see René De La Pedraja, *Wars of Latin America, 1948-1982: The Rise of the Guerrillas* (Jefferson, NC: McFarland, 2013), pp. 229-235.

4. "Chairman of Joint Chiefs of Staff Meets with Honduran Minister of Defense," Cable from U.S. Embassy, Tegucigalpa, 15 August 2003, Wikileaks.

5. Ibid.

6. "Honduras to Withdraw Immediately Troops from Iraq," Secret Cable from U.S. Embassy, Tegucigalpa, 16 April 2004, Wikileaks; "President Maduro Declares Honduras Will Withdraw from Iraq," Cable from U.S. Embassy, Tegucigalpa, 20 April 2004, Wikileaks.

7. Comisión de la Verdad, *Informe*, pp. 100-102; "A Zelaya Administration—What Does That Mean for the U.S. Government on Political Issues?" Cable from U.S. Embassy, Tegucigalpa, 13 December 2005, Wikileaks; Thomas M. Leonard, *The History of Honduras* (Santa Barbara, CA: Greenwood, 2011), p. 172.

8. "Zelaya's Administration First 45 Days-Pro U.S. Zelaya Often Torn in Different Directions by Advisors," Cable from U.S. Embassy, Tegucigalpa, 16 March 2006, Wikileaks; "POTUS June 5 Meeting with Honduran President Mel Zelaya," Cable from U.S. Embassy, Tegucigalpa, 12 May 2006, Wikileaks.

9. This quotation and the rest of the paragraph come from "A Zelaya Administration—What Does That Mean for the U.S. Government on Political Issues?" Cable from U.S. Embassy, Tegucigalpa, 13 December 2005, Wikileaks.

10. Foreign ministries routinely consult with their counterparts in another country whenever a candidate is under consideration for appointment as ambassador to that country. Under international law, the host country always has the right to refuse to accept the proposed candidate for the post of ambassador. But the client states of the Caribbean dared not exercise this right under international law or much less even dream of asking for the recall of the U.S. ambassador. An early sign of attempts to escape the U.S. sphere of influence comes when countries such as Mexico and Venezuela exercised their right to veto appointments or to ask for the recall of U.S. ambassadors, an action completely unimaginable in the client states of the Caribbean. And the issue of ambassadors paled in significance compared to the U.S. authority to approve appointments to high civilian and military positions in the Caribbean countries.

11. "A Zelaya Administration—What Does That Mean for the U.S. Government on Political Issues?" Cable from U.S. Embassy, Tegucigalpa, 13 December 2005, Wikileaks. Comisión de la Verdad, *Informe*, pp. 99-100.

12. "A Zelaya Administration—What Does That Mean for the U.S. Government on Political Issues?" Cable from U.S. Embassy, Tegucigalpa, 13 December 2005, Wikileaks.

13. "POTUS June 5 Meeting with Honduran President Mel Zelaya," Cable from U.S. Embassy, Tegucigalpa, 12 May 2006, Wikileaks; Comisión de la Verdad, *Informe*, p. 101; Leonard, *History of Honduras*, p. 171.

14. "POTUS June 5 Meeting with Honduran President Mel Zelaya," Cable from U.S. Embassy, Tegucigalpa, 12 May 2006, Wikileaks.

15. "Under Attack, Honduran Vice President Troubled by Influence of Certain Leftist Presidential Advisors," Secret Cable from U.S. Embassy, Tegucigalpa, 2 June 2006, Wikileaks.

16. Comisión de la Verdad, *Informe*, pp. 100, 102. "Zelaya's Administration First 45 Days-Pro-U.S. Zelaya Often Torn in Different Directions by Advisors," Cable from U.S. Embassy, Tegucigalpa, 16 March 2006, Wikileaks. U.S. diplomats quickly assimilated the rules of the game of empire and knew that Washington usually did not want to leave a long paper trail of U.S. meddling in the domestic politics of client states. Washington merely received two facts: Vice President Santos wanted to resign but stayed in office. There was no need to explain how Santos went from point A to point C, and I have deduced the ambassador's advice from the general tenor of the cable of 16 March 2006. The last three paragraphs of this chapter provide a blatant and deliberately well-documented example of giving actual orders and not just simply advice.

17. Comisión de la Verdad, *Informe*, pp. 101, 103.

18. "Chief of Defense Romeo Vásquez Velásquez to Stay on Three More Years," Cable from U.S. Embassy, Tegucigalpa, 27 November 2007, Wikileaks; Comisión de la Verdad, *Informe*, p. 204.

19. "Chief of Defense Romeo Vásquez Velásquez to Stay on Three More Years," Cable from U.S. Embassy, Tegucigalpa, 27 November 2007, Wikileaks.

20. "Zelaya Cabinet Shuffle: New Ministers Are Either Loyal or Compliant," Cable from U.S. Embassy, Tegucigalpa, 10 January 2008, Wikileaks. Zelaya was having increasing difficulties finding people to work effectively for him in any area: "In a December meeting with the Ambassador, Zelaya indicated his son Héctor would be the new go-between with the Embassy. In the past, Zelaya has, in turn, appointed the Foreign Minister, the Minister of the Presidency, and a political operative to this role, none of which worked successfully. He is now left with his son as one of those few people he still trusts." Ibid.

21. "POTUS June 5 Meeting with Honduran President Mel Zelaya," Cable from U.S. Embassy, Tegucigalpa, 12 May 2006, Wikileaks; Comisión de la Verdad, *Informe*, p. 118.

22. "Honduras: U.S. Government Should Keep Zelaya Nervous about Chávez, Despite Impending Government of Honduras-Government of Venezuela Oil Deal," Secret Cable from U.S. Embassy, Tegucigalpa, 31 May 2006, Wikileaks.

23. Comisión de la Verdad, *Informe*, p. 103; Leonard, *History of Honduras*, p. 173.

24. "Honduras Signs on to Bolivarian Alternative Despite Wide Opposition," Cable from U.S. Embassy, Tegucigalpa, 26 August 2008, Wikileaks.

25. Comisión de la Verdad, *Informe*, pp. 102–103, 119–120; "Honduran Congress Passes ALBA in Emergency Session," Cable from U.S. Embassy, Tegucigalpa, 11 October 2008, Wikileaks; Leonard, *History of Honduras*, p. 173.

26. "Zelaya and the Fourth Urn," Cable from U.S. Embassy, Tegucigalpa, 9 June 2009, Wikeleaks; Comisión de la Verdad, *Informe*, pp. 103, 119–120.

27. "Zelaya and the Fourth Urn," Cable from U.S. Embassy, Tegucigalpa, 9 June 2009, Wikeleaks; Comisión de la Verdad, *Informe*, p. 125.

28. This extract and the next two come from "Honduran Political Crisis Update and Perspective," Cable from U.S. Embassy, Tegucigalpa, 23 June 2009, Wikileaks.

29. "Update on Fourth Urn," Cable from U.S. Embassy, Tegucigalpa, 1 May 2009, Wikileaks; Comisión de la Verdad, *Informe*, p. 121; Leonard, *History of Honduras*, pp. 172–173.

30. *Los Angeles Times*, 25 March 2009; "Zelaya Calls for Opinion Poll on Constituent Assembly," Cable from U.S. Embassy, Tegucigalpa, 26 March 2009, Wikileaks; Leonard, *History of Honduras*, p. 174. Election-day mechanics confused many outside observers. *Urna* in Spanish means a ballot box and not a ballot. Traditionally in Honduran elections at each voting station voters inserted three distinct ballots in three separate ballot boxes. One ballot box was for candidates for municipal offices, another for candidates for the national unicameral legislature, and the third for presidential elections. The proposal of Zelaya called for installing a fourth ballot box so that voters could cast a fourth ballot on whether to have a referendum on convoking a constitutional convention. Because Honduras could not obtain additional ballots and ballot boxes on such short notice, the government imported them from abroad. Precisely the attempt to rescue these sequestered ballots and boxes on 25 June was the final incident triggering the coup, as the text shows.

31. "Update on Fourth Urn," Cable from U.S. Embassy, Tegucigalpa, 1 May 2009, Wikileaks.

32. "President Zelaya Discusses the Domestic Scene," Cable from U.S. Embassy, Tegucigalpa, 1 April 2009, Wikileaks; Comisión de la Verdad, *Informe*, pp. 122–124.

33. "Fourth Urn Update: President Zelaya Orders Armed Forces to Support," Cable from U.S. Embassy, 1 June 2009, Wikileaks; Comisión de la Verdad, *Informe*, pp. 134, 177.

34. "Zelaya Calls for Opinion Poll on Constituent Assembly," Cable from U.S. Embassy, Tegucigalpa, 26 March 2009, Wikileaks; "President Zelaya Discusses the Domestic Scene," Cable from U.S. Embassy, 1 April 2009, Wikileaks; "Update on Fourth Urn," Cable from U.S. Embassy, Tegucigalpa, 1 May 2009, Wikileaks.

35. "President Zelaya Discusses the Domestic Scene," Cable from U.S. Embassy, Tegucigalpa, 1 April 2009, Wikileaks.

36. "President Zelaya Discusses the Domestic Scene," Cable from U.S. Embassy, Tegucigalpa, 1 April 2009, Wikileaks.

37. "Update on Fourth Urn," Cable from U.S. Embassy, Tegucigalpa, 1 May 2009, Wikileaks.

38. Coverage of pre-coup Honduras in the mainstream English-language media was nearly non-existent. In contrast, television networks such as Univisión and Telemundo had been providing almost constant reporting in their daily news programs since late 2008, and Spanish language periodicals in the United States likewise covered the unfolding crisis in Honduras in great detail.

39. "Zelaya and the Fourth Urn," Confidential/No Forn Cable from U.S. Embassy, Tegucigalpa, 9 June 2009, Wikileaks. My analysis leads me to conclude that at the 7 June meeting the ambassador again repeated the instruction to drop the Fourth Ballot Box proposal, mainly because he had already given that instruction twice before and would do so again afterwards. The ambassador did not include the warning in the cable because it would document the image of a weak U.S. government pleading after Zelaya ignored the two previous warnings. Also, the ambassador had not received any comment from the State Department about his two previous warnings, and he felt it prudent to play it safe and not include similar statements in the record.

40. Among the other implied rules of the imperial game, ambassadors quickly learn that the State Department does not welcome many "Action Requests" and not just because they disrupt the leisurely routines of bureaucrats or because of the dangers of micromanaging from Washington. Too many "Action Requests" sug-

gest that the U.S. Embassy is really not doing its job in keeping problems out of the news and out of Washington's mind. This perception of inadequate performance influences future decisions on promotions, salary increases, and also on transfers to less desirable postings.

41. "Zelaya and the Fourth Urn," Confidential/No Forn Cable from U.S. Embassy, Tegucigalpa, 9 June 2009, Wikileaks.

42. "Guidance on Fourth Urn Issue," Cable from Secretary of State, Washington, D.C., 12 June 2009, Wikileaks.

43. Comision de la Verdad, *Informe*, p. 157; "Zelaya Zig-Zags the Red Lines," Cable from U.S. Embassy, Tegucigalpa, 15 June 2009, Wikileaks; "Honduran Political Crisis Update and Perspective," Cable from U.S. Embassy, Tegucigalpa, 23 June 2009, Wikileaks; "Honduran President Fires Military Joint Chief; Minister of Defense and Other Military Leadership Resign," Cable from U.S. Embassy, Tegucigalpa, 25 June 2009, Wikileaks.

44. Comision de la Verdad, *Informe*, pp. 135-136, 177-178.

45. "Zelaya Zig-Zags the Red Lines," Cable from U.S. Embassy, Tegucigalpa, 15 June 2009, Wikileaks.

46. "Ambassador Warns Honduran Military on Coup Rumors," Cable from U.S. Embassy, Tegucigalpa, 18 June 2009, Wikileaks. Paragraph also draws on Comisión de la Verdad, *Informe*, pp. 123-124.

47. "Ambassador Warns Honduran Military on Coup Rumors," Cable from U.S. Embassy, Tegucigalpa, 18 June 2009, Wikileaks. Paragraph also draws on "Zelaya Zig-Zags the Red Lines," Cable from U.S. Embassy, Tegucigalpa, 15 June 2009, Wikileaks.

48. One of several key areas the Truth and Reconciliation Commission barely touched was precisely the logistics of organizing the poll of 28 June. The commissioners with a minimum of effort could have gathered detailed information on this seemingly non-controversial topic actually driving many of the events.

49. "Honduran President Fires Military Joint Chief; Minister of Defense and Other Military Leadership Resign," Cable from U.S. Embassy, Tegucigalpa, 25 June 2009, Wikileaks; Comisión de la Verdad, *Informe*, pp. 178-179, 205-206; *Wall Street Journal*, 26 June 2009.

50. "Congress Pulls Back from the Brink on Removing President Zelaya," Cable from U.S. Embassy, Tegucigalpa, 26 June 2009, Wikileaks.

51. "Updates on Fourth Urn," Cable from U.S. Embassy, Tegucigalpa, 1 May 2009, Wikileaks; "Congress Pulls Back from the Brink on Removing President Zelaya," Cable from U.S. Embassy, Tegucigalpa, 26 June 2009, Wikileaks.

52. "Honduras Armed Forces Leadership Crisis," Cable from U.S. Embassy, Tegucigalpa, 26 June 2009, Wikileaks; *Wall Street Journal*, 26 June 2009; *New York Times*, 26 June 2009.

53. Comision de la Verdad, *Informe*, pp. 137-138.

54. "Honduras: Political and Legal Standoff Continues," Cable from U.S. Embassy, Tegucigalpa, 26 June 2009, Wikileaks; Comisión de la Verdad, *Informe*, p. 139; *Wall Street Journal*, 26 June 2009.

55. Comisión de la Verdad, *Informe*, pp. 140-142, 166; "Honduras: Political and Legal Standoff Continues," Cable from U.S. Embassy, Tegucigalpa, 26 June 2009, Wikileaks.

56. *New York Times*, 29 June 2009; Comisión de la Verdad, *Informe*, p. 142; *Washington Post*, 29 June 2009.

57. Comisión de la Verdad, *Informe*, pp. 142-144, 194; *Wall Street Journal*, 29 June 2009; *New York Times*, 29 June 2009.

58. Comisión de la Verdad, *Informe*, pp. 144, 150, 194, 207; *Wall Street Journal*, 29 June 2009.

59. For a full description of human rights abuses under the Micheletti presidency, see Comisión de la Verdad, *Informe*, chapter 12 and Human Rights Watch, *After the Coup: Ongoing Violence, Intimidation, and Impunity in Honduras* (New York: Human Rights Watch, 2010), pp. 10-60.

60. Comisión de la Verdad, *Informe*, chapters 9-11. As one of many arguments trying to depict the 28 June 2009 coup as illegal and unconstitutional the U.S. Embassy carefully crafted for public consumption a convenient compilation with the title of "Honduran Coup Timeline." The Embassy deliberately labeled this cable of 2 July 2009 "Unclassified" to make possible its broad distribution to media outlets, foreign governments, and international organizations. This cable is quite remarkable for its omissions, because except for a single direct reference, the role of U.S. diplomats is totally missing. Even though the U.S. government was trying to block the coup this worthy goal had to be kept secret to avoid revealing the informal empire over Honduras.

61. "Updates on Fourth Urn," Cable from U.S. Embassy, Tegucigalpa, 1 May 2009, Wikileaks; Leonard, *History of Honduras*, p. 175.

62. U.S. Government Accountability Office, *Review of U.S. Response to the Honduran Political Crisis of 2009* (Washington, D.C.: Government Printing Office, 2011), pp. 5-8.

63. "Ambassador and President Lobo Discuss the Need to Appoint New Military High Command," Secret Cable from U.S. Embassy, Tegucigalpa, 17 February 2010, Wikileaks. See also for paragraph Comisión de la Verdad, *Informe*, pp. 208-209.

64. "Ambassador and President Lobo Dis-

cuss the Need to Appoint New Military High Command," Secret Cable from U.S. Embassy, Tegucigalpa, 17 February 2010, Wikileaks.

65. "New Minister of Defense," Cable from U.S. Embassy, Tegucigalpa, 27 February 2010, Wikileaks. In an obvious contrast with the reality depicted in the secret embassy cables, the U.S. Government Accountability Office, *Review of U.S. Response*, p. 9 is silent about the pressure U.S. diplomats inflicted upon the hapless President Lobo.

Chapter 10

1. Joint Testimony of Kevin L. Perkins, FBI agent, and Anthony P. Placido, DEA agent, before U.S. Senate Caucus on International Narcotics Control, Washington, D.C., 5 May 2010.

2. "Engaging the New Mexican Administration on Security and Counterterrorism," Cable from U.S. Embassy, Mexico City, 15 June 2006, Wikileaks; Craig A. Deare, *U.S.-Mexico Defense Relations: An Incompatible Interface* (Washington, D.C.: Institute for National Strategic Studies, 2009), p. 4.

3. "Calderón Debuts with a Deft Hand," Cable from U.S. Embassy, Mexico City, 21 December 2006, Wikileaks; Anabel Hernández, *Los señores del narco* (Mexico: Grijalbo, 2010), pp. 472–475; Ernesto Núñez Albarrán, *Crónica de un sexenio fallido* (Mexico: Grijalbo, 2012), pp. 52–53, 56; Ioan Grillo, *El Narco: Inside Mexico's Criminal Insurgency* (New York: Bloomsbury Press, 2011), pp. 112–113.

4. "Calderón Debuts with a Deft Hand," Cable from U.S. Embassy, Mexico City, 21 December 2006, Wikileaks; Alfredo Corchado, *Midnight in Mexico: A Reporter's Journey Through a Country's Descent Into Darkness* (New York: Penguin, 2013), pp. 176–177.

5. "Two-Month Truce Has Drastically Reduced Northern Border Drug Violence," Confidential/No Forn Cable from U.S. Consulate, Monterrey, 23 August 2007, Wikileaks; Hernández, *Los señores del narco*, pp. 486–490.

6. "Two-Month Truce Has Drastically Reduced Northern Border Drug Violence," Confidential/No Forn Cable from U.S. Consulate, Monterrey, 23 August 2007, Wikileaks.

7. "PEMEX Pipeline Attack Causes Economic Damage," Cable from U.S. Embassy, Mexico, 14 September 2007, Wikileaks; Carlos Montemayor, *La guerrilla recurrente* (Mexico: Debate, 2007), pp. 42, 49, 82–84; Jesús López González, *Presidencialismo y fuerzas armadas en México, 1876–2012* (Mexico: Gernika, 2012), pp. 266–267; James F. Rochlin, *Social Forces and the Revolution in Military Affairs: The Cases of Colombia and Mexico* (New York: Palgrave Macmillan, 2007), pp. 125–127.

8. "PEMEX Pipeline Attack Causes Economic Damage," Cable from U.S. Embassy, Mexico, 14 September 2007, Wikileaks; "FARC in Mexico," Secret/No Forn Cable from U.S. Embassy, Mexico City, 28 March 2008, Wikileaks; "EPR-GOM Dialogue Helping to Keep the Peace," Cable from U.S. Embassy, Mexico City, 6 August 2008, Wikileaks; "EPR Mediation Talks Dissolve," Cable from U.S. Embassy, Mexico City, 28 April 2009, Wikileaks; Diego Enrique Osorno, *El Cártel de Sinaloa: Una historia del uso político del narco* (Mexico: Debolsillo, 2011), pp. 84–85.

9. "The U.S.–Mexican Relationship: Meeting Challenges, Tapping Opportunities in 2009—Security and Reform," Cable from U.S. Embassy, Mexico City, 4 February 2009, Wikileaks.

10. Anabel Hernández, *México en llamas: El legado de Calderón* (Mexico: Grijalvo, 2012) pp. 145–146.

11. "Narco-killings Update," Cable from U.S. Embassy, Mexico City, 11 March 2008, Wikileaks; "Weakening of Gulf, Sinaloa Cartels Generates Mexico's Spike in Violence," Cable from U.S. Embassy, Mexico City, 9 June 2008, Wikileaks; Osorno, *El Cártel de Sinaloa*, pp. 282–283; Hernández, *México en llamas*, pp. 207, 232.

12. "Weakening of Gulf, Sinaloa Cartels Generates Mexico's Spike in Violence," Cable from U.S. Embassy, Mexico City, 9 June 2008, Wikileaks.

13. Hernández, *Señores del narco*, p. 467; Hernández, *México en llamas*, pp. 92–97; Corchado, *Midnight in Mexico*, p. 148.

14. Hernández, *Señores del narco*, pp. 471–472.

15. Núñez Albarrán, *Crónica de un sexenio fallido*, pp. 87–107; Hernández, *México en llamas*, pp. 89–92, 105–108, 112–113; Hernández, *Señores del narco*, pp. 527–530.

16. Iñigo Guevara Moyano, *Adapting, Transforming, and Modernizing Under Fire: The Mexican Military 2006–2011* (Carlisle, PA: U.S. Army War College, 2012), pp. 14–15; George W. Grayson, *The Impact of President Felipe Calderón's War on Drugs on the Armed Forces: The Prospects for Mexico's Militarization and Bilateral Relations* (Carlisle, PA: U.S. Army War College, 2013), p. 2; Sylvia Longmire, *Cartel: The Coming Invasion of Mexico's Drug Wars* (New York: Palgrave Macmillan, 2011), pp. 119–120.

17. "Walters Meetings with Mexican Officials," Cable from U.S. Embassy, Mexico City, 13 April 2007, Wikileaks; Guevara Moyano, *The Mexican Military 2006–2011*, pp. 24–25; "Narco-Violence Sparks, Top Cops Targeted,"

Cable from U.S. Embassy, Mexico City, 13 May 2008, Wikileaks.

18. Guevara Moyano, *The Mexican Military 2006-2011*, pp. 19-20

19. Grayson, *Mexico's Militarization*, pp. 55, 57; Mexico, Instituto Nacional de Estadística y Geografía, *Anuario estadístico 2001* (Mexico: Instituto Nacional de Estadística y Geografía, 2002), Table 8.28; Ioan Grillo, *El Narco*, p. 128; Longmire, *Cartel*, p. 103; Roderic Ai Camp, *Generals in the Palacio: The Military in Modern Mexico* (New York: Oxford University Press, 1992), pp. 101-109. Although Camp's findings refer to the twentieth century, the conditions privileging Mexico City as a source of officers remain as strong if not stronger. Instead the complete passing from the scene of the revolutionary officers from northern Mexico has if anything increased the preponderance of officers originating in Mexico City.

20. Grayson, *Mexico's Militarization*, pp. 53-54; Guevara Moyano, *The Mexican Military 2006-2011*, pp. 16-17; Grillo, *El Narco*, pp. 98, 113, 213; Deare, *U.S.-Mexico Defense Relations*, p. 5.

21. Presently those few soldiers who reenlist are automatically promoted to the rank of sergeant with its slightly higher pay. Presumably if all soldiers earn adequate pay, this motivation to acquire rank of sergeant will no longer be as pressing. The tradition-bound and slow to adapt army will then have to develop mechanisms to separate individuals best suited for the rank of sergeant from those who simply make good soldiers. Historically, the weakness of the NCO ranks and the excessive number of officers have been two structural defects hampering the development of the Mexican army.

22. "By the Numbers: Women in the U.S. Military," 24 January 2013, CNN.com.

23. Eileen Patten and Kim Parker, "Women in the U.S. Military: Growing Share, Distinctive Profile," Pew Research Center, September 2011; Rutgers University, Institute for Women's Leadership, "Women in the U.S. Military Services." As another indirect indicator of the behavior of Mexican women, in the U.S. military both male and female Hispanics have almost identical percentages only slightly lower than the Hispanic share of the total U.S. population.

24. Patten and Parker, "Women in the U.S. Military," September 2011. No record seems to exist about any publicity campaign to attract women to the military in Mexico. Recruitment for men tends to take place near military facilities, in remote rural regions particularly in southern Mexico, and in temporary booths set up at public events such as parades. In the United States, a barrage of ads bombards television and other media; however, recruiters, perhaps because of political correctness, have never played the matrimonial card. Without becoming blatant like the match-making ads on television, the appeal could be communicated to American women by subtle messages hinting at opportunities to travel, to meet many people, and to work closely with interesting individuals who will not have all their skin from head to toe covered in full combat gear.

25. Grayson, *Mexico's Militarization*, pp. 43-45; Camp, *Generals in the Palacio*, pp. 133-136, 141-154.

26. Grayson, *Mexico's Militarization*, pp. 44-46.

27. Guevara Moyano, *The Mexican Military 2006-2011*, pp. 27-30; Jorge Luis Sierra Guzmán, *El enemigo interno: Contrainsurgencia y fuerzas armadas en Mexico* (Mexico: Universidad Iberoamericana, 2003), pp. 289-303.

28. Guevara Moyano, *The Mexican Military 2006-2011*, pp. 26, 30-31; "Mexican Navy Operation Nets Drug Kingpin Arturo Beltrán Leyva," Secret Cable from U.S. Embassy, 17 December 2009, Wikileaks; Sierra Guzmán, *El enemigo interno*, pp. 312-313; George W. Grayson, *Mexico: Narco-Violence and a Failed State?* (New Brunswick: Transaction Publishers, 2011), pp. 157-158.

29. "Scene Setter for the Visit of Secretary of Defense Gates to Mexico City," Cable from U.S. Embassy, Mexico City, 10 April 2008, Wikileaks; "Mexico Seeks to Turn the Page on Corruption," Cable from U.S. Embassy, Mexico City, 11 April 2008, Wikileaks.

30. "Scene Setter for the Visit of Secretary of Defense Gates to Mexico City," Cable from U.S. Embassy, Mexico City, 10 April 2008, Wikileaks; Grayson, *Narco-Violence*, pp. 238-239.

31. "Narco-Violence Sparks; Border Affected," Cable from U.S. Embassy, Mexico City, 5 December 2008, Wikileaks; Corchado, *Midnight in Mexico*, pp. 194-195.

32. "Narco-Violence Sparks; Border Affected," Cable from U.S. Embassy, Mexico City, 5 December 2008, Wikileaks.

33. "Violence Escalates, Cartels Take a Stand against GOM," Cable from U.S. Embassy Mexico City, 23 February 2009, Wikileaks.

34. Ibid.

35. Ibid.

36. "Ciudad Juárez at a Tipping Point," Cable from U.S. Embassy, Mexico City, 13 March 2009, Wikileaks.

37. "Mexico's Federal Police to Receive a Facelift," Cable from U.S. Embassy, Mexico City, 14 May 2009, Wikileaks.

38. "Mexican Army Major Arrested for As-

sisting Drug Trafficking Organizations," Secret Cable from U.S. Embassy, Mexico City, 20 January 2009, Wikileaks.
39. "Mexico's Congress Considers Creation of National Guard," Cable from U.S. Embassy, Mexico City, 5 March 2009, Wikileaks; "Mexican Army Major Arrested for Assisting Drug Trafficking Organizations," Secret Cable from U.S. Embassy, Mexico City, 20 January 2009, Wikileaks; "Mexico: More Interagency Cooperation Needed on Intelligence Issues," Secret/No Forn Cable from U.S. Embassy, Mexico City, 10 November 2009, Wikileaks.
40. "Winning or One Step Behind: Government of Mexico's Interdiction Policy for Methlabs," Cable from U.S. Embassy, Mexico City, 31 August 2009, Wikileaks.
41. "Mérida Initiative," website of mexico.usembassy.gov and "Mérida Initiative," www.state.gov; Grayson, *Narco-Violence*, pp. 234–237; Longmire, *Cartel*, pp. 132–133.
42. "Elements of GOM Policy Team Interested in Focusing Together on Improving Security in a Few Key Cities," Cable from U.S. Embassy, Mexico City, 5 October 2009, Wikileaks; Corchado, *Midnight in Mexico*, p. 178.
43. Joint Testimony of Kevin L. Perkins and Anthony P. Placido before U.S. Senate Caucus on International Narcotics Control, Washington, D.C., 5 May 2010; "Elements of GOM Policy Team Interested in Focusing Together on Improving Security in a Few Key Cities," Cable from U.S. Embassy, Mexico City, 5 October 2009, Wikileaks.
44. "Monterrey Shootouts Leave 17 Dead, Expose Shortcomings in State's Crime Fighting Apparatus," Cable from U.S. Consulate, Monterrey, 14 December 2009, Wikileaks; "Scene Setter for the Opening of the Defense Bilateral Working Group, Washington, D.C.," Secret Cable from U.S. Embassy, Mexico City, 29 January 2010, Wikileaks.
45. "Mexican Navy Operation Nets Drug Kingpin Arturo Beltrán Leyva," Secret Cable from U.S. Embassy, Mexico City, 17 December 2009, Wikileaks; J. Jesús Esquivel, *La DEA en México: Una historia oculta del narcotráfico contada por los agentes* (Mexico: Grijalbo, 2012), pp. 188–189; José Reveles, *El Cártel incómodo: El fin de los Beltrán Leyva y la hegemonía del Chapo Guzmán* (Mexico: Grijalbo, 2010), pp. 157–158; Núñez Albarrán, *Crónica de un sexenio fallido*, pp. 159–160.
46. "Mexican Navy Operation Nets Drug Kingpin Arturo Beltrán Leyva," Secret Cable from U.S. Embassy, Mexico City, 17 December 2009, Wikileaks; "Update: SSP to Replace Military as Primary Security Player in Ciudad Juárez," Cable from U.S. Embassy, Mexico City, 8 January 2010, Wikileaks; "GOM Arrests Carlos Beltrán Leyva—An Important Psychological Victory," Cable from U.S. Embassy, Mexico City, 25 January 2010, Wikileaks; "Scene Setter for the Opening of the Defense Bilateral Working Group, Washington, D.C.," Secret Cable from U.S. Embassy, Mexico City, 29 January 2010, Wikileaks.
47. "Monterrey: Money Laundering Moves from Banks to Cash Businesses," Cable from U.S. Consulate, Monterrey, 8 April 2009, Wikileaks; Diego Enrique Osorno, *El Cártel de Sinaloa: Una historia del uso político del narco* (Mexico: Debolsillo, 2011), pp. 29, 37.
48. "Monterrey Sees Bloodiest Day Ever with Nine Assassinations," Confidential/No Forn Cable from U.S. Consulate, Monterrey, Wikileaks; Osorno, *El Cártel de Sinaloa*, p. 29.
49. "Unprecedented Escalation of Narco-Violence in Monterrey," Secret Cable from U.S. Consulate, Monterrey, 23 March 2007, Wikileaks.
50. "New Nuevo León Security Director's Ambitious Plans to Overhaul Corrupt Police Forces," Cable from U.S. Consulate, Monterrey, 7 September 2007, Wikileaks; "Drug Violence Focuses on Small Dealers: Public Yawns," Cable from U.S. Consulate, Monterrey, 12 March 2008, Wikileaks; "Kidnappings Rise in Nuevo León: Federal Forces Decreasing Presence," Cable from U.S. Consulate, Monterrey, 15 May 2008, Wikileaks.
51. "Public Protest Against Mexican Military Likely Organized by Drug Cartels," Cable from U.S. Consulate, Monterrey, 12 February 2009, Wikileaks; Osorno, *Cártel de Sinaloa*, pp. 37, 41; Grayson, *Narco-Violence*, pp. 155–156.
52. "Nuevo León's Efforts to Reform State and Local Police Have Not Been Effective," Secret/No Forn Cable from U.S. Consulate, Monterrey, 4 March 2009, Wikileaks.
53. "Mexican Military Arrests State and Local Police Officers," Secret Cable from U.S. Consulate, Monterrey, 5 June 2009, Wikileaks; "Mexican Military Continuing to Arrest Local Police Officers in Nuevo León," Cable from U.S. Consulate, Monterrey, 25 June 2009, Wikileaks; "Military Presence No Panacea for Nuevo León," Cable from U.S. Consulate, Monterrey, 27 July 2009, Wikileaks.
54. "Intensifying Violence Continues to Shake Public Confidence in Local Authorities," Cable from U.S. Consulate, Monterrey, 21 August 2009, Wikileaks; "Incoming Nuevo León Governor Faces Serious Security Challenges," Cable from U.S. Consulate, Monterrey, 9 October 2009, Wikileaks.
55. "Monterrey Shootouts Leave 17 Dead, Expose Shortcomings in State's Crime Fighting Apparatus," Cable from U.S. Consulate, Monterrey, 14 December 2009, Wikileaks; "Business

Leaders Support Mexican Military and Mérida Initiative," Cable from U.S. Consulate, Monterrey, 7 May 2009, Wikileaks; "Killing of Police Chief Latest in Cartel Violence, Intimidation against Public Officials," Cable from U.S. Consulate, Monterrey, 6 November 2009, Wikileaks.

56. "Monterrey Shootouts Leave 17 Dead, Expose Shortcomings in State's Crime Fighting Apparatus," Cable from U.S. Consulate, Monterrey, 14 December 2009, Wikileaks.

57. "Mexican Army Public Relations Offensive in Nuevo León," Cable from U.S. Consulate, Monterrey, 1 December 2009, Wikileaks; "Border Violence Spreads to Nuevo León," Cable from U.S. Consulate, Monterrey, 26 February 2010, Wikileaks.

58. Grayson, *Militarization of Mexico*, pp. 4–14; "Mexico: More Interagency Cooperation Needed on Intelligence Issues," Secret/No Forn Cable from U.S. Embassy, Mexico City, 10 November 2009, Wikileaks; Grayson, *Narco-Violence*, pp. 104–106.

59. Esquivel, *La DEA en México*, pp. 237–238.

60. *Proceso* No. 1609 of 2 September 2007.

61. "Director of National Intelligence Dennis Blair's Meeting with General Galván Galván," Secret Cable from U.S. Embassy, Mexico City, 26 October 2009, Wikileaks; "Mexico: More Interagency Cooperation Needed on Intelligence Issues," Secret/No Forn Cable from U.S. Embassy, Mexico City, 10 November 2009, Wikileaks; Corchado, *Midnight in Mexico*, p. 249.

62. Esquivel, *La DEA en México*, pp. 241–244.

63. "Elements of Government of Mexico Policy Team Interested in Focusing Together on Improving Security in a Few Key Cities," Cable from U.S. Embassy, Mexico City, 5 October 2009, Wikileaks.

64. "Director of National Intelligence Dennis Blair's Meeting with General Galván Galván," Secret Cable from U.S. Embassy, Mexico City, 26 October 2009, Wikileaks; "Drugs and Downturn on the Border," Cable from U.S. Embassy, Mexico City, 17 November 2009, Wikileaks; "U.S.–Mexico Relations: Progress in 2009, Challenges in 2010," Cable from U.S. Embassy, Mexico City, 21 January 2010, Wikileaks; Hernández, *Los señores del narco*, p. 537.

65. "Mexico Murders at Over 101,000 in Past Six Years, Report Says," 27 November 2012, Fox News Latino.

66. Hernández, *México en llamas*, pp. 223–234.

Chapter 11

1. Estanislao Beltrán, Chief of the Michoacán Self-Defense Forces, Univisión, in "Las Autodefensas ¿Qué hay detrás?" broadcast 16 March 2014.

2. George W. Grayson, *The Impact of President Felipe Calderón's War on Drugs on the Armed Forces: The Prospects for Mexico's Militarization and Bilateral Relations* (Carlisle, PA: Strategic Studies Institute, 2013), pp. 68–69; "U.S. Cooperation Against Cartels Remains Strong," 16 May 2013, *Forbes*; "Security Dominates Talk of U.S.–Mexico Relations," 2 May 2013, CNN.com.

3. Secretaría de la Defensa Nacional, "Comunicados de Prensa," 10 December 2012, 25 October 2013; "U.S. Cooperation Against Cartels Remains Strong," 16 May 2013, *Forbes*; "Maduro's International Support," *The News* (Mexico City), 27 June 2013.

4. Secretaría de la Defensa Nacional, "Comunicados de Prensa," 15 July, 25 November 2013.

5. "Mexico Continues War on Drug Cartel Leaders," *Buffalo News*, 19 August 2013; "Gobernación concentra el presupuesto más importante en seguridad," 8 September 2013, CNNMexico.com.

6. "Con más 'maña' que fuerza, Peña Nieto golpea estructura de cartales," 23 February 2014, CNNMexico.com.

7. Telemundo, Daily TV news broadcasts, February 2014; "Con más 'maña' que fuerza, Peña Nieto golpea estructura de cartales," 23 February 2014, CNNMexico.com.

8. Univisión, "El Chapo Guzmán cayó por su esposa y sus gemelas," 25 February 2014.

9. "Informantes y escuchas de celulares, las claves que llevaron a El Chapo," 24 February 2014, CNNMexico.com.

10. Telemundo, daily TV news broadcasts, February 2014.

11. *New York Times*, 23 February 2014; Univisión, "El Chapo Guzmán cayó por su esposa y sus gemelas," 25 February 2014; "Informantes y escuchas de celulares, las claves que llevaron a El Chapo," 24 February 2014, CNNMexico.com.

12. "El Chapo's Arrest Unlikely to Break Mexican Cartel," *New York Times*, 25 February 2014.

13. George W. Grayson, *La Familia Drug Cartel: Implications for U.S.–Mexican Security* (Carlisle, PA: Strategic Studies Institute, 2010), pp. 13–19.

14. "Los 7 líderes de los Caballeros Templarios," 15 January 2014, *Union* (Jalisco); Grayson, *La Familia Drug Cartel*, pp. 35–37.

15. "La historia de los Caballeros Templarios," 27 August 2013, *El País* (Cali); Grayson, *La Familia Drug Cartel*, pp. 64–65, 76–78.

16. "Cómo operan los Caballeros Templarios de Michoacán," 26 September 2013, LosÁn

gelesPress.org; Grayson, *La Familia Drug Cartel*, p. 55; "Los Caballeros Templarios surgen tras la muerte de La Familia," 15 July 2011, Ágorarevista.com.

17. "Cómo operan los Caballeros Templarios de Michoacán," 26 September 2013, Los Ángeles Press.org; "Ex militares y ex templarios, en autodefensa," 23 January 2014, *El Universal*; "Los Caballeros Templarios surgen tras la muerte de La Familia," 15 July 2011, Ágorarevista.com.

18. Univisión, "Las Autodefensas ¿Qué hay detrás?" broadcast 16 March 2014; "Cómo operan los Caballeros Templarios de Michoacán," 26 September 2013, Los Ángeles Press.org.

19. "Cómo operan los Caballeros Templarios de Michoacán," 26 September 2013, LosÁngelesPress.org; "Ex militares y ex templarios, en autodefensa," 23 January 2014, *El Universal*.

20. As medical doctor José Manuel Mireles explained: "lo que nos levantó el alma fueron las violaciones de nuestras esposas y de nuestras hijas, porque ya en la calle las que pasaban los sicarios les decían a sus pistoleros tráeme esas mujeres," Univisión, "Las Autodefensas ¿Qué hay detrás?" broadcast 16 March 2014.

21. "Las autodefensas festejan un año de su revolución con misa y birria," 12 February 2014, CNNMexico.com; Univisión, "Las Autodefensas ¿Qué hay detrás?" broadcast 16 March 2014. Readers of chapter 17 of my *Wars of Latin America, 1982–2013: The Path to Peace* (Jefferson, NC: McFarland, 2013) will notice striking similarities between the guerrilla insurrections in Latin America and the self-defense revolt in Michoacán. Actually if Knights Templars replace the variable of "hated dictator" the explanations of that chapter are equally valid for Michoacán. The idea is that if rulers earn the hatred of most members of all social classes, the people will rise up in an unstoppable revolt.

22. *New York Times*, 14, 15 January, 25 February 2014.

23. "El Ejército controla la seguridad del puerto de Lázaro Cárdenas, Michoacán," 4 November 2013, CNN Mexico.com. For the revolt against Shining Path, see De La Pedraja, *Wars of Latin America, 1982–2013*, pp. 76–81, 167–172.

24. "Militares y autodefensas se enfrentan en Michoacán," 14 January 2014, *El País*; "Tras llegar a Apatzingán, autodefensas van por el resto de Michoacán," 10 February 2014, CNNMexico.com; "Fausto Vallejo reconoce 'omisiones' en la seguridad en Michoacán," 14 February 2014, CNNMexico.com.

25. "El ejército de las milicias," 30 January 2014, Tierrasdeamerica.com; "10 cosas que debes saber para entender el conflicto en Michoacán," 13 February 2014, CNNMexico.com; *New York Times*, 15 January 2014.

26. *New York Times*, 25 February 2014; daily TV news broadcasts of Telemundo, March 2014.

27. "Militares y autodefensas se enfrentan en Michoacán," 14 January 2014, *El País*; Univisión, "Las Autodefensas ¿Qué hay detrás?" broadcast 16 March 2014; *Proceso*, 21 January 2014.

28. *New York Times*, 25 February 2014; Univisión, "Las Autodefensas ¿Qué hay detrás?" broadcast 16 March 2014; "Las autodefensas entran al bastión de los Caballeros Templarios," 9 February 2014, *El País*; "Tras llegar a Apatzingán, autodefensas van por el resto de Michoacán," 10 February 2014, CNNMexico.com.

29. CNN.Mexico.com, 15, 25 April 2014; *El Universal*, 29 June 2014; Insight Crime, 27 June 2014.

30. "Opponent of Mexico's Cartels Is Detained in Vigilantes' Death," *New York Times*, 13 March 2014; "10 cosas que debes saber para entender el conflicto en Michoacán," 13 February 2014, CNNMexico.com; daily TV news broadcasts of Telemundo, March 2014. The similarities of the self-defense forces in Michoacán with the self-defense committees against Shining Path in Peru come to mind. See De La Pedraja, *Wars of Latin America, 1982–2013*, pp. 166–168, 170–173, 178–179.

Chapter 12

1. Niccolò Machiavelli, *Discorsi sopra la prima deca di Tito Livio* (Milan: Rizzoli Editore, 1984), pp. 543–544. Translation by Beatriz De La Pedraja.

2. "Vice President Biden's March 30 Bilateral Meeting with Costa Rican President Óscar Arias," Cable from U.S. Embassy, San José, 8 April 2009, Wikileaks.

3. "Scene Setter: Codel Burton in Costa Rica," Cable from U.S. Embassy, San José, 3 January 2006, Wikileaks.

4. "Costa Rica: MAMPADs Inventory and Trafficking Controls," Secret/No Forn Cable from U.S. Embassy, San José, 1 August 2005, Wikileaks; "Costa Rica: Visit of Southcom Amb. Trivelli," Cable from U.S. Embassy, San José, 16 October 2009, Wikileaks.

5. "Assessment of Costa Rican Security Requirements," Cable from U.S. Embassy, San José, 30 August 2007, Wikileaks.

6. "Costa Rica: VP Chinchilla on Security Issues," Cable from U.S. Embassy, San José, 14 May 2008, Wikileaks; "Costa Rica: Minister Rodrigo Arias Visits Southcom," Cable from U.S. Embassy, San José, 15 July 2009, Wikileaks.

7. "Costa Rica: Visit of Southcom Amb. Trivelli," Cable from U.S. Embassy, San José, 16 October 2009, Wikileaks; "Costa Rica: Minis-

ter Rodrigo Arias Visits Southcom," Cable from U.S. Embassy, San José, 15 July 2009, Wikileaks.
  8. "Costa Rica: VP Chinchilla on Security Issues," Cable from U.S. Embassy, San José, 14 May 2008, Wikileaks; "Costa Rica: Visit of Southcom Amb. Trivelli," Cable from U.S. Embassy, San José, 16 October 2009, Wikileaks.
  9. "Costa Rica: Minister Rodrigo Arias Visits Southcom," Cable from U.S. Embassy, San José, 15 July 2009, Wikileaks.
  10. "Costa Rican Police Professionalization Program," Cable from U.S. Embassy, San José, 25 May 2007, Wikileaks; "Costa Rica: Visit of Southcom Amb. Trivelli," Cable from U.S. Embassy, San José, 16 October 2009, Wikileaks.
  11. "Assignment of Costa Rican Security Requirements," Cable from U.S. Embassy, San José, 30 August 2007, Wikileaks; "Vice President Biden's March 30 Bilateral Meeting with Costa Rican President Óscar Arias," Cable from U.S. Embassy, San José, 8 April 2009, Wikileaks.
  12. "One Size Does Not Fit All: Enduring Friendship Boats are not the Best Solution for Costa Rica," Cable from U.S. Embassy, San José, 11 July 2008, Wikileaks.
  13. Ibid.
  14. "Assignment of Costa Rican Security Requirements," Cable from U.S. Embassy, San José, 30 August 2007, Wikileaks; "Costa Rican Coast Guard Receives Maintenance Training," Cable from U.S. Embassy, San José, 15 May 2008, Wikileaks; "Costa Rica: Visit of Southcom Amb. Trivelli," Cable from U.S. Embassy, San José, 16 October 2009, Wikileaks.
  15. "Costa Rica and Panama—Cooperation and Competition," Cable from U.S. Embassy, San José, 5 May 2009, Wikileaks.
  16. *New York Times*, 23 December 1989; Corinne Caumartin, "Depoliticisation in the Reform of the Panamanian Security Apparatus," *Journal of Latin American Studies* 39 (2007): 111.
  17. *New York Times*, 24 December 1989, 2 January 1990, 2 February 1990; Ronald D. Sylvia and Constantine P. Danopoulos, "Civil-Military Relations in a Civilianized State: Panama," *Journal of Political and Military Sociology* 33(Summer 2005); Bob Woodward, *The Commanders* (New York: Simon & Schuster, 1991), p. 133.
  18. *New York Times*, 9, 12 January 1990.
  19. Caumartin, "Panamanian Security Apparatus," p. 113; *New York Times*, 30 October 1990; Sylvia and Danopoulos, "Civil Military Relations in Panama."
  20. Caumartin, "Panamanian Security Apparatus," pp. 112, 115; *New York Times*, 30 October 1990.
  21. *Washington Post*, 7 December 1990; *New York Times*, 6 December 1990; Kevin Buckley, *Panama: The Whole Story* (New York: Touchstone, 1991), p. 262; *Boston Globe*, 9 December 1990.
  22. Caumartin, "Panamanian Security Apparatus," p. 112; *New York Times*, 2 February 1990, 21 January 1991.
  23. Caumartin, "Panamanian Security Apparatus," p. 126; *New York Times*, 20 October 1990, 21 January 1991, 1 May 1992.
  24. Caumartin, "Panamanian Security Apparatus," pp. 121–124; *New York Times*, 16 December 1990.
  25. "Panama: Rehabilitating the Dictatorship," Cable from U.S. Embassy, Panama, 12 May 2006, Wikileaks.
  26. *Washington Post*, 16 October 1995; Caumartin, "Panamanian Security Apparatus," p. 125; René De La Pedraja, *Wars of Latin America, 1982–2013: The Path to Peace* (Jefferson, NC: McFarland, 2013), p. 127.
  27. "Panama Government Weathers Fierce Criticism as It Strives to Modernize Its Public Forces," Cable from U.S. Embassy, Panama City, 23 November 2004, Wikileaks; Sylvia and Danopoulos, "Civil Military Relations in Panama"; Caumartin, "Panamanian Security Apparatus," pp. 122–123.
  28. "Panama Government Weathers Fierce Criticism as It Strives to Modernize its Public Forces," Cable from U.S. Embassy, Panama City, 23 November 2004, Wikileaks; "Panama Proposed Merger of Air and Maritime Services—Innovation or a Bad Marriage," Cable from U.S. Embassy, Panama City, 2 November 2005, Wikileaks.
  29. "Panama: Rehabilitating the Dictatorship," Cable from U.S. Embassy, Panama, 12 May 2006, Wikileaks; *New York Times*, 3 May 2004.
  30. "Panama: Rehabilitating the Dictatorship," Cable from U.S. Embassy, Panama, 12 May 2006, Wikileaks.
  31. "Panama Government Weathers Fierce Criticism as It Strives to Modernize Its Public Forces," Cable from U.S. Embassy, Panama City, 23 November 2004, Wikileaks.
  32. "Panama: Militarization Debate Resurfaces," Cable from U.S. Embassy, Panama, 30 May 2008, Wikileaks.
  33. *Washington Report on the Hemisphere*, 26 March 2008; "Panama: Militarization Debate Resurfaces," Cable from U.S. Embassy, Panama, 30 May 2008, Wikileaks.
  34. "Panama Proposed Merger of Air and Maritime Services—Innovation or a Bad Marriage," Cable from U.S. Embassy, Panama City, 2 November 2005, Wikileaks; "Panama's New Security Laws Passed amidst Opposition," Cable from U.S. Embassy, Panama City, 5 September 2008, Wikileaks.

35. "Panamanian Security Cooperation: Irreplaceable?" Secret/No Forn Cable From U.S. Embassy, Panama City, 26 August 2006, Wikileaks; "Plans to Establish Four Maritime Operations Centers," Cable from U.S. Embassy, Panama City, 9 November 2009, Wikileaks.

36. "Panama: Leaders Grapple with Security Reforms," Cable from U.S. Embassy, Panama City, 12 August 2008, Wikileaks; "Panamanian Security Cooperation," Secret/No Forn Cable From U.S. Embassy, Panama City, 26 August 2006, Wikileaks; "Panama: How Mérida Fits Into the Emerging Security Consensus," Confidential/No Forn Cable from U.S. Embassy, Panama City, 3 March 2009, Wikileaks.

37. "Government of Panama Requests Colombian Helicopters to Help Control FARC Incursions in the Darién," Secret/No Forn Cable from U.S. Embassy, Panama City, 16 December 2006, Wikileaks.

38. "Confrontation with FARC Rattles Government of Panama," Secret/No Forn Cable from U.S. Embassy, Panama City, 1 February 2010, Wikileaks.

39. "Darién Activity Update," Secret/No Forn Cable from U.S. Embassy, Panama City, 15 December 2009.

40. "Martinelli Confirms Approval of Cross-Border Attack Plan," Secret/No Forn Cable from U.S. Embassy, Panama City, 17 December 2009, Wikileaks.

41. "Colombian Cross-Border Offer Sows Doubt and Confusion in Government of Panama," Secret/No Forn Cable from U.S. Embassy, Panama City, 11 December 2009, Wikileaks. This sudden discovery of deeply hidden Panamanian fears about Colombian involvement actually was in reality a clever attempt by U.S. diplomats to present as Panamanian the total opposition of the U.S. government to any meddling by Colombia. Precisely President Martinelli had turned to Colombia to try to minimize the massive and smothering presence of the United States in all aspects of Panamanian life.

42. "Chargé Discusses Colombia-Panama Border Cooperation with Police Chief," Secret/No Forn Cable from U.S. Embassy, Bogotá, 16 December 2009, Wikileaks.

43. "Martinelli Confirms Approval of Cross-border Attack Plan," Secret/No Forn Cable from U.S. Embassy, Panama City, 17 December 2009, Wikileaks. The intense activity of the U.S. embassies in Bogotá and Panama City contrasts sharply with the absence of guidance from Washington on how best to deal with FARC and the drug threat in Darién.

44. *Listín Diario*, 28 September 2011; Anthony P. Maingot and Wilfredo Lozano, *The United States and the Caribbean: Transforming Hegemony and Sovereignty* (New York: Routledge, 2005), pp. 54–61; Philippe R. Girard, *Clinton in Haiti: The 1994 U.S. Invasion of Haiti* (New York: Palgrave Macmillan, 2004), pp. 74–75; "Haitian National Police—Past, Present, and Future," Cable from U.S. Embassy, Port-au-Prince, 13 August 2009, Wikileaks.

45. Maingot and Lozano, *The United States and the Caribbean*, pp. 57–62; Girard, *Clinton in Haiti*, pp. 16–18; Bill Clinton, *My Life* (New York: Alfred A. Knopf, 2004), pp. 618–619.

46. *Washington Post*, 21 September 1999; Girard, *Clinton in Haiti*, pp. 8–15, 24–25; Maingot and Lozano, *The United States and the Caribbean*, pp. 56, 62. Estimates put the per capita income of the Dominican Republic in a range from five to ten times larger than that of Haiti.

47. Girard, *Clinton in Haiti*, pp. 123–127.

48. Bob Shacochis, *The Immaculate Invasion* (New York: Penguin Books, 1999), p. 305; *Washington Post*, 21 September 1999; Girard, *Clinton in Haiti*, pp. 124–125, 213n24; *New York Times*, 22 February 2004; "Haitian National Police—Past, Present, and Future," Cable from U.S. Embassy, Port-au-Prince, 13 August 2009, Wikileaks.

49. Girard, *Clinton in Haiti*, pp. 124–126, 213n24; "Haitian National Police—Past, Present, and Future," Cable from U.S. Embassy, Port-au-Prince, 13 August 2009, Wikileaks.

50. "Haiti's Central Plateau Poor and Isolated," Cable from U.S. Embassy, Port-au-Prince, 4 January 2006, Wikileaks; *New York Times*, 22 February 2004.

51. "Ambassador's Call on Dominican President Mejía," Cable from U.S. Embassy, Santo Domingo, 24 February 2004, Wikileaks.

52. *New York Times*, 6, 7 March 2004.

53. "Haitian National Police—Past, Present, and Future," Cable from U.S. Embassy, Port-au-Prince, 13 August 2009, Wikileaks; *New York Times*, 7 March 2004.

54. "Haiti's Central Plateau Poor and Isolated," Cable from U.S. Embassy, Port-au-Prince, 4 January 2006, Wikileaks; "Government of Haiti Budget Moving Forward But Slowly," Cable from U.S. Embassy, Port-au-Prince, 6 September 2006, Wikileaks; "Corruption and Security Void in Hinche," Cable from U.S. Embassy, Port-au-Prince, 1 October 2007, Wikileaks.

55. *Listín Diario*, 30 July 2008; "Ex-Soldiers Occupy Former Army Garrison, Challenge Government," Cable from U.S. Embassy, Port-au-Prince, 1 August 2008, Wikileaks.

56. "Haitian National Police—Status Report and Action Request," Cable from U.S. Embassy, Port-au-Prince, 6 February 2010, Wikileaks; *New York Times*, 20 September 2011.

57. *Listín Diario*, 27, 28 September 2011.

58. *New York Times*, 26 October, 19 Novem-

ber 2011; *Listín Diario*, 28 September 2011, 25 November 2012. Demobilization was not a new problem in the Caribbean, and ample precedents existed. See the chapter "Cuba: Demobilization and War," in René De La Pedraja, *Wars of Latin America, 1899–1941* (Jefferson, NC: McFarland, 2006). It is highly disconcerting to see U.S. officials repeat in Haiti the same mistakes committed one century before.

59. *Caribbean News*, 15 February 2014; *Miami Herald*, 24 March 2014; *Haiti Libre* 2 June 2014.

Chapter 13

1. President Leonel Fernández, quoted in Lauren Mathae, "From Haiti to the Dominican Republic and Back: Disjunctive Pattern of Immigration on Hispaniola in the Aftermath of the Haitian Earthquake," *Washington Report on the Hemisphere*, 17 June 2011.

2. "Dominican Republic: Transparency of Budgets/Military Spending," Cable from U.S. Embassy, Santo Domingo, 7 November 2004, Wikileaks.

3. Ibid.

4. *Listín Diario*, 27 Mary 2011; René De La Pedraja, *Wars of Latin America, 1948–1982: The Rise of the Guerrillas* (Jefferson, NC: McFarland, 2013), pp. 149–155; "Dominican Republic: Transparency of Budgets/Military Spending," Cable from U.S. Embassy, Santo Domingo, 7 November 2004, Wikileaks.

5. "New Dominican Crime Control Initiatives Take Effect Following Spike in Violence," Cable from U.S. Embassy, Santo Domingo, 28 July 2006 Wikileaks.

6. "How Best to Support U.S. Interests in the Dominican Republic—Recommendations," Cable from U.S. Embassy, Santo Domingo, 15 November 2003, Wikileaks. Deploying the Quisqueya Battalion to Iraq at first seemed a brilliant political maneuver to secure U.S. military and economic assistance. But when confirmation of the falsity of the claims of weapons of mass destruction reached Santo Domingo, public outrage forced the government to recall the battalion two months ahead of schedule. See "Dominican Republic: Transparency of Budgets/Military Spending," Cable from U.S. Embassy, Santo Domingo, 7 December 2004, Wikileaks. The return home of the Quisqueya Battalion came too late to save an administration already in difficulties, and the opposition candidate Leonel Fernández easily won the presidential election.

7. "Embassy Officials Visit Dominican Republic-Haiti Border Region," Cable from U.S. Embassy, Santo Domingo, 6 January 2010; Christian Krohn-Hansen, *Political Authoritarianism in the Dominican Republic* (New York: Palgrave Macmillan, 2009), pp. 42, 202n27; Mathae, "From Haiti to the Dominican Republic"; Eugenio Matibag, *Haitian-Dominican Counterpoint: Nation, Race, and State on Hispaniola* (New York: Palgrave Macmillan, 2003), pp. 158–160.

8. "Dominican Deportations of Haitians—Deeply Rooted Problems Remain," Cable from U.S. Embassy, Santo Domingo, 10 June 2005, Wikileaks.

9. Ibid. For recent discussions of anti-Haitian attitudes in the Dominican Republic, see Krohn-Hansen, *Political Authoritarianism*, chapter 7 and more briefly in Eugenio Matibag and Teresa Downing-Matibag, "Sovereignty and Social Justice: The 'Haitian Problem' in the Dominican Republic," *Caribbean Quarterly* 57 (June 2011): 92–117.

10. "Dominican President Faces Violent Demonstrations in Port-au-Prince," Cable from U.S. Embassy, Santo Domingo, 16 December 2005, Wikileaks.

11. "Dominican Politics #14: Dominican President Receives State Deputy Assistant Secretary for Caribbean," Secret Cable from U.S. Embassy, Santo Domingo, 17 January 2006, Wikileaks.

12. "Dominican President Faces Violent Demonstrations in Port-au-Prince," Cable from U.S. Embassy, Santo Domingo, 16 December 2005, Wikileaks.

13. "Haiti's President-Elect Warmly Welcomed by Dominicans," Cable from U.S. Embassy, Santo Domingo, 13 March 2006, Wikileaks.

14. "Dominican President Fernández on Haiti, Border Assessment," Cable from U.S. Embassy, Santo Domingo, 24 February 2006, Wikileaks; "Dominican Deportations of Haitians—Deeply Rooted Problems Remain," Cable from U.S. Embassy, Santo Domingo, 10 June 2005, Wikileaks.

15. "Embassy Officials Visit Dominican Republic-Haiti Border Region," Cable from U.S. Embassy, Santo Domingo, 6 January 2010, Wikileaks.

16. "CESFRONT—Toward More Effective Control of the Dominican-Haitian Border," Cable from U.S. Embassy, Santo Domingo, 11 December 2007, Wikileaks; "CESFRONT—A Perspective from Pedernales," Cable from U.S. Embassy, Santo Domingo, 19 May 2008, Wikileaks; *Listín Diario*, 15 October, 13 November 2007; "Embassy Officials Visit Dominican Republic-Haiti Border Region," Cable from U.S. Embassy, Santo Domingo, 6 January 2010, Wikileaks; Elizabeth Eames Robling, "Dominican Republic: Haitian Vendors Face Daily Military Raids," 21 November 2007, Global Information Network.

17. "CESFRONT—Toward More Effective Control of the Dominican-Haitian Border," Cable from U.S. Embassy, Santo Domingo, 11 December 2007, Wikileaks; "Embassy Officials Visit Dominican Republic-Haiti Border Region," Cable from U.S. Embassy, Santo Domingo, 6 January 2010, Wikileaks; Krohn-Hansen, *Political Authoritarianism*, pp. 38, 43; Elizabeth Eames Robling, "Dominican Republic: Haitian Vendors Face Daily Military Raids," 21 November 2007, Global Information Network; Matibag, *Haitian-Dominican Counterpoint*, p. 199.

18. *Listín Diario*, 1 January 2010; "Kidnapping, Cattle-Rustling Tread on Dominican-Haitian Relations," Cable from U.S. Embassy, Santo Domingo, 29 February 2008, Wikileaks; "Embassy Officials Visit Dominican Republic-Haiti Border Region," Cable from U.S. Embassy, Santo Domingo, 6 January 2010, Wikileaks; *Listín Diario*, 1 January 2010.

19. *Listín Diario*, 11 October 2009; Krohn-Hansen, *Political Authoritarianism*, pp. 38, 201n22; Mathae, "From Haiti to the Dominican Republic."

20. *Listín Diario*, 9 January 2009, 30 September 2009, 14 September 2012; Krohn-Hansen, *Political Authoritarianism*, pp. 64–65, 201n21; Matibag, *Haitian-Dominican Counterpoint*, pp. 198–200. Peru used a similar program of military colonists to bolster its control over disputed territories in the Amazon region. See René De La Pedraja, *Wars of Latin America, 1899–1941* (Jefferson, NC: McFarland, 2006), pp. 90–91.

21. *Listín Diario*, 8 April 2008.

22. "More Government of Dominican Republic Presidential Appointments—in Security Forces," Cable from U.S. Embassy, Santo Domingo, 20 August 2009, Wikileaks.

23. *Listín Diario*, 11 January 2008.

24. *Listín Diario*, 13 March 2009, 1 September 2011. María Isabel Medrano was the first woman to command a platoon, and her father Víctor Medrano was on active duty as major in the army.

25. *Listín Diario*, 9 December 2007, 1 September 2010.

26. "ONDCP Director, DEA Director Call on Dominican President and Senior Officials," Cable from U.S. Embassy, Santo Domingo, 17 March 2008, Wikileaks; *Listín Diario*, 29 October 2010.

27. "ONDCP Director, DEA Director Call on Dominican President and Senior Officials," Cable from U.S. Embassy, Santo Domingo, 17 March 2008, Wikileaks; *Listín Diario*, 24 July, 29 October 2010.

28. "ONDCP Director, DEA Director Call on Dominican President and Senior Officials," Cable from U.S. Embassy, Santo Domingo, 17 March 2008, Wikileaks; *Listín Diario*, 27 February, 28 December 2009, 17 October 2011; "U.S. Assistance and Dominican Commitment to Operation Enduring Friendship," Cable from U.S. Embassy, Santo Domingo, 23 October 2006, Wikileaks.

29. *Listín Diario*, 8 August 2012; "Dominican Republic: Transparency of Budgets/Military Spending," Cable from U.S. Embassy, Santo Domingo, 7 December 2004, Wikileaks.

30. *Listín Diario*, 9 August 2012.

31. Ibid., 24 January 2013.

32. Ibid., 9 August 2012, 24 January 2013.

33. The lists submitted to the president were slates of five candidates (*quinas*) and slates of three candidates (*ternas*). In Latin America *quinas* and *ternas* are quite common, but in the civilian sector the losing candidates do not have to leave the institution. For the military, the rules on mandatory retirement if not selected from a *terna* or a *quina* vary from country to country and usually depend on the rank. *Listín Diario*, 8 March, 21 August 2012.

34. De La Pedraja, *Wars of Latin America, 1948–1982*, p. 141; *Listín Diario*, 11 August 2012.

Chapter 14

1. "Key Trading Partners See No Big Economic Reforms in 2010," Cable from U.S. Interests Section, Havana, 9 February 2010, Wikileaks.

2. Isaac Saney, *Cuba: A Revolution in Motion* (London: Zed Books, 2004), p. 67.

3. Brian Latell, *After Fidel: Raúl Castro and the Future of Cuba's Revolution* (New York: Palgrave Macmillan, 2007), pp. 236–238, 244–245.

4. Saney, *Revolution in Motion*, p. 21; Latell, *After Fidel*, pp. 232–233, 239; Morris Morley and Chris McGillon, *Unfinished Business: America and Cuba after the Cold War, 1989–2001* (New York: Cambridge University Press, 2002), p. 36. The Soviet Union had wanted to phase out its aid to Cuba but under massive U.S. pressure the disintegrating Soviet leadership caved in and agreed to make the cut off as swift, extreme, and painful as possible. Fidel Castro never forgave the Soviet Union for this final betrayal of revolutionary Cuba.

5. Saney, *Revolution in Motion*, p. 22; Hal Klepak, *Cuba's Military 1990–2005: Revolutionary Soldiers During Counter-Revolutionary Times* (New York: Palgrave Macmillan, 2005), pp. 55–56.

6. Latell, *After Fidel*, pp. 233–234.

7. Christopher Marquis, "Pentagon Calls Cuban Forces Weak," *Miami Herald*, 30 March 1998; Hal Klepak, *Raúl Castro and Cuba: A Military Story* (New York: Palgrave Macmillan, 2012), p. 58; Latell, *After Fidel*, p. 233; Klepak,

Cuba's Military 1990–2005, pp. 62–63; George Lambie, *The Cuban Revolution in the 21st Century* (London: Pluto Press, 2010), p. 185; Armando Nova González, *Cuban Agriculture and Necessary Transformations* (Washington, D.C.: Woodrow Wilson Center, 2010), pp. 2–3.

8. Defense Intelligence Agency, *The Cuban Threat to U.S. National Security*, 1998; Klepak, *Raúl Castro*, pp. 58, 63; Latell, *After Fidel*, pp. 233–234; Klepak, *Cuba's Military 1990–2005*, pp. 62–63.

9. Klepak, *Raúl Castro*, p. 59; Lambie, *21st Century*, pp. 196–197, 215; Latell, *After Fidel*, pp. 242–243.

10. Marquis, "Pentagon Calls Cuban Forces Weak," *Miami Herald*, 30 March 1998; Klepak, *Raúl Castro*, p. 61; Dolly Mascareñas, "The Raúl I Know," *Time*, 14 August 2006.

11. Defense Intelligence Agency, *The Cuban Threat to U.S. National Security*, 1998; Klepak, *Raúl Castro*, pp. 64–65; Latell, *After Fidel*, pp. 242–243.

12. Defense Intelligence Agency, *The Cuban Threat to U.S. National Security*, 1998; Saney, *Revolution in Motion*, pp. 29–35; Brenden Carbonell, "FAR from Perfect: The Military and Corporatism," in online publication Cuba Project, *Changing Cuba/Changing World* (New York: Bildner Center, 2009), pp. 178–181.

13. Klepak, *Raúl Castro*, pp. 83–84; Morley and McGillon, *Unfinished Business*, pp. 46–51, 98–113.

14. Morley and McGillon, *Unfinished Business*, pp. 113–130, 137.

15. Christopher Marquis, "Cuba Still No Threat, Pentagon Insists," *Miami Herald*, 8 May 1998; Morley and McGillon, *Unfinished Business*, 137–138.

16. Klepak, *Raúl Castro*, pp. 88–89.

17. G. Alexander Crowther, *Security Requirements for Post-Transition Cuba* (Carlisle, PA: Strategic Studies Institute, 2007), pp. 7, 10, 29; Klepak, *Raúl Castro*, p. 89.

18. Agence France Presse, "Cuba to Hold Military Exercises in November," 26 January 2012; Klepak, *Cuba's Military 1990–2005*, p. 73; *Granma*, 11, 18, 20 December 2004. The recovery of Cuba was so far advanced, that in the wake of the devastation Hurricane Katrina caused in New Orleans on 29 August 2005, the Cuban government was able to offer its skilled rescue teams and wanted to share its expertise on dealing with tropical storms. The Bush administration, rather than save hundreds of lives of the least 1,836 Americans who died in the natural disaster, haughtily rejected the Cuban humanitarian proposal. See Lambie, *21st Century*, p. 244 and *Houston Chronicle*, 30 August 2010.

19. Klepak, *Raúl Castro*, pp. 90, 94–95.

20. Quotation and rest of paragraph from "Raúl Delivers Castro Birthday Speech," Cable from U.S. Interests Section, Havana, 4 December 2006, Wikileaks.

21. Commission for Assistance to a Free Cuba, *Report to the President May 2004*, p. xi. One scholar noted that "from the perspective of international law" the creation of this Commission was "outrageous." See Andreas Pickel, "After Fidel: Mechanisms of Regime Change," in Cuba Project, *Changing Cuba*, pp. 214–215.

22. Commission for Free Cuba, *Report 2004*, p. 15.

23. Marifeli Pérez-Stable, *The United States and Cuba: Intimate Enemies* (New York: Routledge, 2011), p. 92.

24. Jon Lee Anderson, "Letter from Cuba: Castro's Last Battle," *The New Yorker*, 31 July 2006, p. 54; Gary Maybarduk, "The U.S. Strategy for Transition in Cuba," in Cuba Project, *Changing Cuba*, pp. 228–229.

25. Commission for Assistance to a Free Cuba, *Report to the President July 2006*, pp. 1–5; Pérez-Stable, *Intimate Enemies*, p. 92; Anderson, "Castro's Last Battle," *The New Yorker*, 31 July 2006, p. 54.

26. Commission for Free Cuba, *Report 2006*, p. 12. What the two commissions wanted was to have Cuba "resume its historically subordinate position in the Western Hemisphere of the American empire," Pickel, "After Fidel," pp. 205–206.

27. Commission for Free Cuba, *Report 2006*, p. 6.

28. Ibid., p. 5.

29. *Listín Diario*, 3 March, 13 November 2007; *Granma*, 19 February 2008; "Advancing the U.S. Position for Change in Cuba," Cable from Secretary of State, Washington, D.C., 24 February 2008, Wikileaks; *Granma*, 27, 30 November 2009; "Military Exercises Fail to Rouse Cubans," Cable from U.S. Interests Section, Havana, 4 December 2009, Wikileaks. In a stark contrast to the high quality professional reporting of the U.S. Foreign Service, the usually very hostile cables from Havana frequently resort to ridicule, satire, and sarcasm to denigrate in every possible way the Cuban regime. Occasionally the Havana cables received a low security classification or remained unclassified to make easy recycling their contents in the constant propaganda barrages against Cuba.

30. "Cuba-Russia Relationship," Secret/No Forn Cable from U.S. Embassy, Moscow, 2 July 2009, Wikileaks.

31. "Cuba: Russia not Increasing Military Cooperation," Cable from U.S. Embassy, Moscow, 6 August 2008, Wikileaks. The war in Georgia, perceived as an attack on Russia, evoked strong nationalistic feelings among Russians.

But in spite of following U.S. policy slavishly after the collapse of the Soviet Union in 1991, Russia suffered an unexpected outpouring of vitriolic criticism from the United States and Europe over the war in Georgia of 2008. Afterwards a more independent foreign policy served to remind other governments that they could not always take Russia for granted.

32. *Financial Times*, 1 December 2008; *Pittsburg Post Gazette*, 28 November 2008; *Los Angeles Times*, 28 November 2008.

33. Mervyn Bain, "A New Era in Russian-Cuban Relations?" *The International Journal of Cuban Studies* (Spring 2010).

34. "Assistant Secretary of State Shannon and Deputy Foreign Minister Ryabkov Discuss Russia-Latin American Relations," Cable from U.S. Embassy, Moscow, 9 January 2009, Wikileaks; "Castro Visits Moscow to Rejuvenate Relations," Cable from U.S. Embassy, Moscow, 18 February 2009, Wikileaks; "Cuba: Russia Not Increasing Military Cooperation," Cable from U.S. Embassy, Moscow, 6 August 2008, Wikileaks.

35. "Cuban President Meets with Putin and Medvedev in Moscow," 12 July 2012, *Info-Prod Research (Middle East)*.

36. "Russia-Cuba Relations Have Become More Pragmatic—Putin," *Interfax: Russian & CIIS Military Information Weekly*, 13 July 2012.

37. "Washington Response to the Government of Bulgaria on Ukraine-Cuba Arms Deal," Cable from Secretary of State, Washington, D.C., 4 June 2009, Wikileaks.

38. *Los Angeles Times*, 17 July 2013; *Wall Street Journal*, 29 July, 27 September 2013; *Voice of America News*, 28 November 2013; *Asia News Monitor*, 29 November 2013. Most of the details of the discovery, boarding, and seizure of the cargo, the ship, and the crew remain unknown or disputed.

39. Agence France Presse, "Cuba to Hold Military Exercises in November," 26 January 2012; Pérez-Stable, *Intimate Enemies*, p. 137; *New York Times*, 16 July 2014; Al Jazeera America, "Russia Wipes Out Cuban Debt," 11 July 2014; *Moscow Times*, 18 June 2014.

40. For a full discussion of the U.S. experiment in Cuba see the classic study of Louis A. Pérez, Jr. *Army Politics in Cuba, 1902–1958* (Pittsburgh: University of Pittsburgh Press, 1976). For a briefer account, see René De La Pedraja, *Wars of Latin America, 1899–1941* (Jefferson, NC: McFarland, 2006), chapter 4.

41. Commission for Free Cuba, *Report 2006*, p. 14. Most of the ideas in the secret blueprints can be deduced from the Wikileaks cables, in particular, "Cuba's Massive Mobilization Capacity," Secret Cable from U.S. Interests Section, Havana, 26 June 2007, Wikileaks.

42. Pérez-Stable, *Intimate Enemies*, pp. 91–92; Commission for Free Cuba, *Report 2006*, p. 60; "Cuba's Massive Mobilization Capacity," Secret Cable from U.S. Interests Section, Havana, 26 June 2007, Wikileaks.

43. Klepak, *Cuba's Military 1990–2005*, pp. 156–162.

44. Crowther, *Post-Transition Cuba*, pp. 6–7.

45. "Cuba's Massive Mobilization Capacity," Secret Cable from U.S. Interests Section, Havana, 26 June 2007, Wikileaks; Commission for Free Cuba, *Report 2004*, p. 192; Crowther, *Post-Transition Cuba*, pp. 15–17.

46. "Cuba's Massive Mobilization Capacity," Secret Cable from U.S. Interests Section, Havana, 26 June 2007, Wikileaks; Commission for Free Cuba, *Report 2004*, pp. 191–197. Even a free-market zealot admitted that "generating revenue for the purpose of running the new government will prove challenging" yet that did not stop him from insisting that "Cuba must implement a mass giveaway privatization." See Carbonnell, "FAR from Perfect," in Cuba Project, *Changing Cuba*, pp. 177, 189.

47. Crowther, *Post-Transition Cuba*, p. 19; Levi Marrero, *Geografía de Cuba* (New York: Minerva Books, 1966), p. 3.

48. Crowther, *Post-Transition Cuba*, pp. 8, 18.

49. Commission for Free Cuba, *Report 2004*, p. 192.

50. Crowther, *Post-Transition Cuba*, p. 2.

51. Klepak, *Cuba's Military 1990–2005*, pp. 163–166; Crowther, *Post-Transition Cuba*, pp. 18, 22.

52. Coming from a different perspective and analyzing mainly non-military factors, George Lambie in his book likewise concludes that prospects are bleak after the overthrow of the revolutionary regime. He even goes farther and warns Cubans not to be misled by the siren calls of consumer capitalism "It is a vital matter that young people become engaged in shaping Cuba's future, but if their main desire is to be like their counterparts in the developed capitalist nations, they should think again. [...] what many young Cubans may not realize is that such a lifestyle is not available to all, and carries heavy social, physical, and psychological costs." Lambie, *21st Century*, p. 203.

Chapter 15

1. Cable from U.S. Embassy, Managua, 22 May 2009, Wikileaks.

2. George Lambie, *The Cuban Revolution in the 21st Century* (London: Pluto Books, 2010), pp. 241–243. For a scholarly discussion

of the origins of CAFTA-DR, see Mark B. Rosenberg and Lis G. Solís, *The United States and Central America: Geopolitical Realities and Regional Fragility* (New York: Routledge, 2007), pp. 83-96.

3. Arnold August, *Cuba and its Neighbors: Democracy in Motion* (Halifax: Fernwood Publishing, 2013), pp. 36-37.

4. Cable from U.S. Embassy, Tegucigalpa, 1 April 2009, Wikileaks.

5. August, *Cuba in Motion*, pp. 37-39.

6. Cable from Secretary of State, Washington, D.C., 4 June 2009, Wikileaks. If the Obama administration had been serious about improving relations with Cuba, here is how it should have handled the Cuban weapons purchase from Bulgaria. In conversations at an appropriate venue (no easy task), U.S. diplomats should have informed their Cuban counterparts that Pentagon experts had confirmed that all the weapons in the purchase deal were of a purely defensive nature. Despite the well-known U.S. opposition to the spread of SAMs in the region even if older Soviet models, Washington, primarily as a favor to its friends in Bulgaria and Ukraine, was leaning toward not opposing this sale of purely defensive weapons to Cuba. U.S. diplomats should state the official view that Cuba was wasting its money and would be better advised to spend its scarce funds on improving the living conditions of its people. Any similar official response at least was sure to dampen the fear among Cuban officials about a U.S. invasion. The hint would start conversations and might result in a voluntary quid pro quo on issues of mutual interest, but any attempt to link the goodwill gesture with internal changes in Cuba would doom the démarche to failure.

7. Associated Press, "U.S. Covertly Sent Latin Americans to Provoke Political Rebellion in Cuba," 4 August 2014; Al Jazeera America, 4 August 2014.

8. "U.S. Secretly Created Cuban Twitter to Stir Unrest and Undermine Government," *The Guardian*, 3 April 2014; *San Francisco Chronicle*, 11 April 2014; Associated Press, "Inspector General Probing U.S. Cuban Twitter Plan," 17 July 2014.

9. "Reaction to Ambassador's Expulsion," Cable from U.S. Embassy, Caracas, 17 September 2008; "Mixed Signals at Return of Ambassador," Cable from U.S. Embassy, Caracas, 9 July 2009; "Vice Foreign Minister Expresses Interest in 'Normalizing' Relations," Confidential/No Forn Cable from U.S. Embassy, Caracas, 8 January 2010, Wikileaks; *Asia News Monitor*, 25 June 2009.

10. *Listin Diario*, 18 December 2010; *Bloomberg Business Week*, 21-27 May 2012; "Venezuela Stops Efforts to Improve U.S. Relations," *New York Times*, 21 July 2013; *Oxford Analytica Daily Brief Service*, 4 January 2011.

11. "Maduro's International Support," *Mexico City News*, 27 June 2013; "Venezuela Stops Efforts to Improve U.S. Relations," *New York Times*, 6 June, 12, 21 July, 1 October 2013; CNN.com, 30 September, 2 October 2013; daily TV news broadcasts of Telemundo, February-September 2014.

12. René De La Pedraja, *Wars of Latin America, 1982-2013* (Jefferson, NC: McFarland, 2013), p. 109; "Vice President Biden's March 30 Bilateral Meeting with Costa Rican President Óscar Arias," Cable from U.S. Embassy, San José, 8 April 2009, Wikileaks.

13. René De La Pedraja, *Wars of Latin America, 1899-1941* (Jefferson, NC: McFarland, 2006), p. 434.

Annotated Selected Bibliography

This list emphasizes books available in English.

Barbosa Miranda, Francisco. *Historia militar de Nicaragua: Antes del siglo XV al XXI*. Managua: Hispamer, 2010.
Carroll, Rory. *Comandante: Hugo Chávez's Venezuela*. New York: Penguin, 2013. Although not as detailed as Bart Jones, this book covers the presidency of Hugo Chávez until its end in 2013.
Cuba Project/Bildner Center. *Changing Cuba/ Changing World*. New York: Bildner Center, 2009. Online publication.
Davidow, Jeffrey. *The Bear and the Porcupine: The U.S. and Mexico*. Princeton: Markus Wiener Publishers, 2007. This well-organized and very readable book by a seasoned foreign service officer, who served four years as U.S. ambassador to Mexico, provides many insights usually missing in official reports. But see chapter 2, note 12.
De La Pedraja, René. *Wars of Latin America, 1982–2013: The Path to Peace*. Jefferson, NC: McFarland, 2013.
Ellner, Steve. *Rethinking Venezuelan Politics: Class, Conflict, and the Chávez Phenomenon*. Boulder, CO: Lynne Rienner, 2009. The best scholarly guide available to understand the movement of Chávez.
Girard, Philippe R. *Clinton in Haiti: The 1994 U.S. Invasion of Haiti*. New York: Palgrave Macmillan, 2004. Covering the period until 2001, this very careful scholarly study on the U.S. invasion is also a great introduction to Haiti.
Grayson, George W. *La Familia Drug Cartel: Implications for U.S.–Mexican Security*. Carlisle, PA: Strategic Studies Institute, 2010.
———. *The Impact of President Felipe Calderón's War on Drugs on the Armed Forces: The Prospects for Mexico's "Militarization" and Bilateral Relations*. Carlisle, PA: U.S. Army War College, 2013. One of the foremost scholars of Mexico examines the military during the first twelve years of the twenty-first century; indispensable.
Guevara Moyano, Iñigo. *Adapting, Transforming, and Modernizing Under Fire: The Mexican Military 2006–2011*. Carlisle, PA: U.S. Army War College, 2012. This careful and well-crafted study by a Mexican scholar is a fundamental study on the Mexican military in the first eleven years of the twenty-first century; indispensable.
Harriott, Anthony. *Organized Crime and Politics in Jamaica: Breaking the Nexus*. Kingston: Canoe Press, 2008.
Hernández, Anabel. *Los señores del narco*. Mexico: Grijalbo, 2010. This very detailed book contains extensive information on many drug traffickers. An English translation of this book has been announced.
Honduras. Comisión de la Verdad y la Reconciliación. *Para que los hechos no se repitan: Informe*. Tegucigalpa: Comisión de la Verdad y la Reconciliación, 2011. Turgid prose, an excessively legalistic approach, and a frequently chaotic presentation make this report largely inaccessible to all but specialists who will struggle to extract valuable nuggets of information. See also chapter 9, note 48.

Isacson, Adam. *Consolidating "Consolidation."* Washington, D.C.: Washington Office on Latin America, December 2012. A reliable and very careful study.

Johnson, Hume N. *Challenges to Civil Society: Popular Protest and Governance in Jamaica.* Amherst, NY: Cambria Press, 2011.

Jones, Bart. *¡Hugo! The Hugo Chávez Story from Mud Hut to Perpetual Revolutionary.* Hanover, NH: Steerforth Press, 2007. The best and most detailed study of the Bolivarian Revolution until the end of 2006; highly recommended. For the final years of the Chávez presidency see Rory Carroll.

Klepak, Hal. *Cuba's Military 1990–2005: Revolutionary Soldiers During Counter-Revolutionary Times.* New York: Palgrave Macmillan, 2005. This detailed study of the Cuban military based on scholarly research and personal observation is a rare treat. Actually, nothing else of comparable quality exists for the military of any other Latin American country for this recent period.

———. *Raúl Castro and Cuba: A Military Story.* New York: Palgrave Macmillan, 2012. Although less detailed than the previous volume, this easier to read book still contains the essential information and brings the story down to the initial years of the Raúl presidency. For a hostile account, see Latell, Brian.

Krohn-Hansen, Christian. *Political Authoritarianism in the Dominican Republic.* New York: Palgrave Macmillan, 2009.

Lambie, George. *The Cuban Revolution in the 21st Century.* London: Pluto Press, 2010. Highly recommended to understand the resilience of the revolutionary system.

Latell, Brian. *After Fidel: Raúl Castro and the Future of Cuba's Revolution.* New York: Palgrave Macmillan, 2007. This hostile account still contains useful information. For a more sympathetic view, see Hal Klepak.

López Gonzalez, Jesús. *Presidencialismo y fuerzas armadas en México, 1876–2012.* Mexico: Gernika, 2012. Although the bulk of this book covers the period before 2000, the last two chapters dealing with the twenty-first century proved invaluable.

Maingot, Anthony P. and Wilfredo Lozano. *The United States and the Caribbean: Transforming Hegemony and Sovereignty.* New York: Routledge, 2005. This scholarly overview helps put into context the military relations of the island countries.

Morley, Morris and Chris McGillon. *Unfinished Business: America and Cuba after the Cold War, 1989–2001.* New York: Cambridge University Press, 2002.

Pérez-Stable, Marifeli. *The United States and Cuba: Intimate Enemies.* New York: Routledge, 2011. This book is an excellent survey of U.S.–Cuban relations since the collapse of the Soviet Union.

Saney, Isaac. *Cuba: A Revolution in Motion.* London: Zed Books, 2004. Extremely revealing study is the best account on how Cuba survived the collapse of the Soviet Bloc.

Sierra Guzmán, Jorge Luis. *El enemigo interno: Contrainsurgencia y fuerzas armadas en Mexico.* Mexico: Universidad Iberoamericana, 2003. This book by a Mexican scholar covers from the 1960s to the start of the twenty-first century and is unusually strong on the air force and the navy.

Trinkunas, Harold A. *Crafting Civilian Control of the Military in Venezuela; A Comparative Perspective.* Chapel Hill: University of North Carolina Press, 2005. This very meticulous scholarly study is indispensable to understand the military of Venezuela.

Index

A-37 jets 42
Acapulco (Mexico) 79
Acosta Chaparro, Mario Arturo (general) 170
action requests 157, 294*n*40
admirals 84, 217, 237
admissions exam 223
Afghanistan 140, 177, 245, 247, 261
AFI (*Agencia Federal de Investigación* [Mexico]) 74, 76, 78–79, 166, 169, 180
Africa 206
AID (Agency for International Development [U.S.]) 264–265
air force: Colombia 52–53, 85–86; Costa Rica 209–210; Cuba 106, 242, 245–247, 251, 255–256; Dominican Republic 231, 235; Honduras 145; Mexico 105–107, 176; Venezuela 18, 99–104
air-to-air missiles 85
airplanes 52, 85, 101, 103, 171, 209, 215, 235, 249, 255; *see also* air force
airstrips 69
AK-47 202
AK-103 automatic assault rifles 31, 90, 103
ALBA Bolivarian Alternative for America 151–152
all-terrain vehicles 190
Allende, Salvador 10, 17
allies of the United States, unconditional 38–54, 220, 260, 263, 268; El Salvador 140, 143; Guatemala 134; unsuited for 144–145, 207, 238
Aló Presidente (Venezuela) 12, 29–31, 34
aluminum 89
ambassadors 293*n*10, 294*n*40
Americans 185, 196–197, 273; *see also* tourists
Andromeda, operation (Colombia) 115–116, 118–119
anti-militarism 216–217; *see also* pacifism
Apatzingán (Mexico) 200, 203–204
Apiay base (Colombia) 54
Arbenz, Jacobo 95, 131, 147, 244, 261
Arce, Bayardo 129
Arellano Félix brothers 77
Argentina 94, 280*n*22
Arias, Óscar 206–209, 211–212, 222, 270
Aristide, Jean Baptiste 221–226
Armani (suits) 21

arms embargo 95, 102–103, 286*n*24
arms industry 31, 89–90, 95, 133, 138, 234–235
arms purchases and sales 31, 63, 65, 90, 100, 103–105, 107; corruption in 182–183
army: absence of 206–226; Colombia 49–54, 81–88, 108–120, 284*n*7; Cuba 228, 238, 241–242, 251–252, 254; Dominican Republic 227–237, 254; El Salvador 43–49, 137–143; Guatemala 38–43, 133; Haiti 221–226, 231; Honduras 144–146, 149–150; Jamaica 120, 122–123, 125–126, 290; Mexico 55–80, 165–192, 202; Nicaragua 96–98, 129–130; Panama 211–221; Peru 110; Venezuela 26–37, 88–92, 110, 228
arrests 67–68, 123–124, 139, 189; *see also* extraditions
arsenals 93–107, 116, 133, 145, 214; *see also* arms industry; arms purchases
assymmetrical warfare, doctrine of 32–33, 90
ATF (Alcohol, Tobacco, and Firearms Agency [U.S.]) 59
atrocities *see* brutality
attack ads (television) 166
Attorney General's Office (Honduras) 156, 158, 160–161
Attorney General's Office (Mexico) 59–60, 69, 74, 78, 166–167, 171, 178, 180, 194
attrition 145, 287*n*6; *see also* cannibalization
Australia 274
Auto defensas see Self-Defense Forces
automatic weapons 48, 89, 116, 126, 138, 180, 187, 201–202, 214, 217, 224; *see also* AK-103, M-16
avocados 201, 205m
axis of evil 262

Baduel, Raúl (general) 16–17, 20, 23, 37, 88–89
bailout of Mexico 5, 63–64, 258–259, 268–269
Baja California (Mexico) 191
ballot box 153–156, 294*n*39
banks 214
Baralt Avenue (Caracas) 15–16, 276*n*24
Barrero, Leonardo (general) 118–119
barricades 125
Barrio 18 (El Salvador) 141, 143
baseball 227

311

Bastion 2004 (Cuba) 246
Bastion 2009 (Cuba) 249
Bastion 2012 (Cuba) 250–251
Bastion 2016 (Cuba) 251
Batista, Fulgencio 93, 275n8
Bay of Pigs (Cuba) 10, 11, 276n15, 278n12
beauty queen 190
Beijing 27
Belize 133
Beltrán, Estanislao see Papa Smurf
Beltrán Leyva, Alfredo 169, 184
Beltrán Leyva brothers 169–170, 178
Benito Juárez International Airport (Mexico City) 77
Berger, Óscar 41
Betancur, Belisario 115
bicameral and unicameral legislature 6–7
bicycles 242
Biden, Joe 136
Bin Laden, Osama 191
Black Hawk helicopter 209–210, 270
blackouts 240
blacks and dark-skinned 174, 214, 275n8
body counts 109; see also metrics
Bogotá 87, 112, 115–117
Bolívar, Simón 9, 27
Bolivarian Circles (Venezuela) 15, 91–92
Bolivarian National Police 92
Bolivarian Revolution (Venezuela) 26–29, 33–34, 36–37, 81, 88, 91–92
bosses, political 122
Botero, Fernando 119
Brazil 10, 94, 100, 102, 226, 234–235
bribes 118, 123, 133, 139, 208; Dominican Republic 232, 235; Mexico 59–60, 69, 75–76, 167, 169, 170, 182–183; see also corruption
Britain and British 110, 120, 123–124, 253
brutality and savagery 79, 81, 83, 110, 112, 169, 229; see also violence
Buggly Hacker (Bogotá) 115–116
Bulgaria 250, 264, 307n6
Bush, George W. (president and administration) 5, 131–132, 258–270; and Colombia 108; and Costa Rica 209; and Cuba 245–247, 250; and Dominican Republic 228; and El Salvador 44–45; and Haiti 222, 224; and Honduras 145–146, 150–151; legacy 108, 258–267; and Nicaragua 96–99, 127–128, 136; similarities with Obama 140, 250; and Venezuela 5, 7, 9–11, 14, 28–29, 32, 35–36, 99–105, 275n8
businesses and business leaders 214; Central America 130, 132, 136, 152, 163, 255, 263; Dominican Republic 228–229, 233, 236; Mexico 185–188, 201–202; Venezuela 30, 276n9; see also FEDECAMARAS

C-130 cargo airplanes 53, 203
Caballeros Templarios see Knights Templars
Cabello, Diosdado 18, 24

CAFTA-DR (Central American Free Trade Agreement) 43, 45, 135, 142, 148, 152, 258, 260, 268, 272, 307n2
Calderón, Felipe 106, 165–166, 173, 176, 186, 189, 191–193, 196, 199
Cali (Colombia) 49, 116
California 47, 135
Canada and Canadians 124, 224, 273
Canal Zone (Panama) 145
cannibalization 242, 256; see also attrition
Cap Haitïen 224
capitalism 260–262; see also free market
Caracas 9, 13, 15–16, 21, 23, 26–27, 37, 91–92, 275n4; see also Miraflores Presidential Palace
Caracas Metropolitan Police 11, 13–14, 20, 23
Caracazo 26–27
Cárdenas Guillén, Oziel 79
Caribbean 147, 177, 209, 227, 262; and U.S. 95, 105, 155, 239, 251, 265, 293n10
Caribbean Command (Colombia) 53, 54
Carmona, Pedro 8, 12, 14, 17–24
Carrillo Fuentes, Amado see Lord of the Skies
Cartagena (Colombia) 84
Castellanos Trujillo, Reinaldo (general) 53, 81
Castro, Fidel 11, 240–241, 245–246, 248–250, 260–261; fear of 147; and Nicaragua 95–96; and Socialism 278n12; and Soviet Union 304n4; and Venezuela 7, 17, 265
Castro, Raúl 240–243, 245–246, 248–251, 264
Catholicism see Roman Catholic Church
cattle 232
cell phones 139, 141, 196–197
Central America 48, 127–143, 145, 206–208, 253, 255, 262; drug trade 270; and U.S. 95; see also individual countries
Central American Free Trade Agreement see CAFTA-DR
Cerna, Lenin 129
CESFRONT (Cuerpo Especializado Fronterizo [Dominican Republic]) 231–234
Cessnas 209
Chapultepec Military Academy (Mexico City) 172, 175
Chávez, Hugo Frías 5–25, 26–37, 88–92, 95, 99, 249; and Cuba 243, 261; fear of 127, 142, 147, 252; and Honduras 150–151, 162; as model 233, 274; and U.S. 95, 194–195, 252, 258, 260, 265, 266, 269, 275n4
cheeses 232
Chiapas (Mexico) 57, 62, 65, 67, 167
Chihuahua (Mexico) 190
children 112, 125, 232
Chile and Chileans 10, 17, 94, 224
China 27, 104, 242
CIA (Central Intelligence Agency) 59, 95, 115; and Guatemala 10, 127, 130–131, 244, 261; and Venezuela 10, 14; see also intelligence failures

Index

Ciudad Juárez (Mexico) 60, 177–180, 184, 191–192; *see also* Juárez Cartel
Civic Action (Venezuela) 22–23, 27, 29
Civil Patrols (Guatemala) 38–39
class warfare 271; *see also* social classes
Clinton, Bill (president and administration) 50, 99, 258–259, 263–265, 267; and Cuba 5, 243–244; and Haiti 221–222; and Mexico 5, 63–64, 68; and Venezuela 5, 7, 10
Clinton, Hillary 136, 157, 263–265
clubs, private 21, 185
"Coalition of the Willing" 44, 145, 228
coalitions 146; Venezuela 6–8, 19, 21
Coast Guard 210, 216, 218
Cobán (Guatemala) 133–134
cocaine 133, 262; and Central America 46, 208, 210, 218–219; Colombia 182, 208, 259; Dominican Republic 235; Guatemala 133; Jamaica 121–122, 289n35; Mexico 60, 74, 77–79, 168, 198; seizures 133, 178–179, 181–182
coffee 47
Cohen, William 244–245
Coke, Christopher "Dudus" 122–126, 289n35
Cold War 48–49, 93, 95, 102, 128–129, 132, 207, 228, 256, 261, 273
college graduates 84
Colom, Álvaro 43, 130–134, 136, 263
Colombia 44, 46, 49–54, 55, 81–88, 108–120, 272; congress 52, 83; and Costa Rica 208, 210; drug trade 60–62, 69, 73, 75, 77, 121, 182, 235; and Panama 218–221, 302n41; and U.S. 93, 195, 259–260, 268; and Venezuela 32, 90
Colón (Panama) 221
colonels 40, 137, 143, 217, 235–236
colonists 304n20
Comendador (Dominican Republic) 231
Commission for Assistance to a Free Cuba 2004 (U.S.) 247–248, 252, 255
Commission for Assistance to a Free Cuba 2006 (U.S.) 248, 252–253, 255
Committees for the Defense of the Revolution (Cuba) 254, 257
communications systems 86, 182; *see also* computers
Communism and Communists 7, 8, 18, 48, 102, 132, 135, 228, 252
compensation benefits 172, 183
Comptroller General (Panama) 212
computers and computer systems 85–86, 115, 182, 200, 209
Conference of the Armies of the Americas 89, 195
conscripts and conscription 144; Colombia 52, 81–83, 108, 111; Cuba 241–242, 248, 254
Conservative party (Colombia) 109
conservatives 259–261
Consolidación (Colombia) 82, 86
constabularies 1, 41, 120, 129, 227
Constitution of 1917 (Mexico) 55–56, 59, 71

Constitution of 1976 (Cuba) 240
Constitution of 1983 (Honduras) 155
Constitution of 1987 (Haiti) 223
Constitution of 1991 (Colombia) 55
Constitution of 1994 (Panama) 215
Constitution of 1999 (Venezuela) 6–7, 10, 27–28
constitutional convention (Honduras) 153–155, 158, 294n30
Contras (Nicaragua) 95–96, 105, 127, 207, 270
cooperatives, farmer 241
Cops 274
corruption 120, 133, 274; Central America 153, 208; Colombia 111, 116, 118, 120; Dominican Republic 232, 235; Eastern Europe 252; Guatemala 43, 133; Haiti 222; Jamaica 120, 124; Mexico 170, 181–182, 187, 272; Panama 211; *see also* bribes
Costa Rica 162, 206–212, 215, 222, 270; and SOFA 275n4, 283n29
counterinsurgency 29, 32, 39, 62, 93; *see also* guerrillas
Country Club (Caracas) 21
coup of April 2002 (Venezuela) 5–26, 28, 35, 37, 100, 252
coup of June 2009 (Honduras) 144–165
coups 221, 252–253; *see also* military coups
courtesy 98
Creole dialect 232–233
crime waves 273; Costa Rica 207–210; El Salvador 46–48, 135, 138–142; Guatemala 39–40, 42–43, 133–134, 279n4; Honduras 153; Mexico 193, 195; Panama 214, 218–219; Venezuela 91–92
criminals 94, 113–114, 116, 192–193, 214
crops 39; *see also* eradication
CTV (Venezuelan Confederation of Workers) 8, 12, 14, 19
Cuba 121, 136, 227–228, 238–258; fear of 127; and U.S. 5, 147, 258, 264–265, 270, 275n5; and U.S. embargo 95, 273; and Venezuela 7, 11, 32, 130, 96; weapons of 96, 106
Cuban Missile Crisis 249
Cuban Revolution 251–252, 257, 278n12
Cubans *see* Miami Cubans
Cuernavaca (Mexico) 77, 184
Culiacán (Mexico) 191, 197
Curaçao 35, 104

Dajabón (Dominican Republic) 231–233
Darién (Panama) 218–221, 302n43
Darwinian 111
DAS (*Departamento Administrativo de Seguridad* [Colombia]) 116, 288n20
DEA Drug Enforcement Agency (U.S.) 59, 85, 165; and El Chapo Guzmán 189–190; and Mexico 59, 69, 71, 76, 78, 168, 176, 196, 282n12; and Panama 220
death penalty 71, 123
death squads 114

decentralization 147–148
Defense Intelligence Agency 244
Defense Ministry 2, 41–42, 56, 144; *see also* U.S. Department of Defense
degeneration, moral 120, 274
De la Huerta Revolt (Mexico) 63, 282n17
De la Madrid, Miguel 59
demobilization 221–226
democracy 74, 260, 271
dentistry 174
dependency 271
deportation 47, 233
deserters and desertions 70, 79, 172–174, 183
deterrence 240, 243, 245, 251
DFS (*Dirección Federal de Seguridad* [Mexico]) 58–59, 74, 176
dictator 41, 300n21
Dignity Battallions "Ding Bats" (Panama) 217
diplomacy 98
discipline 110, 172, 234, 241, 243, 246, 255
divorces 192–193
dollar, U.S. 43, 135, 142
Dominican Republic 44, 222, 227–237, 254, 257; per capita income 222, 302n46
domino effect 240
dons (Jamaica) 122–126
dress 136, 199, 263–264
drones (UAVs) 86–87, 176
drug cartels 93, 130, 133, 134, 262; Central America 208–210, 218–219, 253; Colombia 93, 259; Mexico 60–62, 64–65, 71–75, 78–80, 168, 171–172, 177–181, 195–196
Drug Enforcement Agency *see* DEA
drug lords 50, 183–186, 189–192, 195–198
drug trade 47, 111, 120–121, 123, 129, 232, 235–236, 270–271, 273; Mexico 59–62, 67–70, 73, 165, 198
drug traffickers 94, 124, 208, 253, 262; Central America 39, 46, 208–210; Colombia 50, 85–86; Dominican Republic 235; Guatemala 39, 43, 133–134; Mexico 59–60, 69, 74–77, 85, 166–167, 176, 179, 184–185
Duddy, Patrick 265–266
"Dudus" *see* Coke, Christopher
Durango (Mexico) 190, 196–197

earthquake 225–226
Eastern Europe 243, 252
economic assistance 95, 163, 206, 259, 303n6
ecstasy 289n36
Ecuador 90, 226
education 46, 84, 144–145, 185, 238
El Chapo Guzmán (Joaquín Guzmán Loera) 75–79, 166–167, 169, 178, 180–181, 189–192, 195–196, 284n41
El Chayo (Nazario Moreno González) 198–202, 205
El Paso (Texas) 191–192
El Salvador 43–49, 132, 135–143, 145, 263–264
elections 261, 269, 271; Colombia 119–120;

Costa Rica 206; Dominican Republic 228–229; El Salvador 135–136, 142–143; Haiti 223; Honduras 146, 152, 155, 162; Jamaica 122, 124, 126; Mexico 73, 80, 165–166, 193, 282n8; Nicaragua 97, 261; Panama 216; Venezuela 6–7, 34–36; *see also* referendums
electricity 240
emails 114–115, 119–120
Embraer (Brazil) 53
employment 228, 272
Endara, Guillermo 211, 213–214
Enduring Friendship Boats 210
engineers 226, 245, 263
English language 120, 137
EPR (*Ejército Popular Revolucionario* [Mexico]) 66–67, 167–168
eradication, crop 60, 171, 178–179, 290n37
Escobar, Pablo 75
Escorcia, Ricardo (general) 168–169
Europe 272–274
executions *see* death penalty; extrajudicial killings
Export-Import Bank 101
extortion 23, 123, 192, 199, 201–202, 231–232, 262
extraditions 68–69, 72, 78, 125–126, 169, 189, 198; *see also* arrests
extrajudicial killings 109–114, 120, 126

F-5Es jets 97, 105–106, 145
F-16s jets 97, 99–102, 106
failed states 144, 223, 226, 257, 259
FAL rifles 31
False Positives (Colombia) 108, 113–114, 118, 120, 288n10, 289n22
La Familia Michoacana (Mexico) 169, 198–200, 204; *see also* Knights Templars
FARC (*Fuerzas Armadas Revolucionarias de Colombia*) 50–52, 54, 81–87, 108–109, 112, 115; and Mexico 168, 192; and Panama 219–221, 302n43; and Venezuela 90, 104
FBI (Federal Bureau of Investigation [U.S.]) 58–59, 74, 165
FEDECAMARAS (Venezuela) 8, 12, 14, 17, 19
La Federación (Mexico) 77–79, 166–167, 169
Federal Agency of Investigation (Mexico) *see* AFI
federalist and federalism 55–56, 58, 271
Fernández, Leonel 227, 229–236
Ferrari 1, 210
flogging 123
Florida 60
FMLN (*Frente Farabundo Martí para la Liberación Nacional* [El Salvador]) 132, 135–143
food 233, 240–241, 250, 255
Fort Tiuna (Venezuela) 14–15, 17–18, 20, 22, 24, 32
49 decree laws (Venezuela) 8, 11, 275n7, 276n8
42nd Paratrooper Brigade (Venezuela) 16, 20, 23

Forward Operating Site (Colombia) 54
four-wheel drive vehicles 171
Fourth Army Brigade (Colombia) 112
Fourth Ballot Box (Honduras) 153–159, 163
Fox, Vicente 73–80, 146, 166–167, 176, 190, 192, 198, 260
Fracica Naranjo, Carlos Alberto (general) 54
France and French 99, 224, 233, 253, 272
free market and free trade 260, 270–271, 306n46, 306n52
"Friends of Mauricio" (El Salvador) 135–138
fuel 116, 240–242
Funes, Mauricio 135–142, 263

gangs 129, 133–134, 262, 273; Central America 253, 262; Costa Rica 209–210; El Salvador 46–48, 139, 141–143; Guatemala 39, 43, 133; Honduras 153; Jamaica 121–126; Mexico 186; Panama 214, 219
García Abreu, Juan 69, 72–73, 283n33
García Carneiro, Jorge (general) 20, 23
García Padgett, Miguel Ángel (general) 162, 164
garrison neighborhoods (Jamaica) 121–126
Garza, Pascual 55
gas, natural 167–168
gasoline 150–151
Gates, Robert 177
gendarmerie 11, 14, 41; *see also* constabulary
General Motors Corporation 171
general strike (Venezuela) 8, 12–14
generals 84; Colombia 53, 84, 119; Cuba 253–254; Dominican Republic 228, 236–237; El Salvador 137, 143; Honduras 143, 150, 158–159, 162; Mexico 55, 63, 69–72, 173, 175, 177, 181, 188, 195
genocide 39, 133, 279n3
George Washington (aircraft carrier) 22, 35
Georgia (Caucasus) 103–104, 249, 305n31
Golding, Bruce 123–126
Gonaïves (Haiti) 224
González del Río, Robinson (colonel) 116–116, 118
González Peña, Óscar Enrique (general) 113
graffiti 199
Great Recession 87, 135, 262, 269
Green Revolution 241
grenades *see* Rocket Propelled Grenades
Guadalajara (Mexico) 75
Guanajuato (Mexico) 167
Guantánamo naval base 145, 273
Guardia Nacional (Nicaragua) 41, 129, 207
Guatemala 38–43, 95, 130–134, 136, 138, 263; and CIA 10, 127, 147, 244, 261; crime wave 279n4; and El Salvador 139; and Mexico 62, 79
Guaviare River (Colombia) 51, 54, 87
Guayabero River (Colombia) 87
Guerra de Todo el Pueblo see nation at arms
Guerrero (Mexico) 57, 66–67, 79, 167

guerrillas 29, 206, 253–254, 287n6, 300n21; Colombia 49–51, 81–88, 93, 109–111, 114–115, 120, 192, 259, 268; El Salvador 44, 46, 135; former 82; Guatemala 38–39, 133; Mexico 59, 62–63, 65–67, 167–168, 176; Panama 214; Venezuela 29, 278n10
Gulf Cartel (Mexico) 61, 69, 72–73, 77–79, 166–167, 169, 180; head of 283n33; in Monterrey 186–187, 189, 196
gun control 94, 116, 138, 232; *see also* smuggling, arms
gunboat diplomacy 35
Gutiérrez, Carlos 248
Gutiérrez Rebollo, José de Jesús (general) 70–72
Guzmán Loera, Joaquín *see* El Chapo Guzmán

hackers 115
Haiti 43, 206, 215, 221–226, 235; and Dominican Republic 227–235; failed state 253, 257, 259, 273; per capita income 222, 302n46
Haitian National Police 222–225
Halleslevens, Omar (general) 129
hatred 129, 216–217, 221, 300n21
Havana 114–115, 119
health system 98, 238
helicopters 63, 69, 96, 100, 123, 212; Colombia 49, 86; Costa Rica 209–210; Cuba 255–256; Dominican Republic 230; Guatemala 42; Mexico 171, 176, 182, 204; and navy 63, 176; Panama 212, 215–216, 219–220
Helms-Burton Act 244
herbicides 60
Herrera Hassan, Eduardo (colonel) 212–213
High Value Targets 54
Hinche (Haiti) 223–225
homicides 46, 91, 120, 122, 139, 143, 141, 228; Mexico 186, 192, 202; *see also* False Positives
Honduran congress 149, 151, 153–154, 156, 158–160, 162
Honduras 44, 95, 97–98, 263; coup of June 2009 144–164
housing 172
hubris 243
human rights 212; abuses 47, 109, 121, 133, 162, 175, 223, 226, 295n59; activists 38–41, 47, 110, 112, 175
human shields 125–126
humanitarian missions 44, 53, 220; *see also* relief efforts
Humvees 171
Hundred Hours' War 145
hurricane 249, 305n18

ICE (Immigration and Customs Enforcement [U.S.]) 196
ideology 259–261, 263
Igla-S 104
Ilyushin cargo planes 103

immigration: to the Dominican Republic 229, 231–233; from El Salvador 45–47; from Mexico 58, 200; to the U.S. 253, 272
imperialism 29
improvised explosive devices 126
income, per capita 222, 302n46
Independence Day (Cuba) 247
Indians 133, 175
inequality, income 46, 238, 252, 257
informants *see* spies
Insulza, José Miguel 141
insurgencies *see* counterinsurgency; guerrillas
intelligence failures 11, 155, 276n15
intelligence services 86, 116; Mexico 59, 67, 69, 168, 176, 181–182, 187, 199; U.S. 184
international law 293n10
Internet café 115
Iraq and Iraq War 96, 101, 177, 245, 253; as distraction 261; model for Cuba 247–248; and SOFA 275n4; troops for 44–45, 127, 136, 140, 145–146, 228–229, 303n6
Israel 85, 107
Izar (Spain) 101–102

Jamaica 108, 120–126, 228, 262
Jamaican Constabulary Force *see* JCF
Jamaican Defense Force *see* JDF
Jamaican Labour Party 122, 124, 126
JCF (Jamaican Constabulary Force) 120–126
JDF (Jamaican Defense Force) 120, 122–123, 125–126
Jesus Christ 9
jet fighters 94, 105–104, 145, 255
Jimaní (Dominican Republic) 231–232
joint commands (Colombia) 53–54
Joint Task Force Omega (Colombia) 54
Juárez Cartel (Mexico) 169, 180
juntas: Haiti 221, 223; Venezuela 17, 19

Kaibiles (Guatemala) 79, 133–134
Katrina, hurricane 305n18
Kerry, John 266–267
Kfir jet fighters 85
kidnapping 186, 192, 262, 195, 199, 262, 273
"Kingpin Approach" (Mexico) 189–192, 195
Kingston (Jamaica) 121–122, 125–126
Knights Templars *Caballeros Templarios* (Mexico) 200–205, 300n21; *see also* La Familia Michoacana

labor aristocracy 8
labor, organized 8, 58, 115, 212; Venezuela 8, 19, 276n9
laboratories 182
Las Vegas (U.S.) 274
laser-guided system 53
Lázaro Cárdenas (Mexico) 198–199
left and leftists 48, 127, 147–148, 261, 263; El Salvador 135–143; Guatemala 43, 130–131, 134; Mexico 67; Venezuela 7–8

Liberal party (Colombia) 109, 149
Liberal party (Honduras) 146, 148, 150–151, 153–154, 156–163
limes 201–202, 204–205
Llanos de la Víbora (Mexico) 60, 71, 168
lobbying 94–95
Lobo Sosa, Porfirio 146, 163
lookouts 190–191, 200–201, 204–205
López Obrador, Manuel 165–166
Lord of the Skies (*El Señor de los Cielos* [Amado Carrillo Fuentes]) 60–61, 69, 71–73, 75, 78, 284n41
Los Cabos (Mexico) 191
Lula (Luiz Inacio Lula da Silva) 102
lynching 39, 134

M-16 rifles 54, 123
M-60 machine gun 116
Machiavelli, Niccolò 271
machine guns *see* M-60 machine gun; 12.7 mm machine guns
macho or maleness 173–174
Madero, Francisco 74
Maduro, Nicolás 194, 266
Maduro, Ricardo 145–146, 149
maffia (Colombia) 118–119
Maiquitía International Airport (Caracas) 37
Major Non-Nato Ally *see* MNNA
management 243
Managua 97, 207, 263
Maneuvers 283n29
manifesto 23
Mara Salvatrucha-13 (El Salvador) 141, 143
Maracaibo (Venezuela) 34
Maracay (Venezuela) 16–17, 20, 23
marihuana: Jamaica 121–123; Mexico 59–60, 74, 178–179, 181, 185, 188
Marines: Colombia 52, 84–85, 256; Cuba 256–257; Mexico 56, 65, 165, 176, 178–179, 183–184, 188–189, 198, 256
marriage and matrimony 175, 297n24
Marshall plan 272
Martelly, Michel, 226
Martinelli, Ricardo 219–221, 302n41
Marx, Karl 271
Marxism 50, 135, 138, 240, 259, 261, 263, 268; *see also* Communism
mass 236; *see also* Roman Catholic Church
Matamoros (Mexico) 167
Mauser rifle 202
Mazatlán (Mexico) 197
McCaffrey, Barry (general) 71
mechanics 245–246
Medellín (Colombia) 112
media 21, 89, 113, 118–119, 261, 276n10; *see also* newspapers; television
medical assistance and medicine 98, 174
Medina, Danilo 236
Medina, Enrique (general) 11, 14
Medvedev, Dmitry 249–250

"mega-elections" (Venezuela) 6-7, 36
Melgar, Manuel 138, 140, 292n35
Mérida Initiative 177, 182-183, 209-210, 262, 268
merit 56
Meta River (Colombia) 54, 87
methamphetamine 182, 198
metrics 120, 288n6; *see also* body count
Mexican congress 106, 170, 180-181
Mexican Revolution 56, 71, 95, 203
Mexico 1, 55-80, 165-193, 194-205, 262, 271-272; bailout 1, 5, 63-64, 258-259, 268-269; and Guatemala 133-134; jet fighters 93, 105-107, 145, 255-256; and SOFA 275n4, 283n29; and U.S. 147, 239, 252-253, 270
Mexico City 59, 75, 167-168, 184-185, 198; source of officers 172, 297n19
Mexico Seguro 80
Miami Cubans 238-239, 241, 244, 248, 257-258, 262
Micheletti, Roberto 151-152, 154, 159-163
Michoacán (Mexico) 166, 171, 177, 198-205
middle classes: Colombia 110; Costa Rica 206; Mexico 56; Venezuela 14, 20, 22
Middle East 228, 245, 269; *see also* Iraq
MIGs 95-96, 99-100, 247, 256
militarism and militarization 206-211, 216
military: Venezuela 26-37
military academies 83-84, 174, 210, 233-234; *see also* police academy
military assistance 65, 83; *see also* U.S. military assistance
military attachés: Colombia 114; France 247; U.S. 266; Venezuela 11, 14
military commissioners (*comisionados militares* [Guatemala]) 38-39
military coups 10, 22, 40-41, 212-213, 252; *see also* coup of April 2002; coup of June 2009
Military Dental School (Mexico) 174
military intelligence 115-116, 183, 189
Military Medical School (Mexico) 174
military zones (Mexico) 56-57, 67, 168-169, 180
militias 30, 254, 256-257
minimum wage 12, 19, 37, 153
Ministry for the Navy (Mexico) 56
Ministry of Defense 2; Dominican Republic 231; El Salvador 137; Guatemala 41; Honduras 144; Mexico 56, 105-106, 165, 171, 195, 203-204, 282n6; Nicaragua 129
Ministry of National Security (Mexico) 74, 180-181, 194
Ministry of Public Security (El Salvador) 137-138, 140-141
Ministry of the Interior (Mexico) 74, 194-195
Miquilena, Luis 6, 8
Miraflores Presidential Palace (Caracas) 13-21, 25, 276n24
Mirage jets 85
Mireles, José Manuel 300n20

missile systems 94; *see also* SAMs
MNNA (Major Non-NATO Ally) 45, 280n22
mobilization 90, 245-246, 251
money laundering 123, 185
Monterrey (Mexico) 69, 77, 184-189
Montoya, Mario (general) 113
morality 120, 172, 199, 274
Moscow 100
Mothers of Soacha (Colombia) 119, 289n24
Mouriño, Juan Camilo 170
Munguía Payés, David (general) 137, 139, 141, 143
municipalities 148, 199, 202, 204
murders *see* homicides

NAFTA North American Free Trade Agreement 62, 268
Naos Island (Panama) 212
Naranjo, Óscar 220
narcotiendas 61, 167, 186, 208
nation at arms (*Guerra de Todo el Pueblo* [Cuba]) 245, 254
nation building 222-223
National Aero-Naval Service (Panama) 218
National Air Service (Panama) 215-216
National Assembly (Cuba) 248-249
National Assembly (El Salvador) 135
National Assembly (Nicaragua) 97, 29
National Assembly (Panama) 213, 218-219
National Assembly (Venezuela) 7, 18, 21, 22, 24, 92
National Civilian Police (El Salvador) 46, 138-141
National Civilian Police (Guatemala) 40, 43, 133-134
National Endowment for Democracy 10
National Frontier Service *see* SENAFRONT
National Guard (Dominican Republic) 227
National Guard (Nicaragua) 41, 129, 216
National Guard (Venezuela) 11, 14
National Institute of Statistics (Honduras) 153-155, 158
National Maritime Service 216-217
National Military Academy (Venezuela) 37
National Party (Honduras) 145-146, 163
nationalism and nationalists 305n31; Cuba 253-254, 257; Dominican Republic 229, 237; Guatemala 42; Mexico 68, 105-106; Nicaragua 98-99, 127, 216-217; Panama 217; Venezuela 34-37, 275n4
nationalization 29-30
NATO (North Atlantic Treaty Organization) 45, 246, 253; *see also* MNNA
naval academy 84
naval intelligence 176, 184, 189
navy: Colombia 84-86, 289n24; Costa Rica 210; Cuba 256; Dominican Republic 231; Mexico 56, 63, 66, 68, 105-106, 165, 172, 176-177, 195-198, 200, 282n6; *see also* Marines

318 Index

NCOs (Non-Commissioned Officers) 289n24, 297n21; Colombia 52, 81, 83–84, 113; Guatemala 40, 42
nepotism 236
Netherlands and Dutch 35, 232, 279n31
New Orleans 305n18
New Republic Viaduct (Caracas) 16, 276n24
New York Times 211
newspapers: Colombia 112–113; Venezuela 9, 18; *see also* press
Ni-Nis ni trabajan ni estudian (Mexico) 172
Nicaragua 41, 44, 48, 93–99, 127–131, 208, 216, 261–263
nickel 249
Nobel Peace Prize 206, 270
Nogales (Mexico) 197
Non-Commissioned Officers *see* NCOs
Noriega, Manuel 211–212, 214–215, 217, 219
North American Free Trade Agreement (NAFTA) 260
North Korea 250
nostalgia 249
Novo-Ogaryovo (Russia) 250
Nuevo Laredo (Mexico) 77–80, 185
nurses 173–174

Oaxaca (Mexico) 67
Obama, Barack (president and administration) 132, 136, 258, 262–266; and Cuba 250–251, 262, 264; similarities with Bush administration 140, 250, 265; and Venezuela 91
obesity 281n41
Obregón, Álvaro 63, 282n17
officers: Colombia 50, 83, 109–111, 113, 116; Cuba 242–243, 252–253; Dominican Republic 234–235; El Salvador 50; Guatemala 40, 42; Haiti 222–226; Jamaica 120; Mexico 72, 171–175, 181, 297n19, 297n21; Panama 211, 215–216; Venezuela 27–28, 33–34, 91; *see also* generals
officers in civilian jobs 33, 89, 149, 243
officers, junior: Mexico 72–73; Panama 215–216; Venezuela 20, 33
oligarchy 6, 214; *see also* social classes
Olney, Richard 93
Operation Caguairán (Cuba) 248
Operation Moncada (Cuba) 248
Operation Restore Democracy (Haiti) 223
Oquist, Paul 129
La Orchila Island (Venezuela) 24–25
Organization of American States 141
Ortega, Carlos 8, 12, 14, 19
Ortega, Daniel 48, 97–98, 127–130, 136, 207, 261
oxen 241

pacifism 206–207, 209; *see also* antimilitarism
paintings 119

PAN (*Partido de Acción Nacional* [Mexico]) 73–74, 78, 165, 193
Panama 206–207, 211–221
Panama Canal 212, 215–217, 250
Panama City 218
Panama Defense Forces *see* PDF
Panamanian National Police (*Policía Nacional Panameña*) 215–218
Panamanian Public Force (*Fuerza Pública Panameña*) 211–212
Papa Smurf (*Papá Pitufo* [Estanislao Beltrán]) 194, 204–205
paper trail and paperwork 10, 155, 183, 293n16
parades 175, 247, 297n24
paramilitaries 112
paratroopers 99
Pastrana, Andrés 50
patrol boats 101–102, 176, 210, 215–216, 256
patrol cars 214
patrols 82, 87, 92
payoff lists 187
PDF (Panama Defense Forces) 211–212, 215–218, 221
PDVSA (Venezuela) 5, 8, 12–13, 31, 34, 37, 277n46
Peace Accords of 1992 (El Salvador) 44, 46–47, 135, 139
Peace Accords of 1996 (Guatemala) 39–40, 42
peace negotiations 115, 119
Pechora air defense system 250
Pedernales (Dominican Republic) 231, 233
PEMEX (Mexico) 58, 167–168
Peña Nieto, Enrique 193–205
penal code (Venezuela) 92
Pentagon 45, 65, 244–247, 253–254, 261, 307n6; *see also* U.S. Department of Defense
People's Revolutionary Army (Mexico) *see* EPR
Pérez, Carlos Andrés 26
Pérez Ballesteros, Ernesto 214–215
Pérez Molina, Otto (general) 134
Perry, William J. 65, 68–69, 177, 283
Peru 93, 110, 199, 203, 300n30, 304n20
Petrocaribe 150–152
petroleum 5, 29, 105–106, 150, 163; Cuba 240, 242–243, 245, 249; *see also* PDVSA; PEMEX
photographs 112, 288n7; *see also* videos
pickup trucks 171
pilots 40, 100, 242, 255
pipelines 167–168
pistols and revolvers 95, 112, 116, 180, 214, 223–224, 277n25
Plan Ávila (Venezuela) 14–15
Plan Bolívar (Venezuela) 27
Plan Colombia 50, 52, 259–260, 268
Plan Guardián (El Salvador) 47
Plan Mano Dura 47
Plan Super Mano Dura 47–48
platoons 82, 234, 236, 304n24
Polanco (Mexico City) 167

police academy 46, 48, 222–223
police forces 1–2, 95; Colombia 50, 52, 83, 86, 108, 111, 220, 259–260, 272, 284n7; Costa Rica 207–210; Dominican Republic 227–228, 235; El Salvador 46–48, 138–142; Guatemala 39–43, 133–134, 138; Haiti 222–225, 230; Jamaica 120–126; Mexico 56, 58, 74, 168, 170–171, 176–182, 186, 191–192, 200, 202, 204, 272, 282n9; Nicaragua 129–130; Panama 211–219; Venezuela 11, 91–92, 276n21, 277n46
Police Reestablishment (Colombia) 268
political parties 109, 115, 121; see also FMLN; PAN; PRI; Sandinistas
polling stations (Venezuela) 36
polls and popularity 44, 56, 171, 207; Dominican Republic 227, 236–237; Honduras 154, 158–159; Venezuela 276n12
polygraph tests 187
poppy plants 59–60, 178–179, 181, 185
Poptún (Guatemala) 133
popular participation 6, 8, 147–148, 274, 276n10
popularity see polls
Port-au-Prince (Haiti) 224–226, 230–232
Portillo, Alfonso 40
Posadas Ocampo, Juan Jesús (cardinal) 75, 284n41
posses (Jamaica) 120–126
Post-Transition Cuba (2007) 255
poverty and poor 120, 153, 214, 222–223, 227, 230, 238, 252, 257, 260–261
Powell, Colin 97, 211, 247, 286n7
PRD (Partido Revolucionario Democrático [Panama]) 214, 216–217, 219
press 112, 218, 246; see also newspapers
Preval, René 230–231
PRI (Partido Revolucionario Institucional [Mexico]) 57–58, 73–74, 78, 193, 195, 282n8
Prime Directive 272
Prince Suazo, Luis Javier (general) 158–159, 161
prisons 68, 75–76, 118, 139, 141, 190, 198
privatization 5, 214, 248, 255, 274, 275n8, 306n46
promotions: Colombia 52, 83–84, 109–112, 281n41; Dominican Republic 234, 236; Guatemala 40; Mexico 56, 60, 172; Venezuela 27–28, 33, 88
propaganda and public relations 71, 87, 91, 112, 168, 244, 248, 252, 305n29
prostitution 272
proxy 85, 91, 97
public relations 71, 109, 171, 177
Puente Grande prison (Mexico) 75–76, 284n41
Puerto Barrios (Guatemala) 133–134
purges 21, 28, 113–114, 204, 225; Dominican Republic 235; Panama 211–212
Putin, Vladimir 103, 249–250, 251

Querétaro (Mexico) 167
quid pro quo 98, 307n6
quinas 304n33
quinceañera 190
Quintero Viloria, Julio (general) 30–31
Quisqueya Battalion (Dominican Republic) 303n6

radio 86, 214
railroads 238
Rangel, José Vicente 100
Rangel Briceño, Gustavo (general) 88
rapes 175, 199, 201, 300n20
Reagan, Ronald 95, 279
referendum 6, 21, 154, 158, 294n30; see also elections
regulation 206
relief efforts 44; see also humanitarian missions
religion 6, 24, 199, 201; see also Roman Catholic Church
Republican party (U.S.) 99, 222
reserves and reservists (military) 43; Cuba 242, 245–246, 254; Dominican Republic 233; Venezuela 30–32, 34, 36, 88–89
revenue 87–88, 231, 255, 306n46; see also taxes
revolvers see pistols
Reyes, Rafael 90
Reynosa (Mexico) 180
rice 231, 232
Rice, Condoleezza 248
riots 230
Riverine Patrol and Action Unit (UMOF [Panama]) 219
robbery and theft 39, 120, 199, 214, 232
Rocket Propelled Grenades 180, 187–188, 217, 289n35
Rodas, Patricia 148
Rodríguez Zapatero, José Luis 101
Roman Catholic Church: Cuba 246; El Salvador 141; Venezuela 6, 9, 20; see also mass
Romero, Daniel 21, 24
Romero, Óscar (archbishop) 136
Roosevelt, Franklin D. 55
Rosales, Manuel 34
Rosenberg, Rodrigo 132
rural areas and countryside 172, 203, 222
rural guards and patrols 204–205, 227
Russia and Russians 130, 249; and Mexico 63, 105–107; and Venezuela 31, 89–90, 99–103; see also Soviet Union

sabotage 215
Saca, Tony 43, 45, 47–48, 136
Sadism 110, 192, 289n24
Saipan 35
salaries 33, 88, 172–173, 183, 212, 236; see also minimum wage
Salinas de Gortari, Carlos 62

Samper Pizano, Ernesto 49–50
SAMs (Surface-to-Air Missiles) 85, 103, 107; Cuba 96, 247, 250, 254, 264, 307n6; Nicaragua 94–99, 107, 127, 129
San Andrés Island (Colombia) 121
San Antonio (Texas) 89
San Cristóbal de las Casas (Mexico) 62
San Diego (California) 191–192
San Francisco (California) 137
San Jorge River (Colombia) 116
San Juan River (Nicaragua) 207
San Miguel (El Salvador) 48, 139
San Salvador (El Salvador) 139, 264
Sánchez Cerén, Salvador 142–143
sanctions 163, 244, 260–262; see also arms embargo
Sandinistas (Nicaragua) 95–99, 127–131, 143, 207, 261–263
Santo Domingo (Dominican Republic) 228, 235
Santos, Elvin 146, 148–149, 152, 163
Santos, Juan Manuel 82, 113, 115, 118–119
scandals 108–118, 288n20
SEALs (U.S. Navy) 191
secret police 58; see also DAS
security guards 123, 228, 236
security rings 190–191
Self-Defense Committees (Peru) 300n30
Self-Defense Forces (*Auto defensas* [Mexico]) 194, 202–205, 300n30
Semana magazine (Colombia) 114, 116, 118
semi-submersible craft 85–86
SENAFRONT National Frontier Service (Panama) 218–221
September 11, 2001, terrorist attacks 7, 51, 96, 245, 267
Seventh of August (Colombia) 50–51
sexual harassment 175
Shining Path (Peru) 199, 203, 300n30
shipyards 101–102, 176
shotguns 202
Shower Posse (Jamaica) 122, 289n34
Sierra Madre Mountain Range (Mexico) 190
Sierra Maestra Mountain Range (Cuba) 254
Silva Luján, Gabriel 114
"Silver" (FARC) 219–221
Simpson-Miller, Portia 124, 126
Sinaloa (Mexico) 190–191, 196
Sinaloa Cartel (Mexico) 169–170, 179, 184, 186, 192, 195–199, 202, 208
Sinú River (Colombia) 116
small arms 95, 116; see also AK-47; M-16
smart bombs 52–53
smuggling 253; arms 94, 116, 121, 125, 208, 219; Dominican Republic 231–232
Snowden, Edward J. 266
Soacha (Colombia) 112–113, 119
social classes 83, 110–111, 146, 201, 271, 300n21; see also middle class; upper class
social cleansing 111
Socialism 29–30, 34, 88–89, 96, 147, 274

SOFA see Status of Forces Agreement
Somalia 226
Somoza, Anastasio 41, 207, 216
Soto Cano U.S. base (Honduras) 145
South Eight Street (Caracas) 276n24
Southern Command (U.S.) 212–213, 218
Soviet Union 50, 95–96, 240, 243, 249, 269, 273, 304n4; see also Russia
Spain 18, 101–102; troops in Iraq 145–146
Spanish language 230, 233
Special Forces 133, 215; see also Kaibiles
Special Period (Cuba) 241, 245–246
speedboats 121, 176
spies and espionage 13, 58–59, 74, 115, 183, 191, 199–200; see also CIA
squatters 233
Star Trek 272–273
starvation 233
state farms 241
statistics 209
Status of Forces Agreement (SOFA) 44, 130, 274, 275n4, 283n29
steel 89
strength of army and navy (Mexico) 66–68
students 58
Suárez, Carlos (general) 113
submarines 85–86
succession 240, 248–249
suffrage and voting rights 27, 233; see also elections
sugar and sugar cane 229, 250
Sukhoi fighter jets 103, 105–106
Supreme Court (Honduras) 153–154, 156, 160–162
Surface-to-Air Missiles see SAMs
surplus warships and weapons 55, 93, 105, 165, 176, 208
surveillance and reconnaissance equipment 50, 86, 209, 215

T-72 tanks 103
tactics 82
takedowns 287n6
Tamaulipas (Mexico) 80, 167, 180
tanks, Venezuela 15, 17, 90, 99, 103, 276n22
taxes 52, 83, 87–88, 132, 255, 271
teachers 145
technology 93, 101–102, 182, 196; see also computers
Tegucigalpa (Honduras) 145, 151
Telemundo (U.S.) 294n38
telephone company 164
telephone conversations 115–116, 118–120, 196, 200
telescopic sights 54
television, broadcasts 12, 153–154, 159, 240, 246–247, 274; see also attack ads
television, private 62, 113; U.S. 294n38, 297n24; Venezuela 8–9, 10, 16–18, 21, 24, 277n25

Index 321

television, state (Venezuela) 12, 15–16, 23–24, 29
Temporary Protected Status (TPS) 45–46
term limits 56
ternas 304n33
terrorism 96, 102, 200–202, 213, 215, 267; *see also* September 11
Tianamen Square (Beijing) 27
Tijuana (Mexico) 178, 191–192
Tijuana Cartel 77–79
Tivoli Gardens (Jamaica) 122, 124–126
Tlatelolco Massacre (Mexico City) 59
Tlaxcala (Mexico) 168
Torricelli Act 244
Torrijos, Martín 216–219
Torrijos, Ómar 214, 216–217, 219
torture 81
tourists, tourism, and travel 123, 228, 242, 255, 262, 273, 289n36
TPS *see* Temporary Protected Status
trade and commerce 228, 232, 240, 244, 260–261, 273
trailers, military 90
training, military 93; Colombia 86, 114; Costa Rica 210; Cuba 242, 245, 253; Dominican Republic 232; Jamaica 123–124; Mexico 55, 64–65, 106, 165, 172, 177, 183; Panama 212; Venezuela 90–91, 101
Trinidad and Tobago 44
truce by gangs (El Salvador) 141–143
trucks 42; *see also* vehicles
Trujillo, Rafael 227, 233–235, 237
Truth and Reconciliation Commission (Honduras) 144, 157, 164, 295n48
Tucano aircraft 53, 85; and Dominican Republic 235; and Venezuela 100, 102
tunnels into U.S. 60, 192, 197
12.7 mm machine guns 145, 188
Twitter *see* Zunzuneo

Ukraine 250
ultimatum 97, 286n7
UMOF *see* Riverine Patrol and Action Unit
unicameral legislature 6–7
uniforms, army 207, 212, 227, 236; Cuba 242, 247; Venezuela 27, 32–33, 37
unions *see* labor, organized
United Nations 223–226
U.N. Commission Against Impunity 132
U.N. Security Council 107
U.S. army 110, 175; *see also* U.S. military
U.S. Attorney General 146
U.S. bases 145
U.S. Border Patrol 231
U.S. Coast Guard 165, 235, 253
U.S. Congress 45, 63, 87, 182, 244, 262, 269, 272
U.S. Department of Defense 45, 65, 106; *see also* Pentagon
U.S. Department of Justice 125, 212

U.S. Department of State 18, 156–158, 266, 294n39, 294n40
U.S. embassies 1, 94, 269–270, 295n40; Bogotá 82, 85–86, 110, 302n43; Caracas 9–11, 18, 34–35, 90, 100, 103, 276n12, 286n24; Guatemala City 41–42, 131–132; Managua 97; Mexico City 105–106, 184; Panama City 216, 220–221, 302n43; Rio de Janeiro 10; San Salvador 44, 138, 140; Santo Domingo 233–234; Tegucigalpa 146–148, 152–153, 155–160, 162–163, 295n60
U.S. Foreign Service 305n31
U.S. invasions 28, 96, 261; Cuba 239–240, 251, 254; Panama 211, 214; Venezuela 26, 28–30, 36, 81; *see also* Bay of Pigs
U.S. Marines 227, 275n4
U.S. military 110, 130, 137, 174–177, 182, 209, 225; *see also* U.S. army
U.S. military assistance 93–95, 97, 123–124, 130, 139, 145, 303n6; Colombia 83, 87, 108, 114, 259; Mexico 65, 165, 194
U.S. Navy 18, 22, 35, 68, 165, 174, 176, 191
U.S. occupations 221–222, 225, 228, 247–248, 259
U.S. Treasury 63–64, 258
U.S. troops 44
Univisión (U.S.) 294n38
Unmanned Aerial Vehicles (UAV) *see* drones
upper class 21, 24, 45, 48, 136, 148, 185–186, 275n8; Guatemala 38–41, 132
Urdaneta Avenue (Caracas) 15–16
Uribe Vélez, Álvaro 52–54, 81–83, 88, 112–113; and Panama 220; and U.S. 44, 108, 260

Valle Hermoso accords (Mexico) 167
Vásquez Velasco, Efraín (general) 16, 17, 23–24, 277n43
Vásquez Velásquez, Romeo (general) 149, 158–164
Vatican (Rome) 17
vehicles 171, 190, 202, 204; *see also* Humvees; trucks
Velasco, Ignacio (Cardinal) 9, 17–18, 21, 24–25
Venezuela 5–25, 26–37, 55, 88–92, 130, 142, 233; arms buildup 99–105; and Colombia 85, 110; and Cuba 243, 261; embassies 90–91, 266; and Honduras 150–151; and U.S. 81, 147, 194–195, 228, 258, 264–266, 269–270, 283n29; and U.S. arms embargo 95
Venezuelan Confederation of Workers *see* CTV
Venezuelan Military Acquisition Office 286n24
Veracruz State (Mexico) 57, 60, 168
vices *see* morality
video 132, 168, 232
Vietnam War 32, 69, 110, 277n25
Villa Ahumada (Mexico) 179
Villavicencio (Colombia) 54
violence 109, 110, 123, 126, 184–186; *see also* brutality

La Violencia (Colombia) 109
visas 138, 163

War College 84
war scares 90–91
wedding 190
West Point (New York) 175
whites 120, 124, 174, 214
Wikileaks 2, 270, 279n33, 286n7
Wilson, Woodrow 260
women 24, 125; Dominican Republic 232, 234, 304n24; Mexico 173–175, 196, 199, 201, 297n23, 300n20; U.S. 297n24
World War II 55, 272

Yance Villamil, Luis Fernando (admiral) 52, 281n41

Zapatista National Liberation Army *Ejercito Zapatista de Liberación Nacional* or EZLN (Mexico) 62, 65–67, 167
Zedillo, Ernesto 63–74, 165, 176–177, 195, 258, 283n29
Zelaya, José Manuel 144–163, 263
Zetas (Mexico) 79–80, 166, 169, 192, 196, 199, 202; and Guatemala 133–134; and Monterrey 186–189
Zona Rosa massacre (El Salvador) 138
Zulia (Venezuela) 34
Zunzuneo (U.S.) 264–265

www.ingramcontent.com/pod-product-compliance
Lightning Source LLC
Chambersburg PA
CBHW051208300426
44116CB00006B/475